By Wes Hayward (W7ZOI)

TriQuint Semiconductor
Beaverton, Oregon

Introduction to Radio Frequency Design

Copyright © 1994-96 by
The American Radio Relay League, Inc.

Copyright secured under the Pan-American Convention.

This work is publication number 191 of the Radio Amateur's Library, published by the League. All rights reserved. No part of this work may be reproduced in any form except by written permission of the publisher. All rights of translation are reserved.

Quedan reservados todos los derechos.

ISBN: 0-87259-492-0

First ARRL Edition
Second Printing

Contents

PREFACE   ix

1   LOW FREQUENCY TRANSISTOR MODELS   1

1.1   Introduction: Basic Models   **1**
1.2   Small-Signal Models for the Bipolar Transistor   **3**
1.3   Biasing and Bypassing the Bipolar Transistor   **10**
1.4   Large-Signal Operation of the Bipolar Transistor   **14**
1.5   Gain, Power, and Negative Feedback   **18**
1.6   Modeling the Field-Effect Transistor   **26**
       References   **30**
       Suggested Additional Readings   **30**

2   FILTER BASICS   32

2.1   Network Analysis in the Time Domain   **33**
2.2   Network Analysis in the Frequency Domain   **36**
2.3   Poles, Zeros, and the Series Tuned Circuit   **39**
2.4   Filter Concepts   **44**
2.5   The Ladder Method   **51**
2.6   Losses in Reactive Components and Quality Factor   **54**
2.7   The All-Pole Low Pass Filter   **59**

2.8 Filter Denormalization, Practical Low Pass,
High Pass, and Bandpass Structures 68
References 73

3 COUPLED RESONATOR FILTERS   75

3.1 Impedance Transforming Networks 75
3.2 Normalized Coupling and Loading Coefficients 82
3.3 The Double Tuned Circuit 90
3.4 Experimental Methods, The Dishal Technique 95
3.5 Quartz Crystal Filters 101
References 108

4 TRANSMISSION LINES   109

4.1 Transmission Line Fundamentals 110
4.2 The Voltage Reflection Coefficient and Standing Waves 115
4.3 The Smith Chart and Its Applications 121
4.4 Practical Transmission Lines 132
4.5 Impedance Transforming Networks 137
4.6 Transmission Line Measurements 151
References 159
Suggested Additional Readings 160

5 TWO-PORT NETWORKS   161

5.1 Two-Port Network Fundamentals 161
5.2 Interconnection of Two-Port Networks 168
5.3 The Hybrid-Pi Transistor Model 172
5.4 Amplifier Design with Admittance Parameters 176
5.5 Design Examples with Admittance Parameters 183
5.6 Scattering Parameters 191
References 201
Suggested Additional Readings 201

6 PRACTICAL AMPLIFIERS AND MIXERS   202

6.1 Noise in Amplifiers 202
6.2 Noise Models and Noise Matching 210
6.3 Distortion in Amplifiers and the Intercept Concept 219
6.4 Mixers 232
6.5 Intermediate Frequency Amplifiers 246

References **259**
Suggested Additional Readings **259**

7 OSCILLATORS AND FREQUENCY SYNTHESIZERS — **261**

7.1 Oscillator Concepts **262**
7.2 The Colpitts Oscillator **265**
7.3 Further *LC* Oscillator Topics **279**
7.4 Crystal Controlled Oscillators **286**
7.5 Noise in Oscillators **292**
7.6 Negative Resistance Oscillators **301**
7.7 Methods of Frequency Synthesis **311**
7.8 Phase-Locked Loop Frequency Synthesis **323**
7.9 Voltage Controlled Oscillators and Improved Synthesizers **330**
References **338**
Suggested Additional Readings **339**

8 THE RECEIVER: AN RF SYSTEM — **341**

8.1 Receiving Systems **342**
8.2 Receiver Evaluation and Measurements **349**
8.3 Intermediate Frequency Amplifier Systems and Gain Control **356**
8.4 Front-End Design **367**
References **372**
Suggested Additional Readings **372**

INDEX — **373**

Preface to the ARRL Edition

Just a few years ago it was common to hear comments among electrical engineers stating that "the future is all digital," or "analog and RF are dead." Even the radio amateur was moving away from a traditional interest in radio frequency methods toward activities based upon digital technology, including packet radio. While digital methods have certainly grown and continue to evolve, RF and analog design has hardly disappeared! Rather, new applications have emerged that effectively combine disciplines. RF and "wireless" communications are more popular than ever.

Workers in the field tend to think of radio frequency design and analog design as being nearly identical. This is not really complete. The traditional analog designer is generally concerned with a system described by input signals, usually as voltages, and the resulting output voltages. RF design is a subset of analog design, with voltage having been replaced by power. The RF designer might characterize an amplifier by the output power that can be extracted when driven by an available input power. The RF designer is compelled to apply conservation-of-energy to his or her analog electronics.

Introduction to Radio Frequency Design was originally published by Prentice-Hall in 1982. The text was intended to present basic RF concepts (with some related analog subjects) to the working engineer. The typical reader often had a foundation in digital hardware methods and was comfortable with basic analog design.

The reader of the original printing might have used a work station for digital design. However, he or she performed most analog design tasks with a hand calculator. The age of the personal computer was still ahead.

This new printing has changed little from the original. There has been no change in the intended use of the text; the primitive information is still needed by the working engineer who may now use RF methods to amplify and condition digital as well as analog data. The text should also be useful to the advanced radio amateur. The reader should have a modest formal background that includes the mathematics and physics usually presented to the university sophomore engineering student.

The major difference with the ARRL printing is the inclusion of a disk containing a collection of computer programs related to and derived from the text. The programs are written for the IBM personal computer* or a suitable "compatible" and run from DOS. A manual is included on the disk. The manual, an ASCII text file named IRFD2MAN.TXT, is short and is easily printed by the user. Installation information is included in the software manual.

Several individuals have contributed to this volume by offering encouragement as well as specific feedback and comment. This assistance is gratefully acknowledged. My colleagues in the Wireless Communications Division at TriQuint Semiconductor have participated in numerous useful discussions and have reviewed and used much of the software included with the book.

Several colleagues deserve special mention. Discussions with Roy Lewallen (W7EL) and Bob Culter (N7FKI) at Tektronix have provided continuing enlightenment. Jeff Damm (WA7MLH) and Terry White (KL7IAK), both from TriQuint Semiconductor, and Dr. Fred Weiss of Analog Devices have all provided encouragement and enlightening discussions. Finally, Joel Kleinman (N1BKE) at ARRL has been especially helpful with the logistic details of this and other projects.

*IBM is a registered trademark of International Business Machines.

Preface to the Prentice-Hall Edition

Modern communications technology is generally segmented into two related bodies of knowledge. One is digital electronics, a subject of widespread interest today. The other is the branch of traditional analog circuit technology known as radio frequency (rf) design. Generally, engineers and scientists with a strong background in one area are not up-to-date in the other.

Numerous books are available at all levels for the student of digital methods, especially those methods concerned with microprocessors. There is less information available for the beginning rf engineer. Most existing texts are sophisticated graduate level treatments, often being rather specialized. This text aims to provide a suitable introduction to modern rf design methods, enabling the reader to move on to more specialized treatments.

This volume, in many ways, is a personalized one. It is the text that the writer would like to have had when making a career change from electron device physics to rf circuit design.

Three specific reader groups were considered during text preparation. The first is the working engineer, scientist, technician, or manager. This reader should have a knowledge of fundamental mathematics through differential equations and of basic circuit theory. Little else is assumed for individual study.

The second group is the student of electrical engineering or computer science. Radio frequency design is a subject rarely covered in today's undergraduate curricula. However, I feel that this volume might find application as an auxilliary text in a traditional communications theory course.

The third group is the electronics hobbiest including the radio amateur. While there are many serious experimenters with an interest in the subject matter, this volume will probably be limited to those "hams" with a moderate formal background.

A central theme emphasized throughout the book is simplicity. I have observed that a thorough and complete analysis of a circuit using the simplest of circuit models is prerequisite to the level of understanding that leads to innovation. More formal treatments of individual subjects are found in the references listed at the end of the individual chapters. Mathematics is used in this text as required to convey intuition to the reader. No attempt has been made toward formal completeness. The text is also practical. Frequent examples are used, most of them based upon designs by the writer.

The first chapter introduces simple, low frequency transistor models. Both bipolar and field effect transistors are considered. The emphasis is on small-signal analysis. However, large-signal models are also presented, for they have important application in some parts of rf design.

The second and third chapters are devoted to filter design. Much of Chap. 2 is a review of basics emphasizing analysis in both the time and frequency domains. Low pass prototype filters are discussed in Chap. 2. The third chapter is devoted to coupled-resonator bandpass filters and their relationship to low pass prototypes. Practical details covered include the Dishal methods for filter alignment and the design of crystal filters.

The fourth chapter covers transmission lines, not only as useful components, but as a different way of viewing an rf circuit or system. Considerable use is made of the Smith chart. Numerous impedance matching networks are presented as are methods for transmission line measurements.

Chapter 5 presents the concepts of two-port network theory. Initial development and amplifier examples use admittance parameters. Scattering parameters are then presented and illustrated. Stability is emphasized throughout the examples.

Chapter 6 deals with the practical details of amplifiers and mixers. The methods for handling noise and distortion presented include information on the intercept concept. Many mixer types are analyzed for basic operation. Numerous amplifiers are presented in connection with low noise and low distortion performance.

Oscillators form the basis of Chap. 7. The chapter has a secondary purpose, that of illustrating what may be realized with a thorough circuit analysis using the most elementary device models. The Colpitts oscillator and its variations are discussed in detail. Other LC and crystal oscillators are presented. Oscillator noise is covered in detail with a simplified and intuitive explanation. S-parameter methods are then presented for the design of microwave oscillators. The chapter concludes with a discussion of frequency synthesizers. The emphasis is on fundamentals and the nature of the compromises presented by the frequency synthesizer.

The final chapter is devoted to receiver design. The primary purpose is as an illustration of an rf system. Hence, the discussion is confined to two receiver types, that used for hf communications and that used for measurement. Included in the later group is the spectrum analyzer.

Many examples used throughout the book involve numerical calculation. Those in Chap. 5 for amplifier design were done on a small digital computer with a program written in BASIC. All other calculations presented in the text were done with an inexpensive, handheld programmable calculator. Programs are not presented. However, many equation sets are given in a form to facilitate easy writing of programs for whatever computing facility the reader might have available.

Numerous individuals have contributed to this volume through discussions and with direct support. This assistance is greatfully acknowledged.

I would like to thank Jeannie MacPhee, Lynn Rasmussen, and Mary Chambers for typing assistance. My wife, Shon, was a great help with editorial chores.

Several colleagues deserve special mention. I would like to acknowledge discussions at Tektronix with R. Bales, R. Brown, L. Gumm, R. Lewallen, G. Long, S. Morton, Dr. E. Sang, and F. Winston. Valuable advice was provided by M. DeMaw (American Radio Relay League, Newington, CT) and by J. Moriarity and F. Telewski (both at John Fluke Co., Mountlake Terrace, WA). W. Sabin (Rockwell-Collins, Cedar Rapids, IA) not only provided advice, but did a very thorough review of the manuscript. I would especially like to thank L. Lockwood and Dr. D. Morton, both at Tektronix. Not only have they provided me with considerable intuition and enlightenment through numerous discussions, but have offered much needed encouragement.

Finally, I would like to acknowledge the support of my wife, Shon, and our sons, Ron and Roger. They have shown great tolerance for the many evenings and weekends I spent in the basement, pounding on a typewriter instead of spending time with them.

Introduction to Radio Frequency Design

1

Low Frequency Transistor Models

1.1 INTRODUCTION: BASIC MODELS

Engineering is an application of science. Circuit analysis and design, the topics covered in this text, are derived from fundamental topics in physics. It is not possible for an engineer to neglect the basics. The more familiar the designer is with them, the more effective he or she will be in design efforts.

While exceedingly important, fundamental device physics is rarely applied directly in circuit design. Instead, models are utilized. A model is a simplified view of the device to be applied. Although models are not complete, they will contain the essential details needed for design. The complexity of the model will grow as the application becomes more sophisticated and where greater accuracy is required. It is generally most informative, however, to use the simplest model that will describe the salient features of the circuit being analyzed.

Simple models are presented in this chapter for the bipolar junction transistor (BJT) and field-effect transistor (FET). They are generally accurate only at low frequencies. However, they may be modified for application at higher frequencies. The simplified low-frequency models are also surprisingly useful in many radio-frequency (rf) applications.

Bipolar and junction field-effect transistors (JFET) are shown schematically in Fig. 1.1. Both NPN and PNP devices and N and P channel JFETs are presented. Our analysis will generally be confined to the NPN or N-channel JFET with the understanding that the others are described by a change in biasing voltage polarities.

The bipolar transistor is regarded as either a current or voltage controlled device.

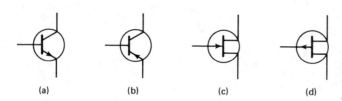

Figure 1.1 Transistor schematic symbols. (a) NPN and (b) PNP bipolar transistors, (c) N-channel and (d) P-channel junction field effect transistors.

If the collector current is controlled by the base current, the transistor is described by

$$I_c = \beta I_b \tag{1.1-1}$$

where β is the common emitter current gain.

The emitter current is related to the base-emitter voltage, V_{be}, by

$$\begin{aligned} I_e &= I_{es}\left[\exp(qV/kT) - 1\right] \\ &\simeq I_{es}\exp(qV/kT) \end{aligned} \tag{1.1-2}$$

where $V = V_{be}$, q is the electronic charge, k is Boltzmann's constant, and T is the temperature in kelvin (K). I_{es} is the emitter saturation current and has a typical value of 1×10^{-13} amp.

Both equations are approximations, models of a generally more complex behavior. Equation 1.1-2 is an approximate form of the first Ebers-Moll model (1). Refined models for the bipolar transistor are summarized by Getreu (2). Equations 1.1-1 and 1.1-2 describe transistor operation for dc biasing and for signal applications.

The N-channel JFET is well described by

$$I_D = I_{DSS}(1 - V_{sg}/V_p)^2 \tag{1.1-3}$$

where I_{DSS} is the drain saturation current, V_{sg} is the potential of the source with respect to the gate, and V_p is the pinch-off voltage. Equation 1.1-3 is limited to V_{sg} between 0 and V_p. The JFET is rarely operated with forward bias on the gate-to-channel diode. I_D is zero for V_{sg} greater than V_p. The equation describes both biasing and signal operation.

The equations presented above will be extended in the following sections. Small-signal approximations will be considered for use in amplifier design. The operation of both device types will be investigated at large signal levels. This will provide limits on the range of application of the small-signal approximations and provide information about the nature of the distortions caused by the nonlinear nature of the equations.

1.2 SMALL-SIGNAL MODELS FOR THE BIPOLAR TRANSISTOR

The models presented in the preceding section describe the operation of a transistor for an arbitrary input voltage. A transistor is usually driven from a combination of inputs: a biasing voltage and a signal. Small-signal modeling examines only the effect of the signal components, ignoring both the bias and nonlinear signal effects.

Figure 1.2 shows a representation of a bipolar transistor amplifier. Operation is described by the Ebers-Moll equation (Eq. 1.1-1). This relationship may be expanded in a Taylor series in the signal variable, V_s, about a bias voltage, V_b

$$I_e = I_{es} \exp(qV/kT) = \sum_0^\infty \frac{q^n I_{es}}{k^n T^n} \exp(V_b q/kT) \frac{V_s^n}{n!} \qquad (1.2\text{-}1)$$

$$= I_0 + g_m V_s + \tfrac{1}{2} g_m^2 V_s^2 + \cdots$$

where g_m is the first derivative of the Ebers-Moll model evaluated at the bias point, $g_m = qI_0/kT$, the small-signal transconductance. The first term represents the bias voltage while the second one, the linear term, represents the effect of a small signal voltage, V_s. The higher order terms are the result of the nonlinear nature of the basic model, negligible if V_s is sufficiently small.

The above analysis is based upon the transistor as a voltage driven device. If viewed as a current controlled element, the defining equation is $I_c = \beta I_b$. This equation is linear. Hence, a Taylor expansion yields

$$I = I_0 + \beta I_s \qquad (1.2\text{-}2)$$

The emitter signal current for a small input signal voltage is given as $I_e = g_m V_s$. The emitter current is the sum of the base and collector currents. Collector current for a small input signal current is given by the linear term of Eq. 1.2-2.

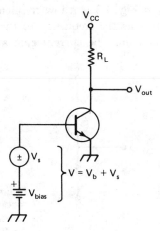

Figure 1.2 Bipolar transistor with voltage bias and input signal.

3

The base current is then $I_e/(\beta + 1)$. The base signal current is thus related to the base signal voltage by

$$I_b = \frac{g_m V_b}{(\beta + 1)} \qquad (1.2\text{-}3)$$

The input resistance for small signals is the ratio of the signal voltage to the signal current

$$R_{in} = \frac{V_b}{I_b} = \frac{(\beta + 1)}{g_m} = (\beta + 1)r_e \qquad (1.2\text{-}4)$$

where $r_e = 1/g_m$ is the emitter resistance. The significance of this term will become apparent below.

An interpretation of this calculation is the small-signal model shown in Fig. 1.3a. It consists of an input resistance and a controlled current source. An equivalent model is shown in Fig. 1.3b. The equivalence may be demonstrated by a nodal analysis. The second version, Fig. 1.3b, is the most common of all transistor small-signal models, the controlled current source with emitter resistance.

Examination of the preceding equations shows that $r_e = kT/qI_0$, or $r_e = 26/I_e$(mA, dc) where I_e is now the dc bias current in milliamperes. The evaluation was performed at a typical ambient temperature of 300 K.

The amplifier output resistance is infinite for it is a pure current source. This is a good approximation for most silicon transistors at low frequency. Output voltage is extracted from a load resistor, R_L. The resistance is assumed small enough that a positive collector bias voltage is maintained for the chosen bias current. The amplifier is termed a common emitter (ce) circuit, for the emitter is common to both the input and the output. The collector supply, V_{cc}, is stable and independent of the signal voltages and is, hence, at signal ground potential.

The model presented in Fig. 1.3 is a low frequency approximation. The most significant change that occurs as the frequency of the input signal increases is an apparent decrease in current gain. The current gain at low frequencies is β_0. This

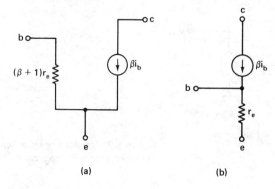

Figure 1.3 Simplified low frequency model for the bipolar transistor, a "beta generator with emitter resistance." $r_e = 26/I_e$(mA, dc).

value is maintained through the audio spectrum, but eventually begins to decrease. At sufficiently high frequency, the small-signal β will drop linearly with frequency, decreasing by a factor of 2 for each doubling of signal frequency. The frequency where current gain is unity is termed the gain-bandwidth product, or F_T, of the transistor. A typical device for lower radio frequency application might have $\beta_0 = 100$ and $F_T = 500$ MHz. The frequency with current gain of $\beta_0/\sqrt{2}$ is termed the F_β and is related to F_T by $F_\beta = F_T/\beta_0$.

The frequency dependence of the current gain is modeled with the small-signal equivalent circuit of Fig. 1.4. This is termed the *hybrid-pi model* (3) and will be discussed in more detail in Chap. 5. The capacitor is chosen to have a reactance equal to the low frequency input resistance, $(\beta + 1)r_e$, at F_β. This is equivalent to a frequency dependent current gain

$$\beta(F) = \frac{\beta_0}{1 + j\beta_0 F/F_T} \qquad (1.2\text{-}5)$$

It is useful to evaluate the magnitude of β for various frequencies. Equation 1.2-5 shows that $|\beta| \simeq F_T/F$ for frequencies well above F_β. It is often useful to assume that a scalar model (Fig. 1.3) is suitable to describe radio frequency operation if the current gain is given by F_T/F. This approximation becomes less accurate as F_T is approached.

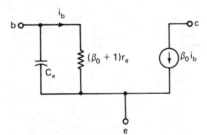

Figure 1.4 The hybrid-pi model for the bipolar transistor.

The small-signal models will be illustrated by analyzing the three common amplifier configurations, those with a common emitter, a common base, and a common collector. The small signal representations are presented in Fig. 1.5.

The ce amplifier is shown in Fig. 1.5a. The nodal equation at the base is, using the model, $V_{\text{in}}/r_e = I_b(\beta + 1)$. Manipulation yields the input resistance, $R_{\text{in}} = r_e(\beta + 1)$ and the base current, $I_b = V_{\text{in}}/[(\beta + 1)r_e]$. The collector current is then

$$I_c = \frac{\beta V_{\text{in}}}{r_e(\beta + 1)} \simeq g_m V_{\text{in}} \qquad (1.2\text{-}6)$$

where the approximate form applies to the case of high β.

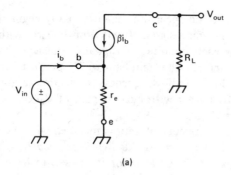

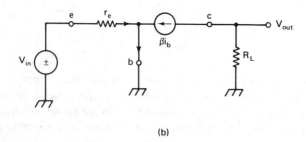

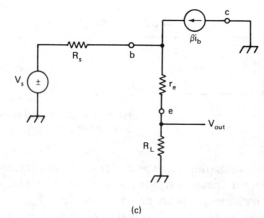

Figure 1.5 Application of small signal models for analysis of (a) the ce amplifier, (b) the cb and (c) the cc bipolar transistor amplifiers.

Noting the direction of current flow predicted from the model, the output voltage is

$$V_{out} = -I_c R_L \qquad (1.2\text{-}7)$$

where R_L is the output load resistance. The voltage gain is then

Sec. 1.2 Small-Signal Models for the Bipolar Transistor

$$G_v = -R_L/r_e \quad (1.2\text{-}8)$$

The output voltage is 180 degrees out of phase with the input as indicated by the minus sign.

The approximate gain given by Eq. 1.2-8 depends upon high beta. Amplifier gain may be controlled by changing r_e, realized by picking the desired operating current or by adding additional resistance in series with the emitter. The latter is a form of feedback termed emitter degeneration.

The cb amplifier (Fig. 1.5b) is similarly analyzed. A voltage generator is again used as the input signal. The nodal equation at the base is identical with that for the ce amplifier. However, the direction of the base and collector currents is different. Manipulation of the nodal equation shows

$$R_{in} = r_e = 1/g_m \quad (1.2\text{-}9)$$

The common base current gain is

$$G_I = \frac{\beta V_{in}}{r_e(\beta+1)} \frac{r_e}{V_{in}} = \frac{\beta}{(\beta+1)} = \alpha \quad (1.2\text{-}10)$$

α is very close to 1, for β is usually large with respect to unity. The cb amplifier, when driven from a voltage source, has a voltage gain of

$$G_V = \frac{R_L \beta}{r_e(\beta+1)} = \frac{\alpha R_L}{r_e} \quad (1.2\text{-}11)$$

This gain is noninverting.

The cc amplifier, often called an emitter follower, is shown in Fig. 1.5c. The driving source for this example contains a source resistance, R_s. Noting the direction of the assumed currents shown in the figure, a nodal equation is written at the base

$$\frac{V_b}{r_e + R_L} = (\beta+1)\frac{V_s - V_b}{R_s} \quad (1.2\text{-}12)$$

This is solved for the base voltage

$$V_b = \frac{V_s(\beta+1)(r_e+R_L)}{R_s + (\beta+1)(r_e+R_L)} \quad (1.2\text{-}13)$$

The output voltage, V_{out}, is related to the base voltage through voltage divider action

$$V_{out} = \frac{V_b R_L}{(r_e+R_L)} \quad (1.2\text{-}14)$$

The emitter current is related directly to the base voltage by

$$I_e = \frac{V_b}{(r_e + R_L)} \qquad (1.2\text{-}15)$$

But the base current is $I_b = I_e/(\beta + 1)$. The input resistance is obtained by combining equations and solving for the ratio of V_b/I_b

$$R_{in} = \frac{V_b}{I_b} = (\beta + 1)(r_e + R_L) \qquad (1.2\text{-}16)$$

The cc amplifier, unlike the ce or cb types, will have a finite output resistance. This is calculated by short circuiting the input voltage source, V_s, and by driving the output port with either a voltage or current source. Figure 1.6 shows the case where the output is driven with a current I_d. The source resistance, R_s, is still contained in the circuit. The nodal equation at the base is

$$-V_b[R_s + r_e(\beta + 1)] + V_{out} R_s = 0 \qquad (1.2\text{-}17)$$

Similarly, the nodal equation is written at the output

$$I_d = \frac{V_{out} - V_b}{r_e} = \frac{V_{out}}{r_e} - \frac{V_{out} R_s}{r_e} \cdot \frac{1}{R_s + r_e(\beta + 1)} \qquad (1.2\text{-}18)$$

Equation 1.2-17 is solved for V_b and substituted into Eq. 1.2-18. The output resistance is then calculated

$$R_{out} = \frac{V_{out}}{I_d} = \frac{R_s}{(\beta + 1)} + r_e \qquad (1.2\text{-}19)$$

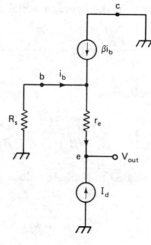

Figure 1.6 The output resistance of a cc amplifier is calculated by driving the output with a current drive, I_d. The input voltage source has been short circuited.

Sec. 1.2 Small-Signal Models for the Bipolar Transistor

The emitter follower voltage gain from the base to the output is evaluated from Eq. 1.2-14, $G_v = R_L/(R_L + r_e)$. This is close to +1 if R_L is large with respect to r_e.

The cc example shows characteristics that are more typical of practical rf amplifiers than the idealized ce and cb amplifiers. Specifically, the input resistance is a function of both the device and the termination at the output. The output resistance is critically dependent upon the input driving source resistance.

The preceding examples have used the simplest of models, the controlled current generator with an emitter resistance, r_e. More refined models will be required for analysis at high frequency. The simplified hybrid-pi of Fig. 1.4 is often a suitable one for approximations.

A more refined small-signal model is shown in Fig. 1.7. It contains an additional capacitance, C_{cb}, from the collector to the base as well as the previously introduced hybrid-pi capacitor. An output resistance and capacitance is included at the collector and a series "base-spreading" resistance, r_b', at the input. Inductors represent the effects of bonding wires connecting the semiconductor to the package. External package capacitances are also taken into account. Clearly, analysis with such a model will be much more complicated than with the simplified ones. This analysis is best done with the aid of a digital computer or with other methods that will be presented in Chap. 5.

The designer should not be intimidated by the complexity of a more complete model. More often than not, surprisingly accurate results may be obtained, even at radio frequencies, from a very simple model. Not only is analysis simplified, but considerable intuition is provided that is buried in the complicated mathematics associated with a more thorough treatment. The model should be the simplest which will describe the salient features of the circuit being studied.

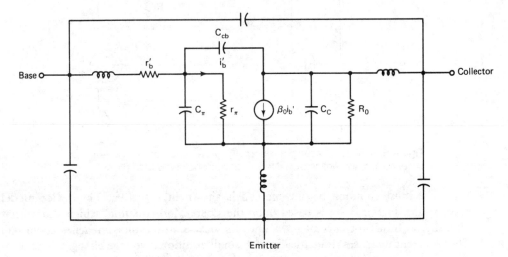

Figure 1.7 A more refined small-signal model for the bipolar transistor. Suitable for many applications near the transistor F_t.

1.3 BIASING AND BYPASSING THE BIPOLAR TRANSISTOR

Proper biasing of the bipolar transistor is more complicated than it might appear. The Ebers-Moll equation would suggest that a common emitter amplifier could be built as shown in Fig. 1.2, grounding the emitter and biasing the base with a constant voltage source. Further examination shows that this presents many problems. Not only does the Ebers-Moll equation predict an exponential dependence on temperature, but the emitter saturation current, I_{es}, is a very strong function of temperature. The power dissipated within the transistor will cause enough heating to alter the biasing, usually toward increased current. The ultimate result is *thermal runaway*. Constant voltage biasing applied to the base is almost never used.

Constant base current biasing, shown in Fig. 1.8a, is sometimes used. R_1 delivers a base current, which is the desired collector current diminished by β. This works reasonably well if the current gain is known. This, however, rarely happens. A transistor with a typical β of 100 might actually have values ranging from 50 to 250. A slightly improved method is shown in Fig. 1.8b where the bias is derived from the collector. As current increases, the collector voltage will decrease. This will, in turn, cause the bias current flowing through R_1 to decrease, ensuring operation in the transistor active region.

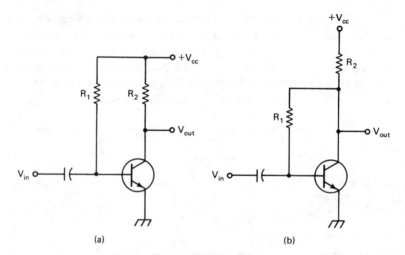

Figure 1.8 Simple biasing methods for a ce amplifier. The scheme at (a) suffers if β is not well known. Negative feedback is used in (b).

The most common biasing method is shown in Fig. 1.9a. The device model used, shown in Fig. 1.9b, is based upon the Ebers-Moll equation, which shows that virtually no transistor current flows until the base-emitter voltage reaches about 0.6. Then, current increases dramatically with small additional voltage changes. The transistor is thus modeled as a current controlled generator with a battery in series with the base. The battery voltage is ΔV.

The circuit is analyzed with nodal equations. The collector resistor, R_5, is initially

Sec. 1.3   Biasing and Bypassing the Bipolar Transistor

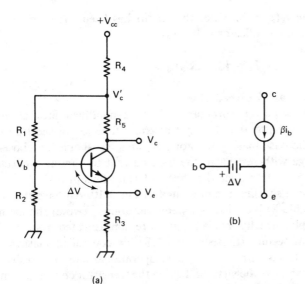

Figure 1.9 (a) Circuit used for evaluation of transistor biasing. (b) The model used for bias calculations.

assumed to be zero. Applying the Kirchhoff current law at the node marked with V'_c yields

$$\frac{V_{cc} - V'_c}{R_4} = \beta I_b + \frac{V'_c - V_b}{R_1} \qquad (1.3\text{-}1)$$

Similarly, at the base

$$\frac{V'_c - V_b}{R_1} = I_b + \frac{V_b}{R_2} \qquad (1.3\text{-}2)$$

The base current is related to the base voltage with

$$\frac{V_b - \Delta V}{R_3} = I_b(\beta + 1) \qquad (1.3\text{-}3)$$

This leaves three equations in the three unknowns V_b, V'_c, and I_b. Simultaneous solution yields

$$V_b = \frac{V_{cc}R_2R_3 + \Delta V R_2[R_4 + R_1/(\beta + 1)]}{R_3R_4 + R_2R_4 + R_2R_3 + R_1R_3 + R_1R_2/(\beta + 1)} \qquad (1.3\text{-}4)$$

$$V'_c = \frac{R_1 V_{cc} + R_4 V_b + \beta R_1 R_4 (\Delta V - V_b)/[R_3(\beta + 1)]}{R_1 + R_4} \qquad (1.3\text{-}5)$$

$$I_b = \frac{V_b - \Delta V}{R_3(\beta + 1)} \qquad (1.3\text{-}6)$$

The emitter current is then $I_b(\beta + 1)$. Once the circuit has been analyzed, R_5 may be taken into account. The final collector voltage is

$$V_c = V_c' - \beta I_b R_5 \qquad (1.3\text{-}7)$$

The solution is valid so long as V_c exceeds V_b.

The previous equations may be programmed for a handheld programmable calculator. Analysis will then show that I_e is not a strong function of the transistor parameters, ΔV and β. The base biasing resistors, R_1 and R_2, should be chosen to draw a current that is large with respect to the base current to eliminate effects of β variation. V_b should be large to reduce the effects of variations in ΔV.

Three additional biasing schemes are presented in Fig. 1.10. All have the virtue of providing bias that is stable with device parameter variations. Two of the methods use a negative power supply, usually available in more sophisticated systems. The circuit of Fig. 1.10c uses a second transistor, a PNP, for control of biasing. The PNP transistor may be replaced with an integrated operational amplifier if desired. All three of the circuits shown have the virtue of having the transistor emitter grounded directly. This is often of great importance in amplifiers operating at microwave frequencies. These circuits may be analyzed using the simple model of Fig. 1.9.

The biasing equations presented may be solved for the resistors in terms of desired operating conditions and device parameters. It is generally sufficient to repetitively analyze the circuit, using standard resistor values.

The small-signal transconductance of a common emitter amplifier was found to be $g_m = qI_e(\text{dc})/kT$ in the previous section. If biased for constant current, the

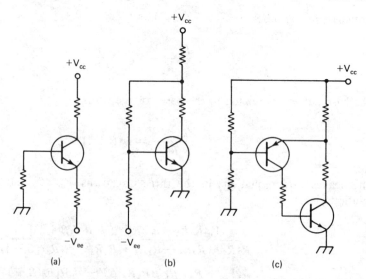

Figure 1.10 Alternative biasing methods. (a) and (b) utilize dual power supplies. (b) and (c) allow the emitter to be at ground while still providing temperature-stable operation.

small signal voltage gain will vary inversely with temperature. Gain may be stabilized against temperature variations with a biasing scheme that causes the bias current to vary in *proportion to absolute temperature*. Such methods, termed PTAT methods, are often used in modern integrated circuits and are finding increased application in circuits built from discrete components.

A complete ce amplifier is shown in Fig. 1.11. The resistors bias the transistor and are part of the small-signal circuit. The parallel combination of R_1 and R_2 will appear in parallel with the input resistance of the amplifier. This component is usually ignored in analysis. Similarly, R_5 will appear in parallel with the load, R_L.

Several capacitors appear in the circuit. They serve three different and distinct functions. C_1 and C_2 are blocking capacitors. They allow a difference in dc voltage to occur across them without altering the circuit operation at ac frequencies. The input from the source resistor is applied directly to the base through C_1 with virtually no attenuation. The capacitive reactance, $X_c = (2\pi f C)^{-1}$, should be small compared with the amplifier input resistance.

C_3 serves a dual role. It is a bypass capacitor that ensures that the supply end of R_5 is at signal ground. The reactance should be low compared with R_5 to serve this function. C_3 is also part of a decoupling network with R_4 being the remaining part. This RC network serves to prevent any stray signals that may be superimposed upon the power supply from reaching the amplifier. The filter should have a cutoff frequency that is low compared with the frequency of any noise that may be anticipated. It is common to find C_3 consisting of a parallel combination of several capacitors of differing value. One might be of relatively small capacitance, perhaps 0.001 μF. Such a value will usually present a very low impedance at high frequency. The other unit might be several μF. While offering low reactance to lower frequency signals,

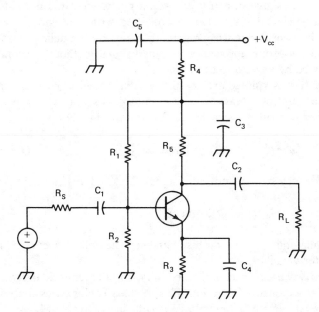

Figure 1.11 Common emitter amplifier with stabilized bias. Capacitors C_1 and C_2 are for dc blocking. C_3 and C_4 are bypass capacitors. C_3 also serves as part of a decoupling network in conjunction with R_4. C_5 is a bypass capacitor that is often redundant, if not undesirable.

the typical large-value capacitor will have enough series inductance to be ineffective at the higher frequencies.

An additional decoupling capacitor, C_5, is sometimes found in practical circuits. More often than not, this unit is redundant. Indeed, it may even cause severe problems. It will couple energy flowing in the ground around the amplifier into the power supply. If a similar amplifier exists elsewhere within a system, the two grounds are coupled together through the power supply, providing a stray signal path beyond that desired. Decoupling is much more effective when there is a series impedance for each capacitor.

C_4 is the emitter bypass element, providing grounding for the ce amplifier. The reactance should be small compared with r_e, the emitter resistance of the model. This is usually much smaller than R_3, the emitter bias resistor. The output resistance of an emitter follower was analyzed in the preceding section and justifies the requirement for the reactance of C_4. Emitter degeneration may be added to the amplifier of Fig. 1.11 without changing the biasing by inserting additional resistance in series with C_4.

Ground symbols appear in many places within the circuit of Fig. 1.11. All of them should be at the same potential. This is commonly realized in rf circuitry through connection to a ground foil on a printed circuit board. Long ground leads should be avoided within a single amplifier stage.

1.4 LARGE-SIGNAL OPERATION OF THE BIPOLAR TRANSISTOR

The models presented in previous sections have dealt with small signals applied to a bipolar transistor. While small-signal design is exceedingly powerful, it is not sufficient for many designs. Large signals must also be processed with transistors. Two significant questions must be considered with regard to transistor modeling. First, what is a reasonable limit to accurate application of small-signal methods? Second, what will the consequences be by exceeding these limits?

Consider an amplifier that is voltage driven such as that shown earlier in Fig. 1.2. The input voltage is the sum of a bias, V_0, and a sinusoidal signal with a peak amplitude of V_p. The emitter current is given by the Ebers-Moll model as

$$I_e = I_{es} \exp[(q/kT)(V_0 + V_p \sin \omega t)] \qquad (1.4\text{-}1)$$

where ω is the angular frequency of the sinusoidal signal. This reduces to

$$I_c = I_0 \exp[(V_p/26)\sin \omega t] \qquad (1.4\text{-}2)$$

where V_p is now in millivolts and I_0 is the dc bias current. The collector current is assumed equal to the emitter value.

The current will clearly vary in a complicated way, for the sinusoidal signal voltage is embedded within an exponential function. The previous Taylor series analysis (Sec. 1.2) predicts a sinusoidal current if the voltage is sufficiently low.

Sec. 1.4 Large-Signal Operation of the Bipolar Transistor

The current of Eq. 1.4-2 may be studied by normalizing the current to the peak value. This produces the relative current

$$I_r = \frac{\exp[(V_p/26)\sin \omega t]}{\exp(V_p/26)} \qquad (1.4\text{-}3)$$

I_r is plotted in Fig. 1.12 for V_p values of 1, 10, 30, 100, and 300 mV. The 1-mV case is very sinusoidal. Similarly, the $V_p = 10$-mV curve is generally sinusoidal with only minor distortions. The higher amplitude cases show increasing distortion. The curve for $V_p = 30$ mV oscillates from a mean relative current of 0.31 up to 1 at the positive peak and down to only 0.1 at the negative peak of the sinusoidal input voltage. The current is completely lacking in the symmetry required of a sinusoid.

Constant base voltage biasing is unusual. More often, a transistor is biased for a nearly constant emitter current. When such an amplifier is driven by a large input signal, the average bias voltage will adjust itself until the time average of the nonlinear current equals the previous constant bias current. Hence, it is vital to consider the average relative current of the waveforms of Fig. 1.12. This is evaluated with an integral

$$\bar{I}_r = \frac{1}{360} \frac{1}{\exp(V_p/26)} \int_0^{360} \exp[(V_p/26)(\sin \theta)] \, d\theta \qquad (1.4\text{-}4)$$

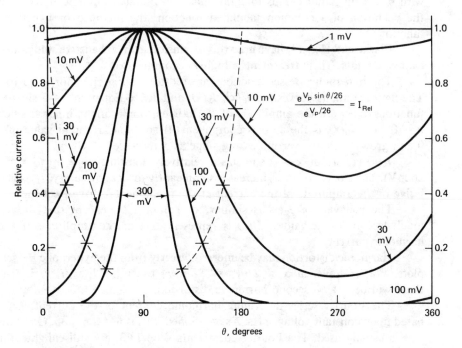

Figure 1.12 Normalized relative current of a bipolar transistor under sinusoidal drive at the base.

The independent variable has been changed to $\theta = \omega t$.

The average relative currents for the cases presented occur at the intersection of the curves with the dotted lines of Fig. 1.12. For example, the dotted curve intersects the $V_p = 300$-mV waveform at an average relative current of 0.12. If an amplifier was biased to a constant current of 1 mA, but was driven with a 300-mV signal, the positive peak current would reach a value greater than the average by a factor of $1/(0.12)$. The average current would remain at 1 mA, but the positive peak would be 8 mA. The transistor would not conduct for most of the cycle.

The analysis presented is formalized by noting that the current, although nonlinear, is periodic. It may thus be expanded in a Fourier series with a fundamental frequency ω. *The original equation is renormalized by replacing $V_p/26$ with a normalized amplitude, x.* The function becomes

$$\exp(x \cos \theta) = I_0(x) + I_1(x) \cos \theta + I_2(x) \cos 2\theta + \cdots \quad (1.4\text{-}5)$$

The cosine form has been used for convenience. The constants are evaluated with the integral

$$I_n(x) = \frac{1}{2\pi} \int_{-\pi}^{\pi} \exp[x \cos \theta] \cos n\theta \, d\theta \quad (1.4\text{-}6)$$

with θ now in radians. This integral must be evaluated numerically. However, it is the definition of a common tabulated function, the modified or hyperbolic Bessel function.

Clarke and Hess (4) have used this expansion in an exhaustive study of communications circuits. Their treatment is highly recommended.

The hyperbolic Bessel functions are plotted in Fig. 1.13 for x up to about 4. The figure contains curves for $I_0(x)$, $I_1(x)$, and $I_2(x)$. Equation 1.4-5 shows that the functions $I_n(x)$ are the amplitude of the nth harmonic of the input frequency. The $n = 0$ case, $I_0(x)$, is the dc bias component. It starts at an initial value of 1 at $x = 0$ and grows with an ever increasing slope as x increases.

$I_1(x)$ is zero at $x = 0$ and has a relatively constant slope up to $x = 1$ ($V_p = 26$ mV). The slope, however, increases significantly for x above unity. $I_1(x)$ is representative of the amplitude of the current at the fundamental drive frequency.

The behavior of $I_2(x)$ is similar to that of $I_1(x)$ except that the values are much less at low x values. $I_2(x)$ is representative of the amplitude of the second harmonic current.

Harmonic distortion may be inferred directly from the hyperbolic Bessel function plots. For example, at $x = 2$ ($V_p = 52$ mV) the ratio $I_2(x)/I_1(x)$ is 0.43. Thus, there will be a 43% second harmonic distortion.

The Fourier series describing harmonic content of the current, Eq. 1.4-5, is based upon constant voltage bias. As mentioned, this is not common. Constant current bias is usually used. The Fourier coefficients, Fig. 1.13, are still enlightening for the constant bias current. Consider a signal of 52 mV ($x = 2$). The value of $I_0(x)$ is

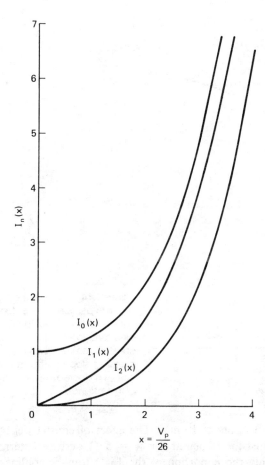

Figure 1.13 Hyperbolic Bessel functions of order 0, 1, and 2. They represent the changes in bias, $I_0(x)$, and the gain at the fundamental drive frequency, $I_1(x)$, of a bipolar transistor with constant base voltage bias. $I_2(x)$ is the magnitude of the second harmonic. The independent variable is $x = V_p/26$ where V_p is the peak sinusoidal amplitude at the base in millivolts.

then about 2.25. The current would have increased by this factor if constant voltage biasing had been used. The current is held constant though. Hence, the amplitude of the fundamental frequency component will be reduced by that factor, 2.25.

This calculation is continued for all values of x to produce Fig. 1.14. The curve is representative of the amplitude of the fundamental frequency component as a function of drive level for the case of constant current biasing. While the curve is increasing with x, it has a decreasing slope. Hence, the gain at the fundamental frequency is a decreasing function of input level. This is vital in determining the operating and limiting characteristics of sine wave oscillators, covered in detail in Chap. 7.

The previous curves have presented data based upon the simplest of large-signal models, the Ebers-Moll equation. Still, the simple model has yielded considerable information. In addition, the behavior of the amplifier circuit has been examined both for time-domain and frequency-domain behavior. Both viewpoints are vital and both are related. The engineer should always consider both during design.

The analysis suggests that a reasonable upper limit for accurate small-signal

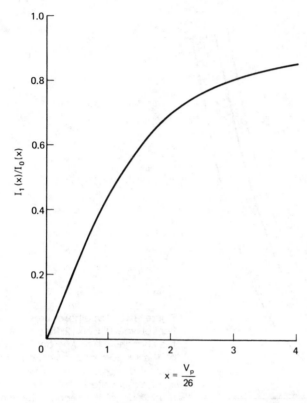

Figure 1.14 Ratio of the first and zero order hyperbolic Bessel functions. The curve is proportional to the current flow as a function of drive for conditions of constant emitter bias current. The decreasing slope with x shows how the gain will decrease with increased amplitude signals.

analysis is a peak base signal of about 10 mV. The effect of emitter degeneration is also evident. Assume a transistor is biased for $r_e = 5\ \Omega$ and an external emitter resistor of 10 Ω is used. Only the r_e portion of the 15 Ω total is nonlinear. Hence, this amplifier would tolerate a 30-mV signal while still being well described with a small-signal analysis.

1.5 GAIN, POWER, AND NEGATIVE FEEDBACK

Both voltage and current gain have been calculated in previous sections for a variety of circuits. These numbers are sufficient for most analog circuit design. The rf engineer needs more information though. Specifically, the power gain of a circuit is of more interest than as merely an algebraic ratio of voltages. In the broadest sense, power gain is the ratio of an output to an input power. There are many different kinds of power gain which will be presented in this section.

A generalized linear network is shown in Fig. 1.15. It is driven by a voltage source with characteristic resistance R_s. The network is represented as having an input resistance, R_{in}, and having an output voltage driving an internal output resist-

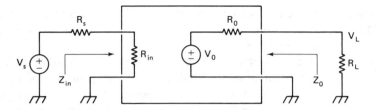

Figure 1.15 General form of a linear network, active or passive.

ance. The network is terminated in a load, R_L. This general form often describes an active circuit, one containing amplifying devices. It could just as well be a passive circuit such as a filter or attenuator. Complex impedances might replace the internal resistors, but the concepts are very general for any network so long as it is linear.

The power gain of the network of Fig. 1.15 is formally defined as the output power delivered to R_L divided by the power delivered to the input, R_i. For the circuit shown

$$G_p = \frac{V_L^2}{R_L} \frac{R_{\text{in}}}{V_{\text{in}}^2} \qquad (1.5\text{-}1)$$

The value of R_{in} may equal that of the source, R_s. This is not always true. If the two values are equal, the source is terminated in a matched impedance and the maximum available power from the source is being delivered to the load. The available generator power is $P_a = V_{\text{in}}^2/4R_s$. An often specified gain figure is the so-called *transducer gain*, G_t, defined by

$$G_t = \frac{P_L}{P_a} = \frac{V_L^2}{R_L} \frac{4R_s}{V_{\text{in}}^2} \qquad (1.5\text{-}2)$$

The values of G_t and G_p will be equal if the input is matched to the source. If the input is not matched, the power gain will exceed the transducer gain.

It is often useful to insert an amplifier in a matched transmission line with the purpose of increasing the signals traveling on the line. The load and source impedances are then equal. The gain measured is then a special case of transducer gain, and is called the insertion power gain.

Another gain parameter of great interest is the maximum available gain from a network, termed G_{max} or MAG. This occurs when both the input and the output of a network are terminated in matched resistances. G_{max} equals both the power gain and the transducer gain of the matched network.

Obtaining the G_{max} value and the terminations required to achieve it would be straightforward if the network was no more complicated than that shown in Fig. 1.15. It would occur merely by evaluating the amplifier in an arbitrary set of terminations and by calculating or measuring the input and output resistances. Matching terminations are then chosen and the procedure is repeated to obtain the value for MAG. If this is possible, the amplifier is said to be unilateral. This is rare in amplifiers used at radio frequencies.

The usual situation is less idealistic. The input resistance, R_{in}, will change as the output termination varies. Similarly, R_{out} is a function of R_s. Formal methods for obtaining the value of G_{max} and the required terminations are presented in Chap. 5. An example of a nonunilateral amplifier is presented later in this section. The emitter follower studied earlier was not unilateral.

The less-than-ideal, nonunilateral nature of most amplifiers has another consequence. When two such networks are cascaded, the net gain is not directly available from simplistic equations. The lack of match between the output of one stage and the input of the next will alter the potential gain. Moreover, the nonunilateral nature will cause change in the output termination to reflect completely through the cascaded chain, altering the net gain and input resistance. Again, the formal methods for calculating the gain of a cascade of two or more networks is presented in Chap. 5.

Gain is usually presented in dB. Formally, $G = 10 \log(P_L/P_{in})$. If the source and load terminations are equal, $G = 20 \log(V_L/V_{in})$. Logarithmic units should be used only with care when applied to a voltage gain.

A dB construct is often used to express powers in absolute units. The most common in rf applications is to express a power in dBm. This is a power in dB with respect to 1 mW.

Amplifier design would be very easy were it not for one phenomenon, feedback. It is feedback that makes an amplifier nonunilateral. It is also feedback, when applied with care, that allows the designer to fabricate amplifiers with desired characteristics that are not critically dependent upon device characteristics. There are four fundamental types of feedback in popular use

1. *Shunt-Voltage Feedback.* The output of a network is sampled to produce a current that is summed in shunt with the input of the network. This is the feedback that is usually used with operational amplifiers.
2. *Series-Current Feedback.* The output current is sampled to produce a voltage that is summed in series with the network input. Emitter degeneration is an example of this, for the emitter current is approximately equal to the collector current.
3. *Series-Voltage Feedback.* The output voltage is sampled with part applied as a voltage in series with the input. The emitter follower is an example.
4. *Shunt-Current Feedback.* The output current is sampled to derive a current which is presented in shunt with the input current.

Feedback will be illustrated with an example. Not only will it show how application of feedback alters the characteristics of the amplifier, but it will present a very practical circuit for rf applications through the microwave spectrum.

The practical embodiment of this circuit is shown in Fig. 1.16a. The amplifier

Sec. 1.5 Gain, Power, and Negative Feedback

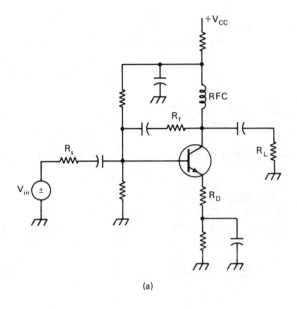

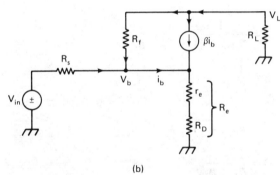

Figure 1.16 Practical amplifier for rf applications. Two forms of feedback are employed, emitter degeneration at R_D and shunt feedback at R_f. The practical circuit is shown at (a) and the small-signal equivalent is presented at (b).

is driven from a source of known internal resistance, R_s, and terminated in R_L. A radio-frequency choke in the collector allows bias to be supplied to the transistor without loading the output except by the desired elements. Emitter degeneration is introduced as R_d while shunt-voltage feedback is introduced with R_f. The small-signal equivalent of the circuit is shown in Fig. 1.16b. The emitter degeneration is combined with the internal transistor emitter resistance to form a net element, R_e. The simplest of transistor models is chosen, the controlled current generator with emitter resistance.

The base current is related to the base voltage by

$$I_b = \frac{V_b}{R_e(\beta + 1)} \tag{1.5-3}$$

Nodal equations may be written using this result. At the base

$$\frac{V_{\text{in}} - V_b}{R_s} + \frac{V_L - V_b}{R_f} = \frac{V_b}{R_e(\beta + 1)} \tag{1.5-4}$$

while at the collector

$$\frac{-V_L}{R_L} = \frac{\beta V_b}{R_e(\beta + 1)} + \frac{V_L - V_b}{R_f} \tag{1.5-5}$$

The equations are rearranged and solved simultaneously for V_b and V_L, yielding

$$V_b = \frac{V_{\text{in}} D / R_s}{BC + AD} \tag{1.5-6}$$

and

$$V_L = -V_b C / D \tag{1.5-7}$$

where

$$A = \frac{1}{R_e(\beta + 1)} + \frac{1}{R_f} + \frac{1}{R_s}$$

$$B = \frac{1}{R_f} \tag{1.5-8}$$

$$C = \frac{\beta}{R_e(\beta + 1)} - \frac{1}{R_f}$$

$$D = \frac{1}{R_f} + \frac{1}{R_L}$$

The amplifier input current is readily calculated as $I_{\text{in}} = (V_s - V_b)/R_s$. This is combined with V_b to yield the input resistance

$$R_{\text{in}} = \frac{R_s}{\left(\dfrac{V_s}{V_b} - 1\right)} \tag{1.5-9}$$

The output resistance may be evaluated using a number of methods. Either a voltage or current source could be used to drive the output with the input voltage source short circuited. This would produce nodal equations that must again be solved. An alternative method is used here for illustration.

Note the general linear network of Fig. 1.15. The output is a voltage source with a series output resistance, R_o. The output resistance may be inferred if two different values are used for R_L and the corresponding V_L voltages are calculated (or measured). Specifically, if the normal termination, R_L, leads to an output V_L and a larger one, $2R_L$, produces V'_L, the output resistance is

$$R_o = \frac{2R_L(V'_L - V_L)}{(2V_L - V'_L)} \tag{1.5-10}$$

This method is messy if a closed-form solution is desired. However, it is easily implemented if numerical calculations are being performed with, for example, a handheld programmable calculator.

This method is much more general and powerful than it might appear. We have managed to infer an output resistance from knowledge of the input voltage, V_{in}, and the resulting output, V_L. This was done by perturbing a terminating impedance. Generalizing, the input and output impedances are

$$Z_i = \frac{(Z_s A_v)|_{Z_s \to \infty}}{A_v|_{Z_s \to 0}} \tag{1.5-11}$$

$$Z_o = \frac{A_v|_{Z_L \to \infty}}{(A_v/Z_L)|_{Z_L \to 0}} \tag{1.5-12}$$

where $A_v = V_L/V_s$, the voltage gain from the unloaded source to the output. Limits are evaluated with respect to the source or load impedances. Note that impedances are used in these theorems rather than resistances. This emphasizes that the theorems apply for complex terminations and complex input or output impedances. These theorems were first presented to the writer by Dr. R. D. Middlebrook (5). The equations may be derived with the use of the *ABCD* matrix of two-port network theory.

The power gains and the input and output resistances of the feedback amplifier (Fig. 1.16) may now be evaluated. A practical example is used. The transistor is biased to $I_e = 20$ mA, leading to $r_e = 1.3$ Ω. R_d, the external emitter degeneration, is set to 8.7 Ω, causing the total emitter resistance, R_e, to be 10 Ω. The feedback resistor is set at 250 Ω and the amplifier is terminated in 50 Ω at both input and output. The transistor is assumed to have a current gain of 30.

A value of $2V$ is used for V_{in}. This would be unrealistically large in practice, but is still suitable for calculations with the small-signal model. The output voltage is $-3.622 V$ with the values assumed. The minus sign indicates that the amplifier is inverting. The transducer gain is 11.18 dB while the power gain is larger by 0.02 dB. The input resistance is 44.1 Ω. The closeness to the 50-Ω source value accounts for the small difference between G_p and G_t. The output resistance, calculated numerically from Eq. 1.5-10, is 56.7 Ω, again a reasonable match to the 50-Ω load.

The effects of feedback are seen by allowing some parameters to vary. Figure 1.17 shows the variation in the output resistance with changes in driving resistance, R_s. A similar inverse relationship is seen if the load is varied while examining the

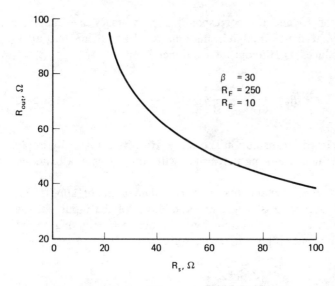

Figure 1.17 Variation of output resistance as a function of input termination for the multiple feedback amplifier of Fig. 1.16.

input resistance. The variation in transducer gain is less than 1 dB for the range of R_s used in Fig. 1.17.

Changes in operating frequency may be evaluated by changing β. Recall that a scalar approximation to the variation in current gain with frequency is $\beta = F_T/F$. This is shown in Fig. 1.18. Note that at low frequency (high β) the gain is less dependent upon β. This is consistent with a traditional analysis of an amplifier with negative feedback where gain is determined by the feedback elements if the

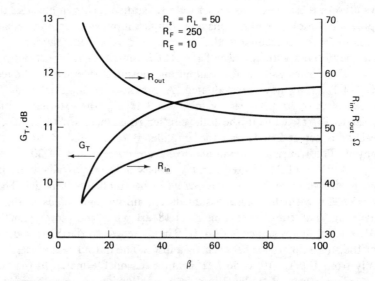

Figure 1.18 Variation of G_t and input and output resistance with current gain for the feedback amplifier. Frequency dependence is inferred from $\beta \simeq F_t/F$.

gain without feedback is high with respect to that with feedback. The impedance match is also much improved with high β.

Single-stage feedback amplifiers will generally be well matched if the net emitter resistance is related to the shunt feedback resistor by $R_e R_f = R_s R_L$. This constraint is applied to plot performance as a function of the feedback resistors in Fig. 1.19.

The model used has been the simplest possible. Still, the results are remarkably accurate for frequencies below about 10% of the transistor F_T.

The preceding analysis described small-signal operating conditions of a specific amplifier, one with two types of feedback. Large-signal operation is also of interest in a practical amplifier. A vital question to ask is what the maximum power output might be. A typical circuit might be that of Fig. 1.16a, perhaps with a minimum of feedback.

Two possible mechanisms will limit the output power depending upon the load resistor and bias conditions. Voltage limiting dominates if the load resistance is high while current limiting dominates if the collector load is small.

Assume R_L is large. Then, the collector voltage may vary from the quiescent value down to a point where $V_c = V_e$. The transistor is then saturated, preventing further voltage swing. Additional input drive will cause a severe departure from linearity. The positive going collector voltage will be equal for a more or less linear operation. The peak-to-peak voltage swing is then $2V_{ce}$, leading to a maximum output power of $V_{ce}^2/2R_L$.

Assume now that the load resistor is relatively low. Current limiting will domi-

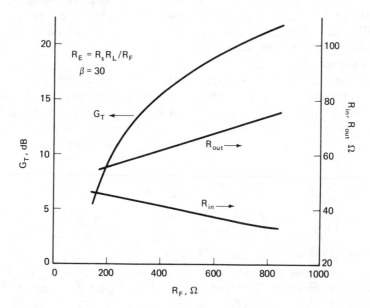

Figure 1.19 Variation in transducer gain, G_t, and input and output resistance with changes in feedback elements for the amplifier of Fig. 1.16.

nate. The collector current may drop from the bias value only down to zero. Further excursion (additional input drive energy) will cause severe nonlinearity. A reasonable peak-to-peak current is then $2I_0$ where I_0 is the quiescent bias value. The corresponding maximum power output is then $\frac{1}{2}I_0^2 R_L$. This power is typical of that level where 1 dB of gain compression occurs with high frequency amplifiers loaded with a relatively low impedance.

The load resistance required to achieve the maximum output power occurs when both voltage and current limiting are present. The load is found by equating the two previously calculated powers, leading to $R_L = V_{ce}/I_0$. Noting that V_{ce} is the peak ac voltage and I_0 is the peak ac current, the maximum output power is then $V_{ce}I_0/2$. The maximum efficiency of a power amplifier is then 50%. Class A operation denotes conduction during the entire drive cycle.

The resistance for maximum output power will contain both the external load and feedback elements. A shunt feedback element will appear in parallel with the external load while an emitter degeneration resistor appears in series. Note that the load required for maximum output power is not related to R_{out}, the value for R_L required for maximum gain.

1.6 MODELING THE FIELD-EFFECT TRANSISTOR

An often used device in rf applications is the field-effect transistor (FET). There are many types, with the JFET being the most common. Metal Oxide Silicon FETs (MOSFETs) including the newer power devices, the so-called vertical MOSFET or VFET, are of increasing utility. Of special interest at microwave frequencies are MOSFETs built from galium-arsenide, usually called GASFETs. The discussion in this section will generally be confined to the JFET with the understanding that other FET types are similar. A JFET symbol was shown in Fig. 1.1.

The bipolar transistor was viewed both as a voltage and as a current controlled device. The JFET, however, is purely a voltage controlled element, at least at low frequencies. The input gate is usually a reverse biased diode junction with virtually no current flow. The drain current is related to the voltage of the source with respect to the gate by

$$I_D = I_{DSS}(1 - V_{sg}/V_p)^2 \qquad 0 \leq V_{sg} \leq V_p$$
$$I_D = 0 \qquad V_{sg} > V_p \qquad (1.6\text{-}1)$$

where I_{DSS} is the drain saturation current and V_p is the pinch-off voltage. Operation is not defined for V_{sg} below zero, for the gate diode is then forward biased. Equation 1.6-1 is a reasonable approximation as long as the drain bias voltage exceeds the magnitude of the pinch-off.

Two virtually identical amplifiers using N-channel JFETs are shown in Fig. 1.20. The two circuits illustrate the two popular methods for biasing the JFET. Fixed

Sec. 1.6   Modeling the Field-Effect Transistor

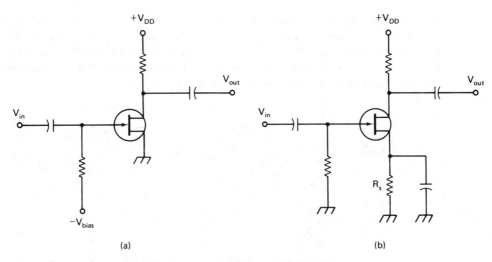

Figure 1.20 Biasing schemes for a common source JFET amplifier.

gate voltage bias, Fig. 1.20a, is feasible for the JFET because of the favorable temperature characteristics. As the temperature of the usual FET increases, the current will decrease, avoiding the thermal runaway problem of bipolar transistors.

A known source resistor, R_s, in Fig. 1.20b, will lead to a known source voltage. This is obtained from a solution of Eq. 1.6-1

$$V_{sg} = \frac{\left[\frac{1}{R_s I_{DSS}} + \frac{2}{V_p}\right] - \left[\left(\frac{1}{R_s I_{DSS}} + \frac{2}{V_p}\right)^2 - \left(\frac{2}{V_p}\right)^2\right]^{1/2}}{\frac{2}{V_p^2}} \qquad (1.6\text{-}2)$$

The drain current is then obtained by direct substitution.

Alternatively, a desired drain current less than I_{DSS} may be achieved with a proper choice of source resistor

$$R_s = \frac{V_p \cdot (1 - \sqrt{I_D/I_{DSS}})}{I_D} \qquad (1.6\text{-}3)$$

The small-signal transconductance of the JFET is obtained by differentiating Eq. 1.6-1

$$g_m = \frac{dI_D}{dV_{sg}} = \frac{-2I_{DSS}}{V_p}(1 - V_{sg}/V_p) \qquad (1.6\text{-}4)$$

A minus sign appears, for the equation describes a common-gate configuration. The amplifiers of Fig. 1.20 are both common-source types and are described by Eq. 1.6-

4 except that g_m is now positive. Small-signal models for the JFET are shown in Fig. 1.21. The simple model is that inferred from the equations while the model of Fig. 1.21b contains capacitive elements which are effective in describing high frequency behavior. Like the bipolar transistor, the JFET model will grow in complexity as more sophisticated applications are encountered.

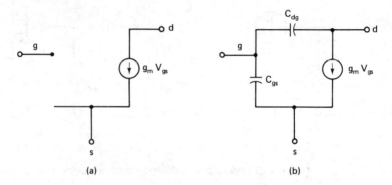

Figure 1.21 Small-signal models for the JFET. (a) is useful at low frequency, (b) is a modification to approximate high frequency behavior.

Large-signal JFET operation is examined by normalizing the previous equation to $V_p = 1$ and $I_{DSS} = 1$, yielding

$$i = (1 - v)^2 \tag{1.6-5}$$

A sinusoidal signal is injected at the source of the form

$$v = v_0 + v_1 \sin \theta \tag{1.6-6}$$

where $\theta = \omega t$. The circuit is shown in Fig. 1.22. Also shown in that figure are examples for a variety of bias and sinusoid amplitude conditions. The main feature is the asymmetry of the curves. The positive portions of the oscillations are further from the mean than are the negative excursions. This is especially dramatic when the bias, v_0, is large, placing the quiescent point close to pinch-off. With such bias and high amplitude drive, conduction occurs only over a small fraction of the total input waveform period.

Consider the average current for these operating conditions. This is given by

$$\bar{i} = \frac{1}{2\pi} \int_0^{2\pi} i(\theta) \, d\theta = \frac{1}{2\pi} \int (1 - 2v_0 - v_1 \sin \theta) \, d\theta \tag{1.6-7}$$

The second integral contains no limits of integration. The integral must be broken into two parts, corresponding to the two regions defined by Eq. 1.6-1, conduction and beyond pinch-off. The voltages are constrained to never cause the source-gate

Sec. 1.6 Modeling the Field-Effect Transistor

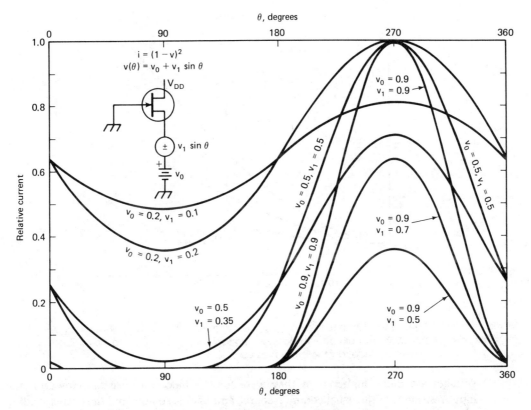

Figure 1.22 Relative normalized drain current for a JFET with constant voltage bias and sinusoidal signals. Relatively "clean" waveforms exist for low signals while large input amplitudes cause severe distortion.

diode to be forward biased. The integral is easily evaluated directly without resorting to numerical methods.

The average current values obtained from Eq. 1.6-7 may be further normalized by dividing by the corresponding dc bias current, $i_0 = (1 - v_0)^2$. The results are shown in Fig. 1.23. The curves show that the average current will increase as the amplitude of the drive increases. This, again, is most pronounced when the FET is biased close to pinch-off.

Although practical for the JFET, the constant voltage operation depicted for the previous curves is not usual. Instead, a resistive bias is usually employed, Fig. 1.20b. With this form of bias, the increased current from high signal drive will cause the voltage drop across the bias resistor to increase. This will then move the quiescent operating level closer to pinch-off, accompanied by a reduced small-signal transconductance. This behavior is vital in describing the limiting found in FET oscillators.

The limits on small-signal operation are not as well defined for a FET as they were for the bipolar transistor. Generally, a maximum voltage of 50 to 100 mV is allowed at the input (normalized to a 1-V pinch-off) without severe distortion. The

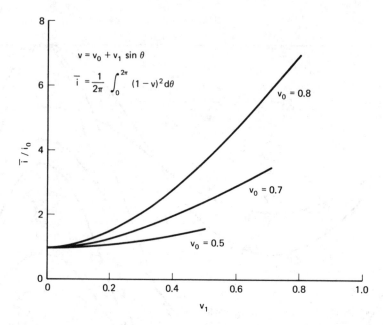

Figure 1.23 Change in average current of a JFET with increasing input signals. The average current with no input signals is i_o, while v_1 is the normalized drive amplitude, and v_0 is the bias voltage.

voltages are much higher than they were for the bipolar transistor. However, the input resistance of the usual common source amplifier is so high and the corresponding transconductance low enough that the available gain is no greater than could be obtained with the bipolar transistor. The distortions are generally lower with FETs, owing to the lack of high order curvature in the defining equations.

Many of the standard circuits used with bipolar transistors are also practical with the FET. This includes the feedback amplifier presented in the previous section. Noting that the transconductance of the bipolar transistor is $g_m = I_e(\text{dc,mA})/26$, the previous equations may be applied directly. The "emitter current" is picked to correspond to the FET transconductance. A very large value is used for current gain. The same calculator or computer program is then used directly. It is generally found in practice that much higher terminating impedances must be used to obtain transducer gain values similar to those of the bipolar transistor.

REFERENCES

1. Ebers, J. J., and Moll, J. L., "Large-Signal Behavior of Junction Transistors," *Proc. IRE,* **42,** pp. 1761–1772, December 1954.
2. Getreu, Ian, *Modeling the Bipolar Transistor,* Elsevier, New York, 1979. Also available from Tektronix, Inc., Beaverton, Oregon, in paperback form. Must be ordered as Part Number 062-2841-00.

3. Searle, C. L., Boothroyd, A. R., Angelo, E. J. Jr., Gray, P. E., and Pederson, D. O., *Elementary Circuit Properties of Transistors,* Semiconductor Electronics Education Committee, Vol. 3, John Wiley & Sons, New York, 1964.
4. Clarke, Kenneth K. and Hess, Donald T., *Communication Circuits: Analysis and Design,* Addison-Wesley, Reading, Mass., 1971.
5. Middlebrook, R. D., *Design-Oriented Circuit Analysis and Measurement Techniques.* Copyrighted notes for a short course presented by Dr. Middlebrook in 1978.

SUGGESTED ADDITIONAL READINGS

1. *The Radio Amateur's Handbook,* American Radio Relay League, Newington, Conn., 1980 or later edition (published annually).
2. Alley, Charles L. and Atwood, Kenneth W., *Electronic Engineering,* John Wiley & Sons, New York, 1962.
3. Terman, Frederick Emmons, *Electronic and Radio Engineering,* McGraw-Hill, New York, 1955.
4. Gilbert, Barrie, "A New Wide-Band Amplifier Technique," *IEEE Journal of Solid-State Circuits,* **SC-3**, *4,* pp. 353–365, December 1968.
5. Egenstafer, Frank L., "Design Curves Simplify Amplifier Analysis," *Electronics,* pp. 62–67, August 2, 1971.

NEW REFERENCES

The following books were published after the first printing of *Introduction to RF Design.* They are offered here as being especially useful to the student of RF methods.

1. Paul Horowitz and Winfield Hill, *The Art of Electronics,* Second Edition, Cambridge University Press, 1989. This is a wonderful general treatment of the field; it is especially good for a program of self study.
2. William Sabin and Edgar Schoenike (editors), and members of the Engineering Staff, Collins Divisions of Rockwell International Corp., *Single-Sideband Systems and Circuits,* McGraw-Hill, 1987.
3. Guillermo Gonzalez, *Microwave Transistor Amplifiers: Analysis and Design,* Prentice-Hall, 1984.
4. Ulrich Rohde and T. T. N. Bucher, *Communications Receivers: Principles and Design,* McGraw-Hill, 1988.

2
Filter Basics

This chapter is devoted to fundamental filter concepts. A filter is a network, passive or active, which passes a desired set of frequencies, the passband, with a minimum of attenuation. Other frequencies, the stopband, are attenuated. Although the fundamental concepts presented are aimed primarily at the analysis and design of *LC* filters, they may be applied equally well to virtually any network. For example, the hybrid-pi bipolar transistor presented in the preceding chapter predicts a low-pass-filter–like response. The gain at low frequencies is high, but decreases with increasing frequency.

The early sections of this chapter are introductory and are intended primarily as a review. These sections are, nonetheless, very important, for they emphasize the dual methods used for network analysis. Either time or frequency may be used as the independent controlling variable. Once an analysis has been done with respect to frequency (in the frequency domain), the circuit performance with respect to time (in the time domain) is also determined. The two domains are related by Fourier or Laplace transforms.

Filters are traditionally characterized in the frequency domain. That is, they may be low pass or bandpass circuits. Time domain behavior has been analyzed, but is often ignored in the design process. It is becoming increasingly important to design filters with a desired time domain characteristic in addition to a suitable frequency response. This is especially critical in modern digital communications systems and in rf instrumentation.

The later part of this chapter deals primarily with low pass filters. The methods presented are extended to the design of simple high pass and bandpass networks.

Chapter 3 is an extension which treats a filter of special interest to the rf designer, the coupled resonator bandpass.

2.1 NETWORK ANALYSIS IN THE TIME DOMAIN

The traditional method for network analysis is with differential equations with time as the independent variable. This is illustrated in this section with a number of simple circuits. The reader is assumed to have a background in differential equations and a familiarity with ac circuit theory. Some texts written specifically for the study of electronic networks will supply both backgrounds. Especially recommended are the fundamental volumes by Van Valkenburg (1) and Skilling (2). The more advanced reader is directed toward the outstanding text by Blinchikoff and Zverev (3).

The first circuit considered is a constant voltage applied to an RC network, Fig. 2.1. The switch is assumed closed until $t = 0$ when it is opened. The charge on the capacitor is related to the voltage by $Q = CV$. But current is the time rate of change of charge. Hence

$$V = \frac{1}{C} \int I \, dt \qquad (2.1\text{-}1)$$

We use Kirchhoff's voltage law to write an equation to describe the total voltage drop in the circuit

$$V = IR + \frac{1}{C} \int I \, dt \qquad (2.1\text{-}2)$$

The equation is differentiated and divided by the resistance

$$\frac{dI}{dt} + \frac{1}{RC} I = 0 \qquad (2.1\text{-}3)$$

The equation is multiplied by an integrating factor, $e^{t/\tau}$, where $\tau = RC$,

$$\frac{dI}{dt} e^{t/\tau} + \frac{I}{\tau} e^{t/\tau} = 0 \qquad (2.1\text{-}4)$$

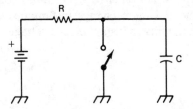

Figure 2.1 Simple resistor-capacitor circuit with a dc drive.

Recalling the rule for differentiation of a product, this reduces to

$$\frac{d}{dt}(Ie^{t/\tau}) = 0 \qquad (2.1\text{-}5)$$

Direct integrations yields

$$Ie^{t/\tau} = K \qquad (2.1\text{-}6)$$

where K is a constant of integration, evaluated from initial conditions. The current prior to opening the switch was V/R. It must be the same just after the switch is opened, for the voltage across the capacitor cannot change instantaneously. The final solution is the familiar decreasing exponential

$$I(t) = \frac{V}{R} e^{-t/\tau} \qquad (2.1\text{-}7)$$

The capacitor voltage is obtained from direct application of Kirchhoff's voltage law.

A similar equation will describe application of a dc voltage to an LR circuit, as shown in Fig. 2.2. The circuit is assumed to be initially in a relaxed condition with no current flowing. The switch is changed at $t = 0$. The differential equation is written directly, noting that the voltage drop across L is $L(dI/dt)$

$$V = L\frac{dI}{dt} + IR \qquad (2.1\text{-}8)$$

This is divided by L and multiplied by the integrating factor, $e^{t/\tau}$, to yield

$$\frac{V}{L} e^{t/\tau} = \frac{dI}{dt} e^{t/\tau} + \frac{I}{\tau} e^{t/\tau} \qquad (2.1\text{-}9)$$

where $\tau = L/R$. Integration and subsequent division by the integrating factor produces

$$I = \frac{V}{R} + Ke^{-t/\tau} \qquad (2.1\text{-}10)$$

The constant of integration, K, is evaluated from initial conditions. $I = 0$ at $t = 0$, for the current in an inductor cannot change instantaneously. The inductor is the dual of the capacitor which inhibits an instantaneous voltage change. The final solution is

$$I(t) = \frac{V}{R}(1 - e^{-t/\tau}) \qquad (2.1\text{-}11)$$

Sec. 2.1 Network Analysis in the Time Domain

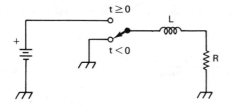

Figure 2.2 Inductor-resistor circuit with dc step input.

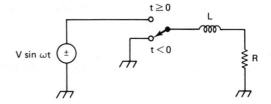

Figure 2.3 Inductor-resistor network with a sinusoidal input.

A more interesting circuit for the rf engineer is one where the dc voltage is replaced with a sinusoidal source. This is shown in Fig. 2.3 and is described by

$$V \sin \omega t = L \frac{dI}{dt} + IR \qquad (2.1\text{-}12)$$

Application of an integrating factor produces

$$\frac{d}{dt}(Ie^{t/\tau}) = \frac{V}{L} e^{t/\tau} \sin \omega t \qquad (2.1\text{-}13)$$

Integration and manipulation yield

$$I(t) = \frac{\frac{V}{L}\frac{1}{\tau}\sin \omega t - \omega \cos \omega t}{\frac{1}{\tau^2} + \omega^2} + Ke^{-t/\tau} \qquad (2.1\text{-}14)$$

The second term vanishes in time and is ignored in this analysis. The first term is simplified by the substitutions

$$\frac{1}{\tau} = K' \cos \phi \qquad (2.1\text{-}15)$$

$$-\omega = K' \sin \phi$$

producing a familiar trigonometric form

$$K' \cos \phi \sin \omega t + K' \sin \phi \cos \omega t = K' \sin(\omega t + \phi) \qquad (2.1\text{-}16)$$

The angle, ϕ, is evaluated as

$$\tan \phi = \frac{\sin \phi}{\cos \phi} = \frac{-\omega/K'}{1/(K'\tau)} = \frac{-\omega L}{R} \qquad (2.1\text{-}17)$$

K' is evaluated with the identity $\sin^2 \phi + \cos^2 \phi = 1$. The final steady state current is

$$I_{ss}(t) = \frac{V \sin(\omega t + \phi)}{(\omega^2 L^2 + R^2)^{1/2}} \qquad (2.1\text{-}18)$$

These results are interpreted by allowing individual components to become small. As L vanishes, I_{ss} becomes the ratio of the ac voltage to the resistance, Ohm's law. Current is in phase with the driving voltage source. If the resistance vanishes, I_{ss} becomes the driving current and is divided by ωL. The tangent of the phase angle becomes infinite, indicating a 90 degree phase difference between the voltage and the current. Intermediate values of L and R produce phase differences between 0 and 90 degrees. The denominator of Eq. 2.1-18 is termed the impedance of the network.

Equation 2.1-18 is termed a transfer function in the time domain if V is set to unity. It is a transfer admittance here, for it is the output current, $I_{ss}(t)$, per unit input voltage. The related output voltage across R of Fig. 2.3 is easily calculated. This results in a voltage transfer function if $V = 1$.

A similar result will be obtained if the RC network (Fig. 2.1) is driven from a sinusoidal voltage source. The inductive term, ωL, is replaced with $1/(\omega C)$ and the phase angle has the opposite sign.

The steady state ac current is the driving voltage divided by a suitable impedance, a characteristic of the network and not of the driving signal.

These methods are easily extended to more complicated networks with many more components. The differential equations become predictably more complicated although the methods for solution are similar.

2.2 NETWORK ANALYSIS IN THE FREQUENCY DOMAIN

Previous circuits were analyzed with time as the independent variable. This traditional approach is reasonable. Length, mass, and time are the three basic units of physics from which all others are derived. Time forms a natural basis for circuit description by differential equations.

A duality is observed when many circuits are analyzed. The frequency dependence is often mathematically similar to the time dependence. It is then natural to examine methods that will emphasize the frequency dependence of a function of time.

Numerous techniques exist to assess the frequency content of a time function. Any periodic function may be expanded in a Fourier series. An example was considered in the previous chapter.

An arbitrary function of time, not necessarily periodic, is transformed to a function only of frequency by the Fourier transform

$$F(\omega) = \int_{-\infty}^{\infty} F(t) e^{-j\omega t} dt \qquad (2.2\text{-}1)$$

The integration removes all time dependence from the result.

A function of frequency is similarly transformed into a time dependent one with the inverse Fourier transform

$$F(t) = \frac{1}{2\pi} \int_{-\infty}^{\infty} F(\omega) e^{j\omega t} \, d\omega \qquad (2.2\text{-}2)$$

These transforms are very useful for many electronics applications. However, they are not well suited to the solution of differential equations. The transforms do not exist for all functions. A better transformation is sought.

Section 2.1 gave the results of RC and RL networks when driven by a dc source. More explicitly, the drive was the unit step function, $u_{-1}(t)$, formally 0 for $t < 0$ and 1 for $t \geq 0$. The results for the networks considered were decreasing exponential time functions. Any transformation that might be picked for analysis should apply well to the unit step. The transformation should, ideally, have physical significance for any arbitrary driving function including a sinusoid.

The desired transformation is the Laplace transform, defined as

$$F(s) = \mathscr{L} f(t) = \int_{0}^{\infty} f(t) e^{-st} \, dt \qquad (2.2\text{-}3)$$

where $f(t)$ is the time domain function and e^{-st} is the multiplying factor. s is the so-called Laplace frequency, and is defined by

$$s = \sigma + j\omega \qquad (2.2\text{-}4)$$

This frequency can be real, imaginary, or complex. Many functions that are not integrable alone, or are not Fourier transformable, have Laplace transforms. This is a result of the $e^{-\sigma t}$ component of e^{-st} which forces the integral to converge at large t. Note that the Laplace transform is dependent on the values of the time function only at times in the future. Earlier history is of no significance as it was for the Fourier transform. Capital letters are used to represent the transformed functions. Hence, $F(s)$ is the transformed, frequency domain version of the time domain function $f(t)$.

There are many details of the Laplace transformation that are both of scholarly interest and vital to its application. These are found in many texts including those referenced earlier. They are not presented here, for our goal is to gain intuition about a specific application, filter design.

The greatest virtue of the Laplace transformation to the electronic designer is the simplicity it affords. It will allow us to study problems directly in the frequency domain. Analysis will be through straightforward algebraic manipulation to yield $F(s)$, the frequency domain solution. The corresponding time function, $f(t)$, is found from the inverse transformation, or from a search of tables of Laplace transforms.

There are two functions of special significance in interpretation of $F(s)$. One is the unit impulse, or Dirac delta function, $\delta(t - x)$. It is zero for all values of time except $t = x$ where it is unbounded. However, it is infinite in a special way such that its integral is unity. In the strictest sense, the unit impulse has significance only when it appears within an integral. As such, it is often termed a functional (4). The Laplace transform of the unit impulse is unity, shown by direct integration.

The other vital function is the unit step, $u_{-1}(t)$, defined earlier. The Laplace transform of the unit step is $1/s$.

The key to application of the Laplace transform in electronics lies in the frequency domain transfer function, $H(s)$. Time domain transfer functions were discussed in the previous section for the case of sinusoidal excitation.

$H(s)$ is written directly in the frequency domain for practical networks through application of Kirchhoff's laws. The impedance of an inductor was shown earlier to be $j\omega L$. In the frequency domain, the impedance is $Z_L(s) = sL$. Similarly, the impedance of a capacitor is $Z_C(s) = 1/sC$. A resistor has an impedance of R in both domains. Examples of $H(s)$ will be presented below.

The casual designer will often use s, the complex frequency, merely as a shorthand notation for $j\omega$. This is valid to the extent that it provides proper results. It is, however, a severely restricted interpretation of Laplace methods. A more general and significantly more profound result is

$$F_{\text{out}}(s) = H(s) F_{\text{in}}(s) \qquad (2.2\text{-}5)$$

Once $H(s)$ is known, the frequency domain output is known for any arbitrary input function, $F_{\text{in}}(s)$, so long as that input function is Laplace transformable.

Recall that the transform of the impulse function was unity. $H(s)$ is then the frequency domain impulse response. Similarly, $u_{-1}(t) = 1/s$. Hence, $(1/s)H(s)$ is the frequency domain response to a step function. Either result is obtained in the time domain with an inverse transformation of the frequency domain calculation.

The response to a sinusoidal excitation is obtained by a direct substitution of $j\omega$ for s in $H(s)$. The result is called the frequency response of the network.

The simplicity and elegance of the Laplace methods are illustrated with an example, the LR circuit analyzed earlier with traditional methods. The circuit is repeated in Fig. 2.4, now with an arbitrary input driving function. The frequency domain impedance of the network is $R + sL$. Hence, the current transfer function is

$$H(s) = \frac{1}{R + sL} \qquad (2.2\text{-}6)$$

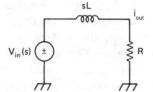

Figure 2.4 Inductor–resistor network with an arbitrary input voltage.

Sec. 2.3 Poles, Zeros, and the Series Tuned Circuit

The frequency response is evaluated initially with direct substitution of $j\omega$ for s

$$H(j\omega) = \frac{1}{R + j\omega L} \quad (2.2\text{-}7)$$

If the input is a voltage at a frequency ω, $V \sin \omega t$, the time domain current is

$$I(t) = \frac{V \sin \omega t}{R + j\omega L} \quad (2.2\text{-}8)$$

or

$$I_{\text{out}}(t) = \frac{V \sin (\omega t + \phi)}{(R^2 + \omega^2 L^2)^{1/2}} \quad (2.2\text{-}9)$$

where $\phi = \tan^{-1}(\omega L/R)$. This is identical to the previous steady state response.

The impulse response is evaluated by obtaining the inverse transform of $H(s)$. A tabulation of Laplace transforms shows that the corresponding time domain function is $e^{-t/\tau}$ where $\tau = L/R$.

Dividing $H(s)$ by s yields the response to a unit step input. This is transformed, again from perusal of a tabulation of Laplace transforms, to

$$I_{\text{out}}(t) = \frac{1}{R}(1 - e^{-t/\tau}) \quad (2.2\text{-}10)$$

Again, this is identical with the earlier result with a driving voltage of $V = 1$. The analysis is easier with Laplace methods and provides considerably more information.

2.3 POLES, ZEROS, AND THE SERIES TUNED CIRCUIT

The next example is the familiar series resonant circuit. This is of importance unto itself and will further illustrate the utility of the Laplace method. The concept of complex frequency will be presented in more detail in connection with the circuit.

The series tuned circuit is shown in Fig. 2.5. The total impedance of the network

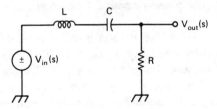

Figure 2.5 Series tuned circuit with arbitrary frequency domain drive.

is $sL + 1/(sC) + R$. Noting the voltage divider action, the frequency domain voltage transfer function is

$$H(s) = \frac{R}{sL + \frac{1}{sC} + R} = \frac{sCR}{s^2LC + 1 + sCR} \qquad (2.3\text{-}1)$$

The usual substitutions are defined

$$\omega_0^2 = \frac{1}{LC} \qquad \frac{\omega_0 L}{R} = Q \qquad (2.3\text{-}2)$$

$H(s)$ then becomes

$$H(s) = \frac{s/(\omega_0 Q)}{s^2/\omega_0^2 + 1 + s/(\omega_0 Q)} \qquad (2.3\text{-}3)$$

The steady state frequency response is first evaluated. Replacement of s by $j\omega$ yields

$$H(j\omega) = \frac{j\omega/(\omega_0 Q)}{1 - (\omega/\omega_0)^2 + j\omega/(\omega_0 Q)} \qquad (2.3\text{-}4)$$

Consider the limiting cases. The denominator of $H(j\omega)$ approaches unity at very low frequencies, while the numerator vanishes. Hence, the frequency response goes toward zero. At $\omega = \omega_0$, $H(j\omega)$ reduces to unity. The response again starts to decrease as the frequency continues to increase, going to zero at very high frequencies. This may be confirmed through detailed calculation, but is generally clear from inspection. The numerator and denominator both increase with frequency. However, the quadratic term in the denominator causes it to dominate.

The inductive reactance equals the reactance of the capacitor at a radial frequency of $\omega = \omega_0$, and f_0 is defined as the resonant frequency where $\omega_0 = 2\pi f_0$. The transfer function is maximum at this frequency because the impedances of the inductor and capacitor, while equal in magnitude, have opposite signs.

The parameter Q was introduced in Eq. 2.3-2. The effect of various Q values is shown in Fig. 2.6. Figure 2.6a shows the relative amplitude plotted on a logarithmic scale while Fig. 2.6b shows the phase response. Curves are presented for $Q = \frac{1}{2}$, 2, and 6.

A number of features of Fig. 2.6 deserve further comment. The curves are not symmetrical about the center frequency, $\omega/\omega_0 = 1$. (They would be symmetrical if the method of graphing had been changed.) The phase changes more rapidly near resonance with higher Q. The phase is $+$ and -45 degrees at the points where the amplitude response is down by 3 dB. Similar results may be obtained for other phases

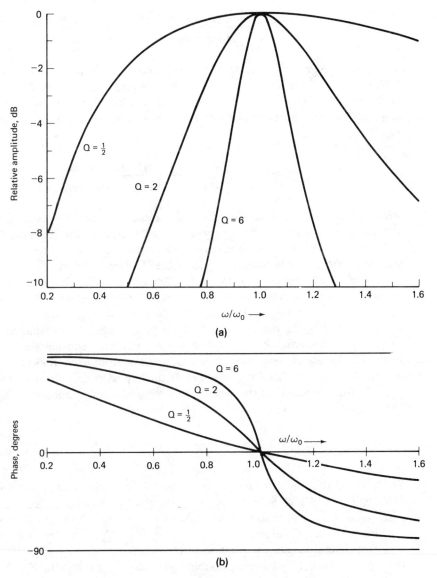

Figure 2.6 Frequency response (a) and phase response (b) for the series tuned circuit with $Q = \frac{1}{2}$, 2, and 6.

and attenuations. This is an example of a more general observation that phase and amplitude responses are not independent of each other. Not all networks have this characteristic, but many do.

Returning to the transfer function, $H(s)$, Eq. 2.3-1, we see that the Laplace frequency s appears in both the numerator and denominator. Generally, the transfer function of an arbitrary linear network is of the form

$$H(s) = \frac{P(s)}{Q(s)} = \frac{b_0 + b_1 s + \cdots + b_m s^m}{a_0 + a_1 s + \cdots + a_m s^m} \qquad (2.3\text{-}5)$$

The transfer function is a ratio of two polynomials in s, the Laplace frequency. The highest power of the denominator polynomial equals the number of reactive elements in the network. The numerator polynomial will have an order that is less than or equal to that of the denominator with the former being more common. The polynomials may be factored to

$$H(s) = K \frac{(s - z_1)(s - z_2) \cdots (s - z_m)}{(s - p_1)(s - p_2) \cdots (s - p_n)} \qquad (2.3\text{-}6)$$

where K is a scalar constant. The numerator vanishes at a complex frequency $s = z_m$. Such a complex frequency is termed a zero of the function. Similarly, at $s = p_n$, the denominator vanishes, causing the transfer function to become infinite. Such a complex frequency is termed a pole of the transfer function.

The transfer function of the series tuned circuit, Eq. 2.3-1, is clearly the ratio of two polynomials. The numerator has no constant frequency independent terms. Hence, the zero occurs at $s = 0$. Recall that the poles and zeros are complex frequencies, $s = \sigma + j\omega$. While the resonant frequency, f_0, might be considered a "natural" frequency of the network, the more significant natural frequencies are the poles and zeros. The frequency response, $H(j\omega)$, occurs not from excitation of the network at a natural frequency, but for a complex frequency with only an imaginary part, $j\omega$.

The denominator of the series tuned circuit transfer function is factored using the quadratic formula

$$p_{1,2} = \frac{-\omega_0}{2Q} [1 \pm (1 - 4Q^2)^{1/2}] \qquad (2.3\text{-}7)$$

The part of this function under the radical is positive if Q is less than ½. The roots will be real and negative. At $Q = ½$, the two roots are identical and have a magnitude of ω_0. The roots become complex for $Q > ½$. The roots are predominantly imaginary at high Q values.

It is useful to plot the position of the poles and zeros of a transfer function in the s plane. Such a plot for the double tuned circuit is shown in Fig. 2.7. Poles are plotted with an X while zeros are shown as a circle. Figure 2-7a shows a case of low Q where both roots are real. The poles lie on the negative real axis. The second part of the figure shows a case of $Q > ½$. The poles are now complex and lie on the circle centered at the origin with radius ω_0. The roots are almost at the imaginary axis with large Q. Note that the roots occur as complex conjugates.

The polynomials are usually much more difficult to factor for the general case, Eq. 2.3-5. A number of methods are available. They are presented in the literature. These range from careful algebra to numerical techniques.

Further meaning of the pole locations is gained from an analysis of the step

Sec. 2.3 Poles, Zeros, and the Series Tuned Circuit

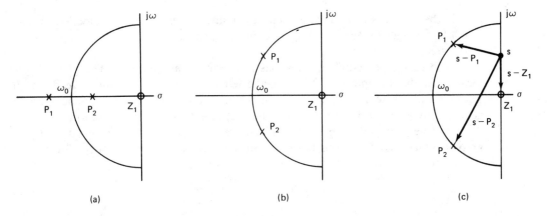

Figure 2.7 Pole-zero plots for the series tuned circuit of Fig. 2.5. (a) shows a low Q example while a higher Q occurs at (b). (c) shows the method of using vectors with a pole-zero diagram to infer network response.

response of the single tuned circuit. Dividing $H(s)$ by s yields the step response in the frequency domain

$$V_0(s) = \frac{1/(\omega_0 Q)}{\dfrac{s^2}{\omega_0^2} + 1 + \dfrac{s}{\omega_0 Q}} \tag{2.3-8}$$

The denominator has already been factored with known pole locations. A table of Laplace transforms shows

$$\mathscr{L}\frac{1}{a-b}(e^{at} - e^{-bt}) = \frac{1}{(s-a)(s-b)} \tag{2.3-9}$$

which leads to the time domain step response

$$V_0(t) = \frac{1/(\omega_0 Q)}{p_1 - p_2} [\exp(p_1 t) - \exp(p_2 t)] \tag{2.3-10}$$

The real part of the exponential functions may be factored from the expression. The imaginary parts are decomposed to a sine function through Euler's identity with the final result for the step response

$$V_0(t) = \frac{2}{\omega_0^2} \frac{\exp(-\omega_0 t/2Q) \sin[(\omega_0/2Q)(4Q^2 - 1)^{1/2} t]}{(4Q^2 - 1)^{1/2}} \tag{2.3-11}$$

This is a sinusoidal oscillation multiplied by a decreasing exponential. It is said to be damped. The frequency of the sine component is less than ω_0 for small Q. The frequency approaches ω_0 as Q becomes large. In addition, the multiplying part in the real exponential component, the damping term, becomes smaller. Hence, it will take longer for the "ringing" to damp to no perceptible signal after application of the step.

The real exponential function has a negative real exponent. The plot of pole location in the s plane, Fig. 2.7, is in the left half plane. A step input leads to an exponentially growing amplitude if a pole occurs in the right half plane. This is generally avoided with most circuits, for it represents an unstable condition.

A pole-zero plot such as the s-plane graphs of Fig. 2.7 are of great utility in graphical analysis. The series tuned circuit has a transfer function of the form

$$H(s) = \frac{(s - z_1)}{(s - p_1)(s - p_2)} \qquad (2.3\text{-}12)$$

where the zero is at the origin, $z_1 = 0 + j0$. If excitation is considered at an arbitrary frequency s, each of the three terms in Eq. 2.3-12 is a vector. An example is shown in Fig. 2.7c where the frequency is a sinusoid and, hence, on the $j\omega$ axis. The magnitude of the frequency response ($s = j\omega$) is proportional to the length of the $(s - z_1)$ vector and inversely proportional to the magnitude of both of the $(s - p_n)$ vectors. The lengths may be measured graphically to infer the response of the network. Similarly, the angles of the individual vectors may be added or subtracted to evaluate the phase response.

The example shown in Fig. 2.7c is for an excitation frequency close to the resonant point, $\omega = \omega_o$. The $(s - z_1)$ and $(s - p_1)$ vectors have approximately the same length, so their ratio is unity. The $(s - p_2)$ vector is longer, so it diminishes the magnitude of the response accordingly.

An excitation frequency close to zero would produce a much different result. The two pole vectors would be of approximately equal length. The $(s - z_1)$ vector is very short, leading to a low output.

If the other extreme is considered, a very high input frequency, the zero vector will effectively cancel with one of the pole vectors. The other pole vector will then dominate. The long length will lead to a small network response.

Pole-zero plots are especially useful for quick graphical analysis of many networks and the subject is covered in many texts. The reader is encouraged to study the methods (5).

Pole-zero plots are equally useful for analysis of active networks. A transistor, modeled at high frequency, will contain reactive terms. These cause poles and zeros to arise in the frequency domain response. The complete amplifier is then characterized by $H(s)$.

2.4 FILTER CONCEPTS

Having acquired some basic tools for analysis, we will now investigate some of the basic filter structures. Figure 2.8 shows several idealized frequency responses.

The idealized low pass filter is shown in Fig. 2.8a. Frequencies below the cutoff

Sec. 2.4 Filter Concepts

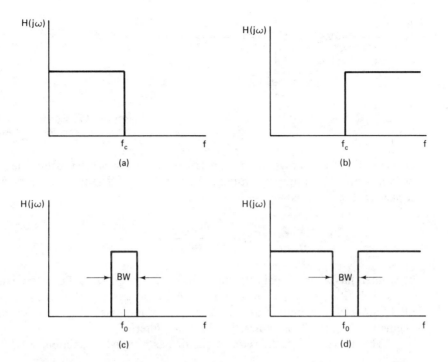

Figure 2.8 Idealized responses for (a) a low pass filter, (b) a high pass filter, (c) a bandpass filter, and (d) a bandstop filter.

frequency, f_c, are transferred through the filter without attenuation. This region is the passband. Frequencies greater than the cutoff are completely attenuated. This region is known as the stopband.

The high pass filter is the opposite of the low pass. Frequencies above the cutoff frequency are passed without attenuation while lower ones are attenuated.

An ideal bandpass filter is presented in Fig. 2.8c. The filter is centered about a frequency f_o, and has a bandwidth shown as BW in the figure. We have already seen one example of a bandpass filter, the single tuned circuit. It was, however, far from ideal.

Finally, Fig. 2.8d shows a bandstop filter. Frequencies outside of the bandwidth of the filter are passed without attenuation. Frequencies within the bandwidth are eliminated from the output.

A final filter type not shown in the figure is the all-pass structure. It passes all frequencies without attenuation. The utility of the all pass is that even though it shows no attenuation, it will shift the phase of signals passing through it. This is of value in numerous applications. One example of an all-pass filter is a matched transmission line.

The high pass and bandpass filters are designed from a low pass prototype filter through suitable transformations. These details will be presented subsequently. First some simple examples will be considered.

Figure 2-9 shows two simple networks. Each contains a resistor and one reactive

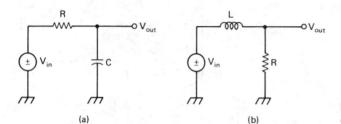

Figure 2.9 Simple low pass filters, each with one reactive element.

element. The frequency domain transfer functions are written using the s plane impedances and by noting the voltage divider action. $H(s)$ for the network with the capacitor, Fig. 2.9a, is

$$H(s) = \frac{1/sC}{R + 1/sC} = \frac{1}{1 + sCR} \quad (2.4\text{-}1)$$

This filter has a single pole in the s plane at $s = -1/RC$. The filter has no finite zeros. Substitution of $j\omega$ for s in Eq. 2.4-1 provides the frequency response. The filter shows no attenuation at dc. As frequency increases, the response immediately begins to decrease. This is clearly a low pass filter.

The network with the series inductor has a transfer function, $H(s)$

$$H(s) = \frac{R}{sL + R} = \frac{1}{1 + \dfrac{sL}{R}} \quad (2.4\text{-}2)$$

This filter has a single real pole at $s = -R/L$. Evaluation of the frequency response is similar to that of the capacitive filter. The inductor presents no impedance at dc; hence the filter has no attenuation. As the frequency increases so does the inductive reactance, leading to increasing attenuation. A low pass response arises.

For both low pass filters, the frequency response is below the dc value by a numerical factor of $1/\sqrt{2}$ at a radian frequency equal to the pole location. The power out of the filter is reduced by the square of this amount to ½. Taking the log of the power ratio and multiplying by 10 shows that the response is down by 3.01 dB, essentially 3 dB. This is defined as the cutoff frequency. The phase of the output then differs from the drive by 45 degrees in each filter.

Neither of the filters has a well-defined stopband. That is, there is never a finite frequency where the attenuation is infinite. This is true of virtually any real filter. The amount of attenuation required in a given application will define what the "stopband" of a filter will actually be. The slope of the response at frequencies well above the cutoff is 6 dB per octave, or 20 dB per decade, for both of the single-pole, low pass filters of Fig. 2.9. The phase of the response is -90 degrees when high frequencies are reached.

Figure 2.10 shows two additional filters. From inspection, we see that they

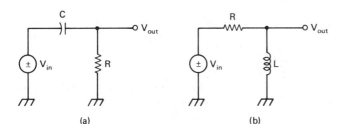

Figure 2.10 Simple high pass filters with (a) a series capacitor and (b) a shunt inductor.

are high pass structures. The capacitor in Fig. 2.10a is an open circuit at dc, producing infinite attenuation there. Similarly, the inductor (Fig. 2.10b) is a short at dc. The frequency domain transfer functions are

$$H(s) = \frac{sCR}{1 + sCR} \qquad (2.4\text{-}3)$$

for the capacitive high pass filter and

$$H(s) = \frac{s}{s + R/L} \qquad (2.4\text{-}4)$$

for the inductive example. The filters each have a single pole. They also exhibit a zero at the origin of the s plane. The 3-dB points occur for a radian frequency, ω_c, equal to the magnitude of the pole location. The slope is also 6 dB per octave at low frequencies. The phase is $+90$ degrees at very low frequencies with respect to the cutoff and is $+45$ degrees at the cutoff.

We have found from the foregoing discussion that low pass characteristics result from series inductors or shunt capacitors. Shunt inductors and series capacitors lead to high pass behavior. We can use this to form more elaborate filters. One example already covered is the series tuned circuit of the last section. An element of each type was placed in the series arm of the filter resulting in a bandpass characteristic. A number of low pass elements may additionally be cascaded to form a more exotic low pass structure. This also applies, of course, to the high pass.

Note that adding two reactances of the same type in the series arm of a filter will not change the general characteristic. For example, two series inductors will act as one inductor. This will lower the cutoff frequency, but will otherwise not alter the response. If a composite filter is to be formed, it must be done by mixing component types.

An example of a composite low pass filter is presented in Fig. 2.11. This filter is doubly terminated. That is, it has a driving source that is not a pure voltage generator, but one with an internal resistance, R_s. The output is terminated in a load resistance, R_L.

The frequency domain transfer function for the filter may be written from inspection, noting the voltage divider action of the composite network. An admittance, Y,

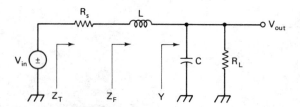

Figure 2.11 A two-pole, doubly terminated low pass filter.

is marked in Fig. 2.11. This is the admittance of the output load resistor in parallel with the capacitor, $Y = 1/R_L + sC$. The impedance of this combination is the reciprocal of the admittance, $Z = R_L(1 + sCR_L)^{-1}$. The total impedance looking in from the voltage generator is $Z_T = Z + R_s + sL$. The transfer function is thus

$$H(s) = \frac{Z}{Z_T} = \frac{1}{s^2 LC + s\left(RC + \dfrac{L}{R}\right) + 2} \tag{2.4-5}$$

where $R_s = R_L = R$.

This filter has no finite zeros, but has two poles. They are evaluated by factoring the denominator polynomial of $H(s)$. The result is

$$P_{1,2} = \frac{-(RC + L/R) \pm [(RC + L/R) - 8LC]^{1/2}}{2LC} \tag{2.4-6}$$

If we allow the source and load resistances to be equal at 1 Ω and assign values of $L = \sqrt{2}$ henry (H) and $C = \sqrt{2}$ farad (F), the poles occur at $p = (1 \pm j1)/\sqrt{2}$.

We will examine the nature of impedance matching when reactive terminations are present before evaluation of the steady state frequency response. Maximum power transfer occurs from a source with characteristic resistance R_s when the load has the same value, $R_L = R_s$. However, this is not the proper condition when the source is a complex impedance.

Assume that the driving generator has a source resistance R_s in series with an inductor. The source impedance at a given angular frequency, ω, is $Z_s = R_s + j\omega L$. If a similar impedance was used as the load, the power transferred to the load is clearly not maximum. If the inductor in the load was removed, the net impedance would decrease, causing an increase in current. Hence, additional power would be dissipated in the load resistor. It is only the power in the resistor that is of significance. The voltage drop in the inductor is always 90 degrees out of phase with the current, leading to no power dissipation.

We ask what the source impedance should be for maximum power delivered to the load. The load would be one with the same resistance as that of the source, with a reactance opposite in sign, but with the same magnitude of the inductor. This reactance is, of course, a capacitor. The combination of inductor and capacitor would form a series resonant circuit. The total reactive impedance is zero. Neither reactive component impedes the flow of current. In the general case, maximum power transfer occurs where the load impedance is the complex conjugate of the source impedance. This is termed a *conjugate match*.

Returning now to the filter of Fig. 2.11, the effective frequency response may be evaluated. We could insert $j\omega$ into Eq. 2.4-5 to obtain the response as was done in earlier examples. This would not be the most useful information though. It is more viable to consider the transducer gain of the filter. Recall that transducer gain was defined as the power delivered to the output termination divided by the available power from the generator.

The frequency response is written directly as

$$H(j\omega) = \frac{1}{(2 - 2\omega^2) + j\omega 2\sqrt{2}} \qquad (2.4\text{-}7)$$

This will be the output voltage for an input of 1 V. Assume that the input is 2 V. Then, the output voltage is $H(j\omega)$ of Eq. 2.4-7 multiplied by 2. The maximum power available from the generator is half of the open circuit voltage across a load equal to the source impedance, or $P_A = 1/R_s$. The equation provides the output voltage, V_{out}. The output power is then V_{out}^2/R_L. Hence, transducer gain is $R_s V_{out}^2/R_L$. We will use this definition for filter frequency response throughout the rest of the text.

The frequency response of the filter is presented in Fig. 2.12 for the filter of

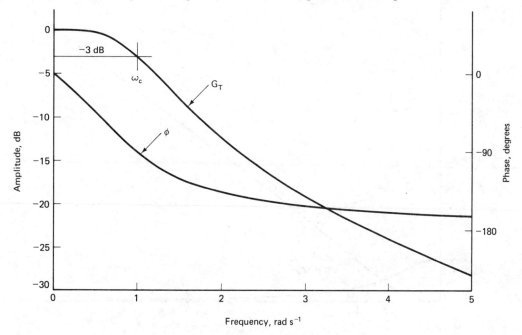

Figure 2.12 Frequency response, or transducer gain, G_t, of the two-pole low pass filter. Phase response is also shown.

Fig. 2.11. The phase with respect to the driving voltage generator is also presented. We have assumed a filter with equal terminations of 1 Ω and $C = \sqrt{2}$ F and $L = \sqrt{2}$ H.

This filter shows no attenuation near dc. However, $G_T = -3$ dB at a radian frequency of $\omega = 1$ rad s^{-1}. This is the cutoff frequency. Also, the phase is -90 degrees at the cutoff. The attenuation slope approaches 12 dB per octave and the phase nears -180 degrees at high frequencies.

The filter input impedance is also evaluated, shown in Fig. 2.11 as Z_f. The impedance is easily calculated in the frequency domain

$$Z_f(s) = sL + \frac{R_L}{1 + sCR_L} = \frac{(R_LLC)s^2 + sL + R_L}{1 + sCR_L} \qquad (2.4\text{-}8)$$

The input impedance, like the transfer function, $H(s)$, is a ratio of polynomials. It has two zeros and one pole, which are evaluated using methods already presented.

The input impedance, $Z_f(j\omega)$ is calculated by substitution of $j\omega$ for s in Eq. 2.4-8. The results are plotted in Fig. 2.13 showing both the resistive and reactive portions. The imaginary part of the impedance is positive. This represents an inductive input. This is intuitively reasonable—at high frequencies the capacitor at the output of the filter tends to behave as a short circuit, leaving the inductor to dominate.

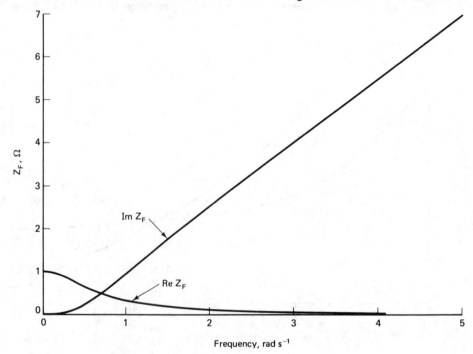

Figure 2.13 Input resistance and reactance for the two-pole, doubly terminated low pass filter.

The resistive portion of Z_f is 1 Ω and there is no reactance at dc and frequencies well below the cutoff. Maximum power is transferred and the attenuation is zero. The impedance changes when the frequency increases. Not only does the resistive portion tend toward zero, but the reactive part increases. This is characteristic of all *LC* filters. The filtering action is a result of mismatch between the source or load and the filter.

2.5 THE LADDER METHOD

The previous filter examples have been simple. They have contained at most two poles, representing two reactive components. Each time a reactive component is added, the degree of the polynomials describing the filter increases accordingly. Modern communications circuitry often requires filters with a dozen or more reactive components. The polynomials can, of course, be factored. However, the degree of algebraic cleverness required of the designer is at least commensurate with the degree of the polynomial. Other methods are required for the rest of us.

Our interest is not confined to the filter transfer function. We are also interested in the immittances (impedances or admittances) associated with the terminals of the filter. The filter must be properly terminated to function properly. Amplifiers at each end must be designed so that they not only function properly with the impedances presented at frequencies in the passband, but tolerate the impedances seen in the stopband.

A method is presented in this section that is especially suited to filters built in a ladder configuration. A ladder structure is one built from combinations of shunt and series elements. Some of the more complicated structures, such as the lattice, are not allowed. Fortunately, the ladder filter is a large category—nearly all of the typical filters in routine use are of this type. The method presented will not only evaluate the transfer function, but will provide impedance data. The method will be illustrated through the study of a third order low pass filter.

The first variation of the ladder method is sometimes termed the "tack hammer" approach. Any practical filter has a transfer function. It may not be known, but we can still write equations of a generalized form. A transfer function in the frequency domain, *H(s)*, has a corresponding steady state response, $H(j\omega)$. The ac output voltage at a given frequency is related to the input voltage by

$$V_{\text{out}} = H(j\omega) V_{\text{in}} \qquad (2.5\text{-}1)$$

Solving for the input voltage yields

$$V_{\text{in}} = V_{\text{out}} H(j\omega)^{-1} \qquad (2.5\text{-}2)$$

This suggests a method for filter analysis.

Figure 2.14 shows a third order low pass filter. We would speculate from the intuition gained in previous examples that this filter would have no finite zeros and three poles in the transfer function, *H(s)*.

Two critical nodes, *x* and *y*, are labeled in Fig. 2.14. Assume that the output voltage across the load resistor, R_L, is 1 V. Knowing this, the voltage across the capacitor, C_3, is also known. Hence, the current out of node *x* is known. Specifically, the total current out of the node is the voltage times the admittance looking into the node from the left, or $I = G_L + j\omega C_3$. Arrows representing the direction of assumed current flow are included in the diagram.

The current out of node *x* equals that flowing in. Hence, the current flowing into the node, which is that flowing through the inductor, is also known. The voltage drop across the inductor is calculated from this. The voltage at node *y* is evaluated as $V_y = 1 + Y_x j\omega L_2$ where $Y_x = G_L + j\omega C_3$. Now that the voltage at node *y* is known, the current in C_1 is also specified. This defines the total current into node *y* from the generator.

The procedure is continued until the voltage at the generator is known. Note that the calculations all involve vectors, for the impedances and admittances are all complex. The transfer function is found from calculated input voltage, $H(j\omega) = V_{in}^{-1}$.

In addition, once the current into node *y* and the voltage at that node are calculated, the input immittance is defined.

This method is easily extended to an arbitrarily long ladder. It is only necessary to attack each node, one at a time in tack hammer fashion, until the filter is analyzed.

The tack hammer version of the ladder filter analysis involves the calculation of numerous immittances. Study suggests a slightly modified method, equivalent to the original, but with the virtue of being perhaps somewhat more ordered. Both are easily programmed with a handheld calculator.

Consider the vertical lines in Fig. 2.14. These are labeled with the letters "a" through "e" and represent different "planes" in the ladder filter. A different plane occurs whenever an additional component is added to the circuit. From each plane, an immittance (*Z* or *Y*) is well defined looking to the right toward the output termination.

The admittance is G_L at plane "a." Y_b, the admittance looking into plane "b," is $Y_a + j\omega C_3$. Knowing the admittance at plane "b," the corresponding impedance is calculated, the reciprocal of the admittance. When the admittance is complex, the conversion is performed by transforming from polar to rectangular coordinates.

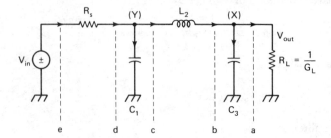

Figure 2.14 A three-pole low pass filter used to illustrate the ladder method of analysis.

Sec. 2.5 The Ladder Method

The magnitude is replaced with its reciprocal and the sign of the angle is changed. The result is then converted back to rectangular coordinates.

The impedance looking to the right into plane "c" is that at plane "b" plus the inductive reactance of L_2. The result is then converted to an admittance and the admittance of C_1 is added. Another conversion results in Z_d. Finally, Z_e is obtained by addition of the source resistor. Y_e results from a final conversion.

The power of this calculation becomes evident with additional algebra. This is based upon having calculated all of the impedances or admittances at each plane in the filter. First, a value for V_{in} is specified. Typically, this will be 2 V, for we are interested ultimately in evaluation of filter transducer gain. The input current is now known, $I_e = V_{\text{in}} Y_e$. But this is also the current into plane "d." Knowing the current, the voltage at plane "d" is calculated. The procedure is repeated as needed. The method is summarized in the following set of equations

$$I_e = V_{\text{in}} Y_e$$
$$I_d = I_e$$
$$V_d = I_e Z_d \qquad (2.5\text{-}3)$$
$$I_c = V_d Y_c$$
$$V_b = I_c Z_b$$

But V_b is the output voltage, V_{out}.

Examination of the family of equations, Eqs. 2.5-3, shows that each contains a current or a voltage on the right-hand side. However, that parameter is defined in the next equation above. A continued substitution will eventually terminate at the input voltage which was defined. The result is

$$V_{\text{out}} = Z_b Y_c Z_d Y_e V_{\text{in}} \qquad (2.5\text{-}4)$$

This method appears formidable. It would be if it were done with a slide rule or with a very simple handheld calculator. However, modern programmable calculators make it a relatively simple task.

Each stage in the calculation involves the conversion of an impedance to an admittance, or the opposite. It is necessary to perform a rectangular to polar conversion at each step. Upon taking the reciprocal of the magnitude and changing the sign of the angle, the desired result is available, the needed new immittance. The magnitude can be multiplied into a storage register and the angle summed into another. The polar result is converted back into a rectangular format. This is needed to add the next series impedance or shunt admittance. The transfer function is generated "for free" during the calculation of the input immittance seen from the driving generator.

An order is present in Eq. 2.5-4 that may not be immediately apparent. The product of immittances is alternating in type. That is, the products are an impedance followed by an admittance followed by an impedance, etc. Also, admittances correspond to planes looking into series elements while impedances correspond to filter planes looking into parallel elements. This allows a program to be written quickly.

2.6 LOSSES IN REACTIVE COMPONENTS AND QUALITY FACTOR

The parameter Q was introduced in a preceding section dealing with the series tuned circuit. No explanation was presented other than observing that the bandwidth of the circuit decreased as Q increased. This section will present a more formal definition and rationale for the parameter and will show how it is used in network analysis.

We have assumed in our studies that components are ideal. This implies that they show no loss. This is, of course, not realistic. It is only a model that serves for analysis. Real components do have loss.

There are numerous mechanisms that will produce loss in capacitors and inductors. Dielectric heating and surface resistance on the conducting elements often introduce loss in capacitors. Inductors show loss from motion of magnetic core material and from wire resistance. The physics surrounding these phenomena is interesting and of vital concern to rf designers, especially those working at microwave frequencies.

The inherent loss mechanisms resulting from component imperfections are usually beyond the control of the designer. They must be taken into account during analysis though. Sometimes loss is purposefully introduced into a circuit. The most common example is the termination that is found with any practical network. If there was no termination, it would not be possible to extract energy from the network and the circuit would have no purpose. All of these losses are represented with the parameter Q, which stands for the quality factor.

Formally, the Q of a circuit is defined as 2π times the energy stored in the circuit divided by the energy lost per cycle of oscillation. The Q concept is usually applied to a resonant or tuned circuit, which is often called a resonator. When applied to a single component the Q is that which would result from resonating the component at the frequency of interest with an ideal lossless reactance of the opposite type. That is, the Q of an inductor would be the Q of a tuned circuit using that inductor with a lossless capacitor.

It may be shown that the Q of an inductor can be modeled by a series resistor (6). The resistor value is

$$R_s = \frac{\omega L}{Q} \qquad (2.6\text{-}1)$$

This resistor represents either the internal losses of the nonideal inductor or the external loss associated with a load combined with internal losses. If the Q value represents only the internal losses, the nonideal nature of the component, the Q is termed the "unloaded Q" and is signified by a subscript, Q_u. It is termed Q_L for "loaded Q" if the Q value represents external loads as well as the internal losses.

The higher the Q, the smaller the series resistance and hence the smaller the loss. This is seen from the definition, Eq. 2.6-1.

We have seen in our study of networks that impedance and admittance are used interchangeably. Admittance is more convenient than impedance when evaluating a shunt component. Similarly, Q may be represented by a parallel instead of a series resistance.

The impedance of a real inductor, as modeled by Q, is $j\omega L + R_s$. The admittance is then

$$Y = \frac{1}{j\omega L + R_s} \qquad (2.6\text{-}2)$$

Multiplying the top and bottom by the complex conjugate of the denominator and inserting Eq. 2.6-1 for R_s yields

$$Y = \frac{(\omega L/Q) - j\omega L}{\dfrac{\omega^2 L^2}{Q^2} + \omega^2 L^2} \qquad (2.6\text{-}3)$$

The admittance is of the form $G + jB$ where G is a conductance, $G = 1/R_p$, and B is a susceptance, $B = 1/X_p$. Separating Eq. 2.6-3 into real and imaginary parts yields, upon multiplication by Q^2

$$Y = \frac{1}{R_p} - j\frac{1}{X_p} = \frac{Q\omega L}{\omega^2 L^2(1 + Q^2)} - j\frac{\omega L Q^2}{\omega^2 L^2(1 + Q^2)} \qquad (2.6\text{-}4)$$

For large value of Q, $(1 + Q^2) \simeq Q^2$. Hence

$$R_p \simeq Q\omega L \qquad X_p \simeq \omega L \qquad (2.6\text{-}5)$$

Both forms of modeling are used almost interchangeably in practice. The choice is usually one of convenience. The difference is only significant for Q values of less than 10.

Practical inductors show a wide range of Q_u values. They are always a function of frequency. Q increases with frequency, usually in proportion to the square root of the frequency for a given fixed size air core inductor. Inductors wound on a powdered iron core typically have a peak in Q_u as frequency is changed. The best inductors are thus built with a core picked especially for the frequency range of interest. The highest Q_u values are usually obtained with toroidal cores for the hf region. Larger Q_u values result from air cores, but the physical size is often unreasonable. A typical high Q inductor in the hf range would have $Q_u = 250$ or higher.

If large wire is used at vhf, air core inductors become practical. It is easily possible to achieve Q_u greater than 400 if large wire is used. This is best realized if the inductor is shielded, owing to radiation losses. The Q of the components should be measured or obtained from tables provided by manufacturers of powdered iron cores if careful work is to be done (7).

Capacitors also suffer from losses which may be represented by a Q_u. Assume a capacitor has a loss that is modeled by a parallel resistance, R_p. If this capacitor was placed in parallel with a perfect, lossless inductor, the resonator would have an unloaded Q_u equal to that of the capacitor. If Q_u was associated with the inductor, R_p would be $Q_u \omega L$. The inductive susceptance equals that of the capacitor in the resonant circuit. Hence

$$R_p = \frac{Q_u}{\omega C} \qquad (2.6\text{-}6)$$

Following the same arguments as were used with the inductor, the Q_u value of a capacitor may be related to a series resistance, R_s, by

$$R_s = \frac{1}{Q_u \omega C} \qquad (2.6\text{-}7)$$

As was the case with the inductor, Eqs. 2.6-6 and 2.6-7 are not exactly related to each other. The two depend upon high Q_u values. We will use $R_p = Q_u \omega L = Q_u/\omega C$ as a formal relationship in most of our work.

Most capacitors used in tuned circuits have a very high Q_u in the hf region. Values well over 1000 are typical. Hence, it is valid for most applications to assume that parallel capacitors are lossless and that the loss of a circuit is related to the inductor.

This ideal behavior is not always true. At vhf and higher, capacitors often exhibit Q values that are low enough that they must be taken into account in analysis. This is true even in the hf spectrum with some capacitors. An example of a lossy capacitor is the varactor diode often used as a voltage controlled variable capacitor. The manufacturers of such diodes usually specify the Q_u at a stated frequency with the understanding that Q_u will decrease linearly with increasing frequency. A better model is that of an ideal diode with a frequency independent series resistance.

Consider the parallel tuned resonator of Fig. 2.15 as an example of the effect of Q in a tuned system. Assume that the source equals the load resistance. Using methods developed earlier, the transfer function is

$$H(s) = \frac{sL/R}{s^2 LC + 1 + 2sL/R} \qquad (2.6\text{-}8)$$

We find that $H(j\omega) = \frac{1}{2}$ at $\omega^2 = 1/LC$. Hence, all of the power available from the voltage generator and source resistance is being delivered to the load resistor. The resonator is transparent at this frequency, that of resonance.

The transfer function may be compared with that of the series tuned circuit,

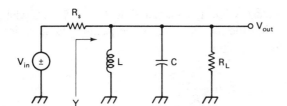

Figure 2.15 A parallel tuned circuit, doubly terminated.

Eq. 2.3-3. Some manipulation then shows that the Q equals half that of the series tuned circuit. However, the series tuned circuit studied contained only one resistor while the parallel one of Fig. 2.15 has two. The circuit Q is thus given by $Q_L = R_p/\omega L$ where R_p is now the parallel combination of the source and load resistors. Note that the same form of transfer function describes both the series and the parallel tuned circuit.

Assume now that the components in the filter of Fig. 2.15 are not ideal, but have a loss described by measurable Q_u values. The equivalent parallel resistance associated with the inductor is $R_{pL} = Q_{uL}\omega L$ while the loss in the capacitor is $R_{pC} = Q_{uC}/\omega C$. But $L = 1/\omega C$. The equivalent loss resistance is the parallel combination of the two. The final result is the equivalent Q_u

$$Q_{net} = (Q_{uL}^{-1} + Q_{uC}^{-1})^{-1} \quad (2.6\text{-}9)$$

This equation is quite general. When a number of lossy reactances are connected in parallel or in series, the equivalent Q is obtained in the same way that the equivalent resistance of parallel resistors is obtained.

Assume that the Q_u value of the capacitor is arbitrarily high. The Q_u of the tuned circuit is then that of the inductor alone, resulting in $R_p = Q_u\omega L$. The effects of the reactive components disappear at resonance, leaving the equivalent circuit shown in Fig. 2.16. The transducer gain of this network will be evaluated. We assume that $R_s = R_L = R$ and the voltage generator has an amplitude of 2 V. The output voltage is written from inspection. Algebraic reduction produces the result

$$V_{out} = \frac{2\left(\dfrac{R_p R}{R + R_p}\right)}{R + \dfrac{R_p R}{R + R_p}} = \frac{1}{1 + R/2R_p} \quad (2.6\text{-}10)$$

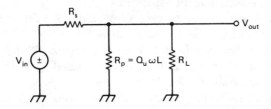

Figure 2.16 The equivalent of the parallel tuned circuit at resonance. R_p is the resistance representing the losses in the resonator.

The unloaded Q is $Q_u = R_p/\omega L$ while the loaded Q is $Q_L = R_e/\omega L$ where R_e is the equivalent resistance loading the inductor. R_e is the parallel combination of the load, the source, and R_p. Hence

$$R_e = \frac{R_p R}{2R_p + R} \tag{2.6-11}$$

for the case of equal load and source resistances.

Consider the ratio of Q_L/Q_u

$$\frac{Q_L}{Q_u} = \frac{R_e}{R_p} = \frac{R}{2R_p + R} \tag{2.6-12}$$

This simplifies to the relationship

$$\frac{R}{2R_p} = \frac{Q_L}{Q_u - 1} \tag{2.6-13}$$

which is then inserted into Eq. 2.6-10 to produce

$$V_{out} = \left(1 - \frac{Q_L}{Q_u}\right) \tag{2.6-14}$$

Because the source and load resistances are equal, the insertion loss of the filter of Fig. 2.15 is the negative of the transducer gain, or

$$\text{Insertion loss (IL)} = -20 \log (1 - Q_L/Q_u) \text{ db} \tag{2.6-15}$$

This filter is a doubly terminated single resonator.

The bandwidth of the single resonator, terminated or not, is analytically related to the Q. Recalling that the bandwidth is defined as the frequency where the output power is down by half, or 3 dB, bandwidth is related to center frequency and Q by

$$Q = f_0/\text{BW} \tag{2.6-16}$$

where F_0 and BW are both measured in the same units.

This is well illustrated by an example. Assume a resonator has a center frequency of 100 MHz and $Q_u = 400$. The unloaded bandwidth is $100/400 = 0.25$ MHz (Q is a dimensionless number). A bandwidth of 1 MHz is measured if the resonator is then placed between an equal source and load. The loaded Q is then $Q_L = 100/1 = 100$. The insertion loss from Eq. 2.6-15 is 2.5 dB.

Examination of the equations reveals a method for measuring the Q of a resonator. Note that as the loaded Q of a filter approaches the unloaded value, the insertion loss becomes very large. If a resonator is terminated equally by both the generator and load and the values are adjusted so that the insertion loss is very large, the measured 3-dB bandwidth will produce a loaded Q according to Eq. 2.6-16. Measuring the insertion loss will allow calculation of Q_L/Q_u. The loaded Q is very close to Q_u if the insertion loss is high enough, typically 30 to 40 dB.

Q has been related to a second order network in our discussion. That is, Q is a parameter of a network containing two reactive elements, described by a transfer function of second order. Sometimes a parameter Q appears in design equations for third or even fourth order networks. The meaning for this Q is only loosely related to the Q parameter we have been discussing. The actual energy storage in the network and the related bandwidth properties are sometimes completely unrelated to Q.

2.7 THE ALL-POLE LOW PASS FILTER

Previous sections have presented background information and analysis methods. Now, the problem of filter design is finally approached. The type of filter considered in this section is shown in Fig. 2.17. It contains only series inductors and shunt capacitors. The filter is a ladder configuration and may be analyzed using the ladder method presented earlier.

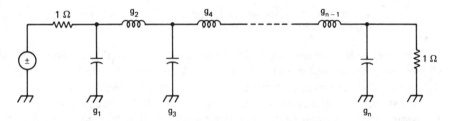

Figure 2.17 An nth order low pass filter. The text has equations for evaluating the component values and the location of the poles.

We find that the transfer function, $H(s)$, of the filter of Fig. 2.17 contains no finite zeros. The transfer function is the reciprocal of a simple polynomial in the variable s. As such, the filter is described completely by factoring of the denominator polynomial to extract the location of the complex poles. This filter, an *all-pole* type, represents many of the practical low pass filters used in routine rf design.

Most design work with all filters is done with low pass prototypes like that of Fig. 2.17 where the source and load resistance are 1 Ω and the cutoff frequency is 1 rad. This has a number of consequences. First, analysis is simplified. Second, the normalized filters are in a form that may be easily scaled to other terminations and cutoff frequencies. The final rationale is somewhat less obvious, though. At a frequency of 1 rad s^{-1}, a 1-H inductor and a 1-F capacitor have the same immittance, 1 Ω or 1 Siemen (S). This duality allows us to treat them more easily than we could if another frequency of normalization were chosen. One consequence is that one form of filter may be transformed into another with no change in numerical values.

This duality is illustrated by the two filters in Fig. 2.18. One has two capacitors of 1 F and an inductor of 2 H. The other has two inductors of 1 H each and a single capacitor of 2 F. Both of these filters are identical. The cutoff frequency (−3 dB) is 1 rad s^{-1} and the transfer functions, *H(s)*, are identical. Only one set of normalized parameters is required for tabulation of normalized filter components. These are usually given by g_k as shown in the component labeling of Fig. 2.17.

The defining polynomial has a degree that is numerically identical to the number of reactive components in the prototype filter. The filters of Fig. 2.18 are third order filters with the highest power of *s* in *H(s)* being 3. The filter g_k values are chosen for a cutoff of 1 rad s^{-1}. However, there are no other restrictions. An infinite number of component values are still allowed.

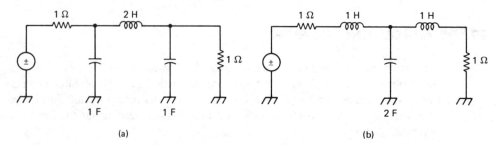

Figure 2.18 Two different, but equivalent versions of the three-pole normalized Butterworth low pass filter.

The chore of picking useful components is a complicated one and is the essence of much of the subject of filter synthesis. Approximation methods are usually applied. An ideal filter serves as a starting point. Then, an attempt is made at choosing filter polynomials which approximate one characteristic or another of the ideal filter. The term "ideal" is itself something of a puzzle. A filter that has a perfect "brick wall" shape in the frequency domain, such as the low pass filter of Fig. 2.8a, may have a horrible step response.

The nature of the approximations used in synthesis is covered well by many writers and will not be dealt with here. The curious reader should refer to the text by Blinchikoff and Zverev.

This section will present two popular and practical polynomial types and the corresponding filters, the Butterworth and Chebyshev. The Butterworth filter is optimized near the lowest frequencies and is often called a maximally flat response. It has reasonable time domain behavior although not as ideal as some other types.

The Chebyshev filter is optimized for frequency domain response. It has excellent differentiation between the passband response and that of the stopband. The price for these virtues is some ripple, or variation, in attenuation within the passband and poor time domain response.

The frequency response of the normalized Butterworth filter is

$$H(j\omega) = \frac{1}{(1 + \omega^{2n})^{1/2}} \qquad (2.7\text{-}1)$$

Sec. 2.7 The All-Pole Low Pass Filter

where ω is the angular frequency and n is the degree of the polynomial. Using this frequency response, the parameter values, the location of the complex poles, and the frequency domain transfer function may all be derived. The pole locations in the s plane are given by

$$\sigma_k = -\sin(2k-1)\frac{\pi}{2n}$$

$$\omega_k = \cos(2k-1)\frac{\pi}{2n} \qquad k = 1, 2, 3, \ldots, n \tag{2.7-2}$$

where σ_k is the real part and ω_k is the complex part of the kth pole. The pole locations may be combined with s to form the factored form of $H(s)$. Multiplication will produce the usual form. The transfer functions for $n = 2$ and $n = 3$ are

$$H(s) = \frac{1}{s^2 + 1.414s + 1} \qquad n = 2 \tag{2.7-3}$$

$$H(s) = \frac{1}{s^3 + 2s^2 + 2s + 1} \qquad n = 3$$

For larger values of n, the polynomials become more complex.

Noting the pole locations, Eqs. 2.7-2, we see that the real and complex parts are described by a sine and a cosine function with the same argument. Hence, the poles of the Butterworth low pass filter lie on a circle in the s plane.

The component values of the normalized Butterworth low pass are given by

$$g_k = 2\sin(2k-1)\frac{\pi}{2n} \qquad k = 1, 2, 3, \ldots, n \tag{2.7-4}$$

for the case where the source and load resistances are 1 Ω. As it turns out, this restriction is not a firm one. The end terminations need not be equal with proper renormalization. The reader should refer to the literature for details (8).

The function for component values, Eq. 2.7-4, is an especially simple relationship. Consider $n = 3$. Substitution then yields $g_1 = 1$, $g_2 = 2$, and $g_3 = 1$. The resulting filters are those of Fig. 2.18.

The analytical representation of the Chebyshev filters is considerably more complex than the Butterworth. Where the Butterworth filter was described by circular functions, the Chebyshev is dominated by hyberbolic functions. These are somewhat more complicated, but may be manipulated easily with a programmable handheld calculator. The analog to the Euler identities can be used:

$$\sinh x = \frac{e^x - e^{-x}}{2} \tag{2.7-5}$$

$$\cosh x = \frac{e^x + e^{-x}}{2}$$

The frequency response of the Chebyshev low pass filter is given by

$$H(j\omega) = \frac{1}{[1 + \epsilon^2 C_n^2(\omega)]^{1/2}} \tag{2.7-6}$$

where $C_n(\omega)$ are the Chebyshev polynomials defined by

$$\begin{aligned} C_n(\omega) &= \cos(n\cos^{-1}\omega) & \omega < 1 \\ C_n(\omega) &= \cosh(n\cosh^{-1}\omega) & \omega > 1 \end{aligned} \tag{2.7-7}$$

The parameter ϵ is the ripple in the passband and is given by

$$\epsilon = (10^{0.1A} - 1)^{1/2} \tag{2.7-8}$$

where A is the depth of the ripple in dB.

Figure 2.19 will shed additional light on the nature of the Chebyshev filter. The figure shows the $H(j\omega)$ plot for a filter with $n = 5$ and $A = 1.5$ dB. The frequency response of the filter oscillates by a peak-to-peak value of 1.5 dB throughout

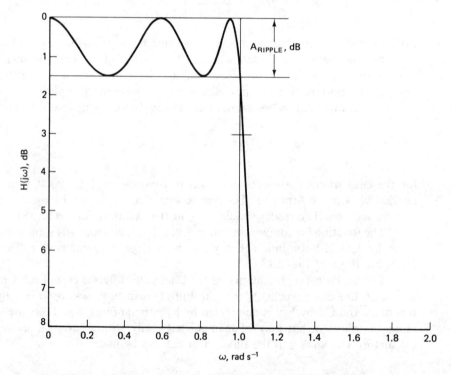

Figure 2.19 Frequency response of a five-pole low pass filter of the Chebyshev type. The passband ripple is $A = 1.5$ dB. The attenuation equals the ripple at the cutoff frequency of 1 rad s^{-1}. Note that the number of half cycles of passband ripple equals the order of the filter.

Sec. 2.7 The All-Pole Low Pass Filter

the passband. One other feature is apparent in Fig. 2.19. The response at $\omega = 1$ rad s^{-1} is equal to the value of A, the ripple, and is not -3 dB as is usually the case. The frequency where the response is down by 3 dB is given by

$$\omega_{3dB} = \cosh\left(\frac{1}{n}\cosh^{-1}\frac{1}{\epsilon}\right) \quad (2.7\text{-}9)$$

for $A < 3$ dB.

The number of poles in a Chebyshev filter is equal to the number of half cycles of oscillation in the ripple passband response. N was 5 for the filter used for the curve of Fig. 2.19. Hence, at dc, the response was zero dB, or unity, for evaluation of Eq. 2.7-6. This cannot be true for n an even number and preserve the Chebyshev response of Eq. 2.7-6 unless the source and load reactances are unequal. If the load is kept at 1 Ω, the required source resistance to preserve the desired filter shape is given by

$$R_s = \begin{cases} 1, & n \text{ odd} \\ \tanh^2\left(\frac{B_0}{4}\right), & n \text{ even} \end{cases} \quad (2.7\text{-}10)$$

where the parameter B_0 is defined below.

The equations for the component values, g_k, of the Chebyshev low pass filter are standard and are presented in many texts. They are rarely presented in a closed form. Instead, several equations of related parameters are given that are then used to calculate the g_k values. A modified version of these equations is presented below where the hyperbolic functions have been replaced by their exponential equivalents. This facilitates evaluation with a handheld calculator.

The set of equations appears formidable. It is not. A program is easily written for the handheld calculator to evaluate the functions. The following set of equations also contains the adaptations of Eqs. 2.7-9 and 2.7-10 in an exponential form

$$d = \frac{A}{8.68589} \quad (2.7\text{-}11)$$

where A is the passband ripple in dB.

$$B = \frac{1}{2n}\ln\left[\frac{e^d+1}{e^d-1}\right] \quad (2.7\text{-}12)$$

$$B_0 = 2nB \quad (2.7\text{-}13)$$

$$N = \tfrac{1}{2}(e^B - e^{-B}) \quad (2.7\text{-}14)$$

$$a_k = \sin\left[\frac{(2k-1)\pi}{2n}\right] \qquad k = 1, 2, \ldots, n \qquad (2.7\text{-}15)$$

$$b_k = N^2 + \sin^2\left(\frac{k\pi}{n}\right) \qquad k = 1, 2, 3, \ldots, n \qquad (2.7\text{-}16)$$

The g_k values are calculated using these parameters

$$g_1 = \frac{2a_1}{N} \qquad (2.7\text{-}17)$$

$$g_k = \frac{4a_{k-1}a_k}{b_{k-1}g_{k-1}} \qquad (2.7\text{-}18)$$

The 3-dB down frequency is evaluated using

$$\omega_{3\text{dB}} = \tfrac{1}{2}(e^Y + e^{-Y}) \qquad (2.7\text{-}19)$$

where Y is given by

$$Y = \frac{1}{n}\ln\left[\frac{1}{\epsilon} + \left(\frac{1}{\epsilon^2} - 1\right)^{1/2}\right] \qquad (2.7\text{-}20)$$

for $A <$ 3-dB ripple. The source required for proper termination to produce the Chebyshev response in an even order filter is

$$R_s = \left[\frac{e^{2x} - 1}{e^{2x} + 1}\right]^2 \qquad (2.7\text{-}21)$$

where

$$x = \frac{B_0}{4} \qquad (2.7\text{-}22)$$

The parameters a_1, b_1, and g_1 are calculated. Then these values are used, along with a_2 and b_2, to evaluate g_2. This continues. The evaluation of the kth component depends upon earlier calculations associated with $k-1$.

Table 2.1 is a tabulation of some results of these equations. The g_k values are presented for filters from order 2 through 5. The cases shown are for the Butterworth filter, shown as $A = 0$, and Chebyshev filters with passband ripple of $A = 0.1$, 0.25, 0.5, 0.75, 1.0, and 1.5 dB. The frequencies where the attenuation is -3 dB are also shown as well as the source resistances needed for achieving the Chebyshev response.

Sec. 2.7 The All-Pole Low Pass Filter

Table 2.1 g_k Values for Chebyshev and Butterworth ($A = 0$) Low Pass Filters. (Normalized to $R_L = 1\ \Omega$ and a ripple bandwidth of 1 rad s^{-1})

A, dB	n	g_1	g_2	g_3	g_4	g_5	R_s^{-1}	ω, -3 dB
0	3	1.000	2.000	1.000	—	—	1.000	1.000
	4	0.7654	1.8478	1.8478	0.7654	—	1.000	1.000
	5	0.6180	1.618	2.000	1.6180	0.618	1.000	1.000
0.1	2	0.8430	0.6220	—	—	—	0.7378	1.9432
	3	1.0316	1.1474	1.0316	—	—	1.000	1.3890
	4	1.1088	1.3062	1.7704	0.8181	—	0.7378	1.2131
	5	1.1468	1.3712	1.9750	1.3712	1.1468	1.000	1.1347
0.25	2	1.1132	0.6873	—	—	—	0.6174	1.5981
	3	1.3034	1.1463	1.3034	—	—	1.000	1.2529
	4	1.3782	1.2693	2.0558	0.8510	—	0.6174	1.1398
	5	1.4144	1.3180	2.2414	1.3180	1.4144	1.000	1.0887
0.5	2	1.4029	0.7071	—	—	—	0.504	1.3897
	3	1.5963	1.0967	1.5963	—	—	1.000	1.1675
	4	1.6703	1.1926	2.3661	0.8419	—	0.504	1.0931
	5	1.7058	1.2296	2.5408	1.2296	1.7058	1.000	1.0593
0.75	2	1.6271	0.7002	—	—	—	0.4304	1.2852
	3	1.8243	1.0436	1.8243	—	—	1.000	1.1236
	4	1.8988	1.1243	2.6124	0.8172	—	0.4304	1.0689
	5	1.9343	1.1551	2.7833	1.1551	1.9343	1.000	1.0439
1.00	2	1.8219	0.6850	—	—	—	0.3760	1.2176
	3	2.0236	0.9941	2.0236	—	—	1.000	1.0949
	4	2.0991	1.0644	2.8311	0.7892	—	0.3760	1.0530
	5	2.1349	1.0911	3.0009	1.0911	2.1349	1.000	1.0338
1.50	2	2.1688	0.6470	—	—	—	0.2983	1.1307
	3	2.3803	0.9069	2.3803	—	—	1.000	1.0574
	4	2.4586	0.9637	3.2300	0.7335	—	0.2983	1.0322
	5	2.4956	0.9850	3.4017	0.9850	2.4956	1.000	1.0205

Other characteristics of the Chebyshev filter are apparent from the table. Looking at the -3-dB frequencies, it is seen that the rate of attenuation increases both for increasing n and for increased passband ripple. If passband ripple is allowed, but as much stopband attenuation as possible is needed, the Chebyshev low pass filter with high ripple is called for. If the passband must be "clean" with a minimum of ripple, the Butterworth, or a low ripple Chebyshev, is the choice.

Evaluation of the 3-dB frequencies gives some indication of the rate of attenuation as the frequency moves from the passband to the stopband. However, it is less than analytical. Knowledge of the stopband characteristics is often of great importance in filter specification. As often as not, the desired goal is to provide a given level of attenuation at a specific frequency that is out of the passband of a filter. Table 2.2 presents this data.

Table 2.2 shows the attenuation values in the stopband for $\omega = 1, 2, 3, 4, 5, 6, 8,$ and 10. All of the filters presented in Table 2.1 are considered. The data in the stopband attenuation table is extended to filters up to the seventh order.

Table 2.2 Attenuation values in dB for Butterworth and Chebyshev filters for orders from 2 to 7. $\omega = 1$ corresponds to the edge of the ripple band for Chebyshev filters and the 3-dB point for the Butterworth filter.

Table 2.2a Butterworth Filter

n	1	2	3	4	5	6	8	10
2	3.0	12.3	19.1	24.1	28.0	31.1	36.1	40.0
3	3.0	18.1	28.6	36.1	41.9	46.7	54.2	60.0
4	3.0	24.1	38.2	48.2	55.9	62.3	72.2	80.0
5	3.0	30.1	47.7	60.2	69.9	77.8	90.3	100.0
6	3.0	36.1	57.3	72.2	83.9	93.4	108.4	120.0
7	3.0	42.1	66.8	84.3	97.9	108.9	126.4	140.0

Table 2.2b 0.1-dB Chebyshev Filter

n	1	2	3	4	5	6	8	10
2	0.1	3.3	8.9	13.7	17.6	20.7	25.8	29.7
3	0.1	12.2	23.6	31.4	37.4	42.2	49.8	55.6
4	0.1	23.4	38.9	49.3	57.3	63.7	73.8	81.6
5	0.1	34.8	54.2	67.3	72.2	85.3	97.9	107.6
6	0.1	46.3	69.5	85.2	97.1	106.8	121.9	133.6
7	0.1	57.7	84.8	103.1	117.0	128.3	146.0	159.0

Table 2.2c 0.25-dB Chebyshev Filter

n	1	2	3	4	5	6	8	10
2	0.25	5.9	12.6	17.6	21.6	24.8	29.8	33.7
3	0.25	16.1	27.6	35.5	41.4	46.3	53.9	59.7
4	0.25	27.5	43.0	53.4	61.4	67.8	72.9	85.7
5	0.25	38.9	58.3	71.4	81.3	89.3	101.9	111.7
6	0.25	50.3	73.6	89.2	101.2	110.8	126.0	137.7
7	0.25	61.8	88.9	107.2	121.1	132.4	150.0	157.3

Table 2.2d 0.5-dB Chebyshev Filter

n	1	2	3	4	5	6	8	10
2	0.5	8.4	15.6	20.7	24.7	27.9	32.9	36.8
3	0.5	19.2	30.8	38.6	44.6	49.4	57.0	62.8
4	0.5	30.6	46.1	56.5	64.5	70.9	81.0	88.8
5	0.5	42.0	61.4	74.5	84.4	92.5	105.1	114.8
6	0.5	53.5	76.7	92.4	104.3	114.0	129.1	140.8
7	0.5	64.9	92.0	110.3	124.2	135.5	153.2	166.8

Sec. 2.7 The All-Pole Low Pass Filter

Table 2.2e 0.75-dB Chebyshev Filter

n	1	2	3	4	5	6	8	10
2	0.75	10.1	17.4	22.6	26.6	29.8	34.8	38.7
3	0.75	21.1	32.7	40.5	46.5	51.3	58.9	64.7
4	0.75	32.5	48.0	58.4	66.4	72.8	82.9	90.7
5	0.75	43.9	63.3	76.3	86.3	94.3	107.0	116.7
6	0.75	55.4	78.6	94.3	106.2	115.9	131.0	142.7
7	0.75	66.8	93.9	112.2	126.1	137.4	155.1	168.7

Table 2.2f 1.0-dB Chebyshev Filter

n	1	2	3	4	5	6	8	10
2	1.0	11.4	18.8	24.0	27.9	31.6	36.2	40.1
3	1.0	22.5	34.0	41.9	47.8	52.7	60.3	66.1
4	1.0	33.9	49.4	59.8	67.8	74.2	84.3	92.1
5	1.0	45.3	64.7	77.7	87.7	95.7	108.4	118.1
6	1.0	56.7	80.0	95.6	107.6	117.2	132.4	144.1
7	1.0	68.2	95.3	113.6	127.5	138.8	156.4	170.1

Table 2.2g 1.5-dB Chebyshev Filter

n	1	2	3	4	5	6	8	10
2	1.5	13.3	20.8	26.0	30.0	33.2	38.2	42.1
3	1.5	24.5	36.1	43.9	49.9	54.7	62.3	68.1
4	1.5	35.9	51.4	61.8	69.8	76.2	86.3	94.1
5	1.5	47.3	66.7	79.7	89.7	97.7	110.4	120.1
6	1.5	58.8	82.0	97.7	109.6	119.3	134.4	146.4
7	1.5	70.2	97.3	115.6	129.5	140.8	158.5	172.1

Care must be exercised in using the data in Table 2.2, for the standard Chebyshev equations have been used in evaluation. Hence, for a Chebyshev filter with a passband ripple of 0.25 dB, the attenuation at $\omega = 1$ will be 0.25 dB. Data on the -3-dB frequencies given in Table 2.1 will allow the data of Table 2.2 to be normalized to -3 dB as needed. Similar data are presented in graphical form by Zverev (9) where all information has been normalized to a 3-dB cutoff frequency.

Use of these tables is illustrated by an example. Assume a low pass filter is to be built with the specification that the stopband attenuation should be greater than 60 dB. The stopband, in this case, will be defined as frequencies greater than twice the edge of the ripple band. We find upon examining $\omega = 2$ that attenuations greater than 60 dB occur for the seventh order Chebyshev filters with a ripple of 0.25 dB or more.

The Butterworth filter has an advantage that is not apparent from the calculations. It has the virtue that it is generally insensitive to small changes in component values. This is demonstrated by differentiating the salient equations with respect to component value. The Chebyshev response, however, is much more dependent upon careful component selection.

The symmetry of the filters is interesting. The Butterworth filters are always symmetrical with respect to component value. That is, equal g_k values appear about a centerline drawn through the filter. The odd ordered Chebyshev filters are also symmetrical. This is not true for the even order Chebyshev prototypes though.

The location of the poles of the Chebyshev filter are given as

$$\sigma_k = \sinh N \sin(2k-1)\frac{\pi}{2n} \qquad (2.7\text{-}23)$$

$$\omega_k = \cosh N \cos(2k-1)\frac{\pi}{2n}$$

where N is given by

$$N = \frac{1}{n}\sinh^{-1}\frac{1}{\epsilon} \qquad (2.7\text{-}24)$$

It may be shown that the poles of the Chebyshev filter lie on an ellipse. They are, of course, in the left half of the s plane. The ellipse is inside the circle of the Butterworth filter. This is significant. It indicates that the real part of the poles of the Chebyshev filter have a smaller magnitude than the Butterworth does. Hence, damping is not as great—the poles have a higher Q than do those of an equal ordered Butterworth filter. This is reflected in the time domain responses. The Chebyshev filter shows more overshoot in both the impulse and step responses than does the Butterworth. For this reason, the Chebyshev is not recommended for applications where very high transition rates are encountered, such as pulses.

The Butterworth filter has a somewhat improved response to an impulse and a step than the Chebyshev does. However, it is still not optimum. Other polynomials are suggested for improved time domain response. These include the Bessel, the Gaussian, and the minimum phase ripple filters. These are covered in detail in the work of Zverev (9).

2.8 FILTER DENORMALIZATION, PRACTICAL LOW PASS, HIGH PASS, AND BANDPASS STRUCTURES

The previous section showed how normalized components were calculated for low pass filters. It is rare indeed when filters are built for a cutoff frequency near 1 rad s^{-1} and a 1-Ω termination. Impedances are usually higher and cutoff frequencies are arbitrary. The equations for denormalization of the previous filters will be presented in this section.

Sec. 2.8 Filter Denormalization, Practical Low Pass, High Pass, and Bandpass Structures

A low pass filter may begin with the first component being either a capacitor or an inductor. The next component in the ladder must, however, be of the opposite type. Each additional part must be opposite from the previous.

Intuition will generally prevail in the denormalization process. If a termination impedance higher than 1 Ω is desired, the reactances are multiplied accordingly. The reactances are then scaled so that they have the same value at the new desired cutoff frequency, ω_c. The stopband attenuation values presented in Table 2.2 will scale with ω replaced by the ratio of the actual cutoff to the frequency where a given value of attenuation is observed.

The high pass filter is the dual of the low pass. Series inductors are replaced with series capacitors while shunt capacitors are replaced by inductors. The reactances are the same at the normalized cutoff of 1 rad s^{-1}.

The components for a low pass filter based upon the normalized results, Table 2.1, are

$$C_k = \frac{g_k}{R_0 \omega_c} \qquad (2.8\text{-}1)$$

and

$$L_k = \frac{g_k R_0}{\omega_c} \qquad (2.8\text{-}2)$$

The high pass filter is described by the following denormalization of the normalized low pass tables

$$C_k = \frac{1}{g_k R_0 \omega_c} \qquad (2.8\text{-}3)$$

and

$$L_k = \frac{R_0}{\omega_c g_k} \qquad (2.8\text{-}4)$$

In all of the equations above, $\omega_c = 2\pi f_c$ where f_c is the cutoff frequency in hertz (Hz), and R_0 is the termination in ohms.

Assume that low pass and high pass filters with a Butterworth response are to be built with a cutoff frequency of 5 MHz and terminations at each end of 50 Ω. Assume that the attenuation should be 30 dB or more at frequencies separated by a factor of 2 from the cutoff. We see from Table 2.2 that a fifth order filter is good enough to provide the needed attenuation. The normalized component values are $g_1 = g_5 = 0.618$, $g_2 = g_4 = 1.618$ and $g_3 = 2.000$. The filters are designed using this data with the preceding equations.

The resulting filters are shown in Fig. 2.20. The low pass is designed with a

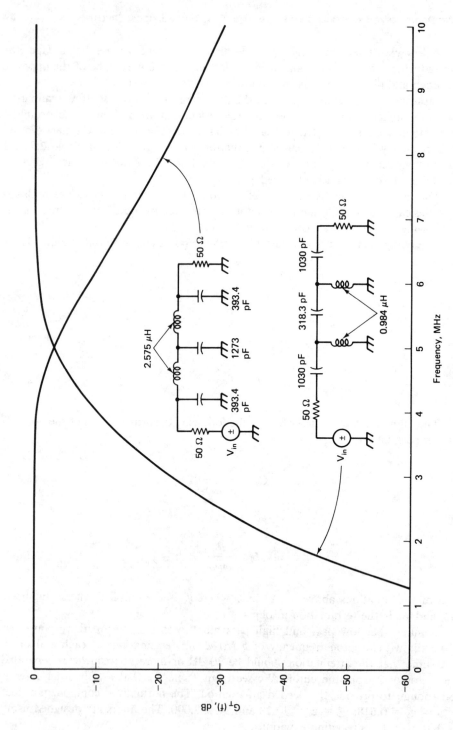

Figure 2.20 Circuits and frequency responses for five-pole low pass and high pass filters, each with a 5-MHz cutoff frequency. The frequency response curves would be absolutely symmetrical about the cutoff frequency if a logarithmic frequency scale was used.

shunt capacitor as the input element while the high pass uses a series capacitor at the input. The transducer gain of each is also presented in the figure. They were obtained with the ladder method of analysis.

The duals of the filters of Fig. 2.20 are presented in Fig. 2.21. The choice of the type to build is purely arbitrary. It will depend upon the application and the practicality of the components.

The low pass filter is the mirror image of the high pass although it may not appear so from the curves of Fig. 2.20. Recall that normalized data are presented in terms of frequency ratios. The cutoff of each of the filters in Fig. 2.20 is 5 MHz. The low pass is 30 dB down at twice this value, 10 MHz. Similarly, the high pass is 30 dB down at half the cutoff, or 2.5 MHz.

The transfer functions of the two filters will differ significantly. The low pass will be an all-pole function. However, the high pass filter will show zeros. Still, the high pass structure is usually termed a five-pole filter, for it resulted from denormalization of a five-pole low pass.

A simple bandpass filter is designed with little more effort than a low pass. The procedure is straightforward. First, the two band edges are determined. The difference is the bandwidth. The center frequency is the geometric mean of the two edges

$$f_{center} = (f_1 f_2)^{1/2} \qquad (2.8\text{-}5)$$

where f_1 and f_2 are the passband edges. A low pass filter is first designed with a bandwidth equaling that of the desired bandpass filter. Then, the resulting low pass elements are resonated with the addition of components at the center frequency of the filter as given by Eq. 2.8-5.

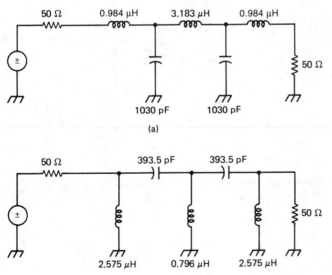

Figure 2.21 The dual forms of the 5-MHz cutoff filters of Fig. 2.20.

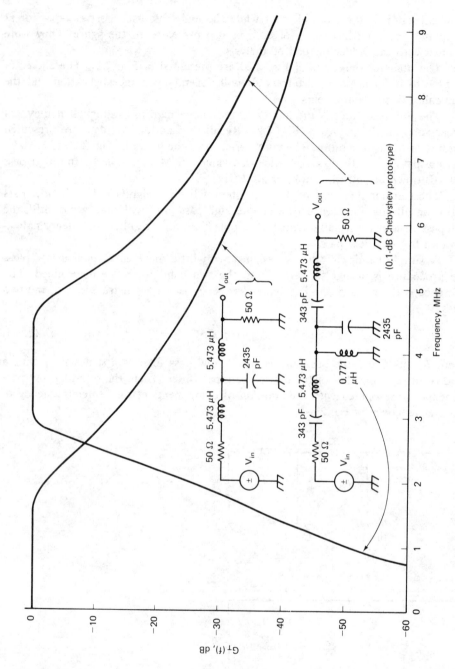

Figure 2.22 Circuits and frequency responses for a 1.5-MHz cutoff low pass filter and a bandpass filter with a 1.5-MHz bandwidth. The bandpass circuit is realized by resonating the elements of the low pass at the desired center frequency.

Assume that a bandpass filter is to be built to cover the range from 3 to 4.5 MHz and that the variation in response within the passband should not vary more than 0.1 dB. Assume also that stopband requirements are consistent with a third order Chebyshev filter and a 50-Ω termination is needed. The resulting filter design is shown in Fig. 2.22. From Table 2.1, a low pass filter with an 0.1-dB Chebyshev response is designed. The series inductors are then resonated at 3.674 MHz with series capacitors while the parallel capacitor is tuned with an appropriate parellel inductor. The result is the desired bandpass filter. Both low pass and bandpass responses and both circuits are given in Fig. 2.22.

This bandpass filter lacks arithmetic symmetry. If needed, this is approximated by cascading a suitable low pass filter with a cutoff equal or just higher than the upper cutoff of the bandpass structure.

The transfer function of the bandpass filter will contain both poles and zeros. Where the low pass prototype had three poles, this filter has six. They will occur in two bunches arranged as points on ellipses. The ellipses are centered about the positive and negative points in the s plane corresponding to the center frequency of the filter.

Even though this filter has six poles in its response, it is still a three-pole bandpass filter. It has three pole pairs and is derived from a three-pole low pass prototype. Generally, bandpass filters described by the number of poles are really being described by the number of resonators in the filter.

This method of bandpass synthesis is generally limited to wide bandwidths, usually 10% of the center frequency or more. In addition, the component values that are encountered are sometimes much less than practical. Methods for the design of narrower bandpass filters are presented in the next chapter.

The Chebyshev examples presented have been denormalized on the basis of the edge of the ripple band. Assume that it is desired to build a filter with a Chebyshev response, but with a well-defined 3-dB bandwidth. This is done by modifying the normalized component values. The key is the 3-dB frequency, which has been evaluated for the g_k values shown in Table 2.1. The g_k values are all multiplied by the normalized 3-dB frequency. The resulting filter will then have the same response as the original, but will have 3 dB of attenuation at 1 rad s^{-1}. The new normalized g_k results are then used in the denormalization equations.

The designer using any of the many collections of tabulated data should be careful to ascertain the form of that data. The extensive tables and graphs presented by Zverev are all normalized to a 3-dB cutoff frequency of 1 rad s^{-1}, making comparisons more direct.

REFERENCES

1. Van Valkenburg, M. E., *Network Analysis*, 3rd ed., Prentice-Hall, Englewood Cliffs, N.J., 1974.
2. Skilling, Hugh Hildreth, *Electric Networks*, John Wiley & Sons, New York, 1974.

3. Blinchikoff, Herman J. and Zverev, Anatol I., *Filtering in the Time and Frequency Domains*, John Wiley & Sons, New York, 1976.
4. Band, William, *Introduction to Mathematical Physics*, D. Van Nostrand, Princeton, N.J., 1959.
5. Johnson, D. E., Hilburn, J. L., and Johnson, J. R., *Basic Electric Circuit Analysis*, Prentice-Hall, Englewood Cliffs, N.J., 1978.
6. Ramo, Simon, Whinnery, John R., and Van Duzer, Theodore, *Fields and Waves in Communication Electronics*, John Wiley & Sons, New York, 1965.
7. DeMaw, M. F., *Ferromagnetic-Core Design and Application Handbook*, Prentice-Hall, Englewood Cliffs, N.J., 1981.
8. See reference 3.
9. Zverev, Anatol I., *Handbook of Filter Synthesis*, John Wiley & Sons, New York, 1967.

3

Coupled Resonator Filters

The previous chapter introduced filter concepts. This led to normalized low pass filters. Methods were given for denormalization of these into practical low and high pass filters and bandpass filters of 10% or greater width.

The study of filters continues in this chapter. The emphasis is now on narrower bandwidth bandpass filters. They are of great importance in all rf design.

The wide bandwidth bandpass filters studied earlier were limited in many ways. The design methods work well only for wide filters. The components were often less than practical, especially at higher frequencies. Finally, little flexibility was open to the designer. Virtually all filter elements were determined during the synthesis process for a desired response shape and number of poles.

The filters described in this chapter offer more flexibility to the designer. Starting with inductors of known Q_u, but a more or less arbitrary inductance, a filter of arbitrary shape and complexity may be designed through control of coupling between the resonators and loading of the end sections of the filter.

3.1 IMPEDANCE TRANSFORMING NETWORKS

Most filters in common rf usage are doubly terminated. That is, they require a resistive termination at each end of the filter. The terminations are mandatory to achieve the desired response. A common value at all rf frequencies is 50 Ω. This allows the circuit to be interconnected with others through a 50 Ω transmission line.

As we will find later, great flexibility may be achieved by terminating a filter in immittances that are different than 50 Ω. Yet, if the filter terminals are to be connected to a 50 Ω load, some means of transforming impedances will be required.

The first transformation element considered is the ideal transformer, shown in Fig. 3.1. Assume that the secondary of the transformer is left open with the load,

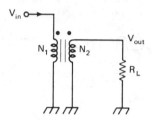

Figure 3.1 Ideal transformer with an N_1/N_2 turns ratio.

R_L, infinite. A voltage source is applied to the primary of the transformer from an ac generator.

The primary will have some inductance related to the nature of the core and the number of turns. From this inductance, L_p, a frequency domain impedance, sL_p, is determined which will allow evaluation of the current flowing in the primary. The current will be small if the inductance is large. The presence of this current will establish a magnetic field in the core. It is this field and the energy stored in it that leads to the inductive effects.

The secondary of the transformer "sees" the magnetic field from the primary current. A voltage appears across the secondary, although no current can flow owing to the open circuited termination.

Note the dots in Fig. 3.1. They indicate the polarity of the voltages. If the instantaneous voltage on the primary dot is positive, the voltage on the secondary dot is also positive. The output voltage is related to the input by $V_{out} = V_{in} N_2/N_1$. This is a result of the fundamental physics of the device. Note that this relationship is independent of the current flowing in the primary inductor. The primary inductance might be so large that the current is vanishingly small. While no current flows in the open circuited secondary, the voltage is still well defined by the turns ratio.

Assume now that a finite load resistance, R_L, is connected to the secondary. A voltage is again applied to the primary. The magnetic field within the transformer core begins to increase at the instant primary current begins to flow. The secondary voltage follows, but now with a current flowing. If the primary voltage is initially positive, that at the secondary will also be positive. Hence, a current flow into the primary dot causes a current flow out of the secondary dotted end of the transformer.

Examine the resulting magnetic field within the core. The net field has two components, one resulting from current flowing in the primary and the other from secondary current. The fields are of opposite polarities. The net result is that they cancel. The inductive nature of the original core is gone with no net magnetic field present. The output voltage merely follows that of the input according to the turns ratio of the transformer, N_2/N_1. The current flowing in the primary is in phase with the driving voltage—hence, the input impedance appears as a resistor. This assumes that the primary inductance was high enough that negligible current would flow when the secondary was open circuited.

Conservation of energy may be applied to the transformer to infer the impedance transformation. The transformer is assumed lossless, for it is ideal. The output voltage is $V_{out} = V_{in} N_2/N_1$ for a given input, V_{in}. The current in the secondary is that flowing in the load, $I = V_{in} N_2/(N_1 R_L)$. The energy dissipated in the resistor is the VI product, $P = V_{in}^2 N_2^2/(N_1^2 R_L)$. The same energy must also be flowing into the primary. Hence, the primary current is $I_p = P/V_{in}$. The resistance seen looking into the primary is the ratio of the input voltage to the primary current, $R_{in} = R_L(N_1^2/N_2^2)$. The impedance transformation provided is in accordance with the square of the turns ratio.

There would be a phase difference between the secondary current and the voltage if the output termination was reactive instead of a pure resistance. However, it is the current flow that leads to the canceling magnetic fields within the core. The ideal transformer will have a primary current phase difference which is identical to that of the secondary. Hence, a reactive termination is also transformed by the square of the turns ratio.

Both primary and secondary currents were calculated in evaluating impedance transformation. Consider the ratio of the currents, I_s/I_p. This is equal to N_p/N_s. That is, while the voltage changes according to the turns ratio, the currents in the primary and secondary change according to the reciprocal of the turns ratio. The sum of the currents into the dots multiplied by the number of turns associated with each winding is zero. Analytically

$$\sum i_k N_k = 0 \qquad (3.1\text{-}1)$$

This relationship is of greatest significance when evaluating a transformer with more than two windings.

The transformer of Fig. 3.1 is ideal. While this is often an adequate model for rf circuit design, many effects have not been considered. The primary inductance is not always arbitrarily large. Sometimes transformer construction and low core permeability cause leakage inductance to appear. This results from not all of the primary magnetic field passing through the secondary winding. Also, transformers show loss.

A more complete model for the rf transformer is shown in Fig. 3.2. It consists of the ideal transformer with added components. L_p is the inductance of the primary.

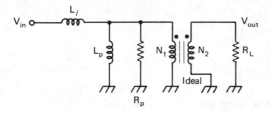

Figure 3.2 Model for a broadband transformer. R_p represents core losses. L_p is the primary inductance while L_l is the leakage inductance.

This inductance presents a decreasing impedance as the input frequency is lowered, shunting some of the available drive current from the ideal primary. This is the factor that usually limits low frequency performance. The inductive reactance of L_p should be at least 5 or 10 times larger than the impedance that is desired at the input.

The resistor represents the losses in the core material. It is usually large and of little significance at low frequencies. However, losses become more important as the vhf and uhf ranges are reached. It is common for ferrite cores to specify the resistance of a single turn, with the total parallel resistance, R_p, then being proportional to the square of the number of primary turns. L_1 represents the leakage inductance. This element usually limits the high frequency performance of a real transformer.

It is necessary to consider only effects in the primary owing to the impedance transforming properties of the ideal transformer. The nonideal nature of the secondary is reflected through the transformer to the primary.

Transformers are often wide band devices. They have frequency ranges from one to several decades. The widest bandwidth rf transformers are those using a combination of a ferrite core and transmission line windings. They will be discussed later in the text.

A series or shunt reactive element will perform a significant impedance transformation over a limited narrow frequency range. Consider the effect of a series capacitor as shown in Fig. 3.3. This is equivalent to the parallel combination of a larger parallel resistor, R_p, and a somewhat smaller parallel capacitor, C_p.

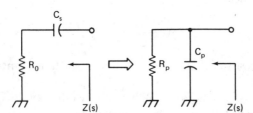

Figure 3.3 Impedance transformation with a series capacitor.

The impedance, $Z(s)$, looking into the series combination is

$$Z(s) = \frac{1}{sC} + R_0 = \frac{1 + sCR_0}{sC} \tag{3.1-2}$$

The admittance is the reciprocal of $Z(s)$

$$Y(j\omega) = \frac{j\omega C}{1 + j\omega CR_0} = \frac{j\omega C(1 - j\omega CR_0)}{1 + \omega^2 C^2 R_0^2} \tag{3.1-3}$$

A complex admittance is always of the form $Y = G + jB$ where G is a conductance, the reciprocal of a parallel resistance, and B is a susceptance, the reciprocal of a parallel reactance. Taking the reciprocal of the real part of Eq. 3.1-3

$$R_p = \frac{1+\omega^2 C^2 R_0^2}{\omega^2 C^2 R_0} \qquad (3.1\text{-}4)$$

A value of parallel resistance, R_p, is usually required while R_0 is known. The capacitor value must be calculated. Noting that $1/(\omega C)$ is the capacitive reactance, Eq. 3.1-4 may be solved for X_C

$$X_C = (R_p R_0 - R_0^2)^{1/2} \qquad (3.1\text{-}5)$$

The imaginary part of Eq. 3.1-3 provides the susceptance, B. Note that a parallel capacitor has a susceptance, ωC. Hence, the parallel capacitance, C_p, is obtained by dividing the imaginary part of Eq. 3.1-3 by ω

$$C_p = \frac{C}{1+(R_0^2/X_C^2)} \qquad (3.1\text{-}6)$$

C_p is very close to C for even moderate impedance transformations.

Assume that a terminating resistance of 1000 Ω is required from a 50-Ω characteristic impedance, R_0. From Eq. 3.1-5 the series capacitor must have a reactance of 217.9 Ω. The parallel capacitor, C_p, will have a value of 95% of the series C.

The transformed parallel impedance resulting from a series capacitor is reactive. However, the extra reactance is usually absorbed in following circuitry. A parallel inductor may be placed in shunt with the series RC combination if it is desired to see a purely real impedance after transformation. This is one form of the popular L-network shown in Fig. 3.4a. The inductor may be made variable if adjustment is required. Alternatively, a slightly smaller inductance may be used with an adjustable parallel capacitor. This form is shown in Fig. 3.4b.

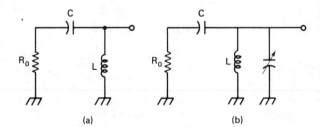

Figure 3.4 Impedance transformation with a series capacitor and (a) a shunt inductance to eliminate reactance. The inductance is replaced with a resonator in (b).

A similar impedance transformation may be performed by placing an inductor in series with R_0. The equations are similar and easily derived.

Another popular form of impedance transformation is the tapped inductor. This is shown in Fig. 3.5 along with the model used for evaluation. Also shown is the essentially identical form where a link, or small winding over a larger inductor, is used. The inductor has a value L with a total number of turns N_t with the load tapped at a point N_L turns from the grounded end. This is modeled by the parallel

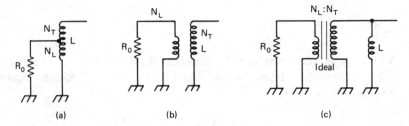

Figure 3.5 Tapped (a) or link coupled inductor (b) as an impedance transforming element. A model with an ideal transformer is shown at (c).

combination of the inductor and an ideal transformer with the same turns ratio as the tapped inductor.

The model of Fig. 3.5 works well for inductors where the total inductance is proportional to the square of the number of turns. Examples are inductors wound on toroid cores of reasonable permeability (six or greater), pot cores and solenoid inductors with a length small with respect to the diameter. Solenoidal inductors with a length greater than the diameter do not increase in inductance as the square of the turns. This is because some of the magnetic flux leaks out and does not pass through turns at the opposite end of the coil. Tapped impedance transformations may still be used. However, the ratio of the transformation must be determined empirically or from more elaborate transformer models.

The tapped or link coupled inductor has one virtue that is not shared by the series capacitor, a significantly extended bandwidth. The loading on the inductor from the tap is constant over a wider frequency range than would result from a series capacitor.

The series capacitor transformation of Fig. 3.3 is limited to increased resistance levels. It is required sometimes to transform in the other direction. This is done with a parallel capacitor as shown in Fig. 3.6. The analysis is similar to the series

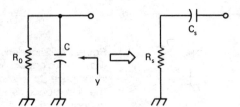

Figure 3.6 Use of a shunt capacitor for impedance transformation.

capacitor. Admittance is calculated, transformed into an impedance and separated into real and imaginary parts. The real part is the equivalent series resistance, R_s

$$R_s = \frac{R_0}{1 + \omega^2 C^2 R_0^2} \quad (3.1\text{-}7)$$

This equation may be solved for the case where a given series resistance is desired from an available termination, R_0, yielding the required capacitive reactance

$$X_s = R_0 \left(\frac{R_s}{R_0 - R_s}\right)^{1/2} \tag{3.1-8}$$

The equivalent series capacitance is

$$C_S = \frac{1 + \omega^2 C^2 R_0^2}{\omega^2 C R_0^2} \tag{3.1-9}$$

As an example of application of these transformation methods, the simplest of bandpass filters, the single tuned circuit, will be designed. Assume that the filter has a center frequency of 5 MHz and a bandwidth of 200 kHz. Hence, the loaded Q of the resonator will be $5/0.2 = 25$. The inductor will be wound on a toroid fabricated from powdered iron. The core has an inductance constant $K = 2.9$ nHt^{-2}, the inductance of a toroid being well approximated by $L = Kn^2$. In addition, this core will show a Q_u at 5 MHz of 200, relatively insensitive to the number of turns of the coil. Lossless capacitors are assumed.

The insertion loss is already determined even though the inductance has yet to be chosen. It is given by IL $= -20 \log(1 - Q_L/Q_u) = 1.16$ dB.

The filter will be inserted in a 50-Ω system with equal loading by the input and the output. The loaded Q accounts for both internal losses and external loading. The external loading is described by Q_e with the relation $Q_L^{-1} = Q_e^{-1} + Q_u^{-1}$. Hence, $Q_e = 28.57$.

Assume the toroid is wound with 45 turns, producing an inductance of 5.87 μH. The nodal capacitance, the value required to resonate the inductor at the center frequency, is thus 172.5 pF. The inductive reactance is 184.4 Ω. The equivalent parallel resistance that must load the resonator externally is $Q_e \omega L = 5270$ Ω. However, half of the loading must come from each side of the doubly terminated filter. Each end loading must appear as 10,540 Ω in parallel with the resonator.

One end will be loaded through a link while a series capacitor is used at the other. The ratio of the required resistance to 50 Ω is 210. The turns ratio must be 14.5. This will be closely approximated by a three-turn link on the 45-turn inductor. Equation 3.1-5 is used at the other end to determine that a capacitor with a reactance of 724.3 Ω is needed for the series connection. This is 43.9 pF at 5 MHz. The equivalent parallel capacitance is 43.7 pF. Hence, the additional capacitance required to tune the resonator, C_T, is the nodal capacitance less 43.7 pF, or 128.8 pF. The final circuit is shown in Fig. 3.7. Note that little error is involved in using the actual series capacitance in place of the equivalent parallel value when evaluating C_T. This is commonly done, for in practice C_T is usually adjustable.

The use of a series capacitor as an impedance matching element is common. However, it does present a problem. It introduces additional zeros in the frequency domain transfer function. This forces filters using this method to take on high pass characteristics away from the passband. The stopband attenuation at high frequencies is compromised.

If series resonators are used instead of parallel ones, the end matching is often done with parallel capacitors. This leads to a low-pass–like behavior away from the

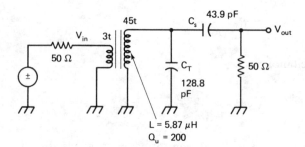

Figure 3.7 Single tuned, doubly terminated resonator. Input coupling from a 50-Ω source is through a link while output coupling is from a series capacitor to a 50-Ω load. The inductance is resonated with a combination of capacitances.

passband. The best filters, where stopband attenuation is critical, will probably use a combination of both methods.

3.2 NORMALIZED COUPLING AND LOADING COEFFICIENTS

Figure 3.8 shows the normalized low pass filter discussed in Chap. 2. The component values, g_k, were derived from polynomials which approximated a desired characteristic in either the frequency or time domain. A given set of g_k values for the inductors and capacitors will produce a unique $H(s)$.

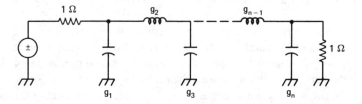

Figure 3.8 Normalized all-pole low pass filter prototype.

The role of the individual L and C values can take on a different meaning. The relationship between the elements and the terminating impedance will determine the Q of the end elements of the filter. The ratio of the individual elements will determine the energy proportion that is stored in each element. While the results are the same, the new viewpoint will allow a different form of denormalization, one that is especially suited to the narrow bandwidth bandpass filter.

Consider the filter shown in Fig. 3.8 with a capacitor as the input element. Assuming that only the end capacitor is loaded by the external termination, the q is defined by Eq. 2.6-6 as $q = R\omega C$. The lower case q is used to signify that a normalized value is being evaluated. The q value is g_k for the typical case of an end termination of 1 Ω. For special cases, such as the even ordered Chebyshev filter, $R \neq 1$ and a slightly different value is needed. Evaluation always occurs at the low pass cutoff frequency.

Generator voltage and resistance along with filter input impedance will determine the current injected into the filter. However, input immittance is a function of the output termination and of each of the filter elements. A more pertinent characterization is the way each pair of elements shares current, or energy, when the pair is isolated. For example, the coupling of energy between C_1 and L_2 (g_1 and g_2) in the filter of Fig. 3.8 is evaluated by shorting C_3. This results in the circuit of Fig. 3.9.

The admittance, Y, of Fig. 3.9 is evaluated directly. An impedance results upon

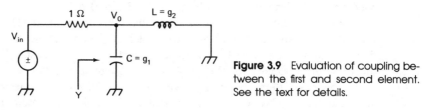

Figure 3.9 Evaluation of coupling between the first and second element. See the text for details.

inversion. Assuming a 1-V source, the voltage at the first node is given from voltage divider action as

$$V_{\text{out}} = \frac{Z}{Z_t} = \frac{sL}{1 + sL + s^2LC} \qquad (3.2\text{-}1)$$

The energy stored in the capacitor and inductor is, respectively, $\tfrac{1}{2}CV^2$ and $\tfrac{1}{2}LI^2$ where the V and I values are peaks. The inductor current is

$$I_L = V_{\text{out}} Y_L = \frac{V_{\text{out}}}{sL} = \frac{1}{1 + sL + s^2LC} \qquad (3.2\text{-}2)$$

Hence, the stored inductor energy is

$$E_L = \frac{\tfrac{1}{2}L}{(1 + sL + s^2LC)^2} \qquad (3.2\text{-}3)$$

The energy stored in the capacitor is similarly

$$E_C = \frac{\tfrac{1}{2}Cs^2L^2}{(1 + sL + s^2LC)^2} \qquad (3.2\text{-}4)$$

Taking the ratio of the two

$$\frac{E_L}{E_C} = \frac{L}{Cs^2L^2} = \frac{1}{s^2LC} \qquad (3.2\text{-}5)$$

This is the square of a resonant frequency related to the two components when evaluated at $\omega = 1$.

Using this as a basis, the normalized coupling coefficient between the n and $n + 1$ components is defined as

$$k_{n,n+1} = (g_n g_{n+1})^{-1/2} \qquad (3.2\text{-}6)$$

The square of this value is indicative of the energy transfer between the two.

Consider the third order Butterworth filter where $g_1 = g_3 = 1$ and $g_2 = 2$. The end resistances are 1. Hence, the end q values are 1 at both ends. The coupling coefficients, k_{12} and k_{23} are both, from Eq. 3.2-6, 0.707.

Extensive tables of k and q values have been tabulated (1). Most are what is termed "predistorted" which will be explained below. These tables, being derived from normalized low pass filter data, are essentially dual data sets. They are generally more useful, especially for the design of narrow bandpass filters.

Low pass filters are easily designed from k and q value tables. The end elements are picked to have a Q equal to q_1 and q_n for a desired cutoff frequency and the new terminating resistance. The k values provide a relation between the elements. Hence, the complete normalized filter is designed.

Denormalization of the k and q values for a bandpass filter is different than we found with the earlier bandpass design. There, a low pass filter with a bandwidth equal to that of the bandpass was designed. Components were then added to resonate each of the low pass elements.

Using k and q values, each element in the low pass prototype filter will transform into a resonator in the final bandpass circuit. A key parameter is the filter Q, or Q_f. This is the ratio of the center frequency of the desired bandpass circuit to the desired bandwidth. The normalized q values are multiplied by Q_f while the normalized k's are divided by Q_f. This results in denormalized end-section Q values and denormalized K values. The end resonators are then loaded as required, using the impedance transformations of the previous section, to produce the proper Q_L values. Similarly, the resonators are coupled to each other to the desired K levels.

A two-pole bandpass filter will be designed for 10 MHz with a bandwidth of 400 kHz as an initial example of design with k and q values. A Chebyshev response with an 0.1-dB ripple will be used.

The normalized low pass data from Table 2.1 is obtained, $g_1 = 0.843$ and $g_2 = 0.622$. The low pass terminations are not equal since this is an even ordered Chebyshev filter. One is 1 Ω while the other is $(0.73738)^{-1} = 1.355$ Ω.

First, the g_k values must be normalized to a cutoff corresponding to the 3-dB frequency rather than the edge of the ripple passband. The 3-dB frequency is 1.9432 rad s^{-1}. Hence, the normalized g_k values are multiplied by this value to realize a normalized low pass filter with a 1 rad s^{-1} cutoff. The new values are $g_1 = 1.638$ and $g_2 = 1.2087$. The prototype filter is shown in Fig. 3.10.

The nonunity termination is associated with the smaller g value. Hence, $q_1 = L/R = 1.638$ and $q_2 = RC = 1.2087 \times 1.355 = 1.638$. The coupling coefficient, k_{12}, is $(g_1 g_2)^{-1/2} = 0.7106$.

The normalized k and q values may now be denormalized to obtain the desired bandpass filter. The filter Q is given as $Q_f = 10/0.4 = 25$. The denormalized Q is then $Q_1 = Q_2 = 25 \times 1.638 = 40.96$. The denormalized coupling factor is $K_{12} = k_{12}/25 = 0.0284$.

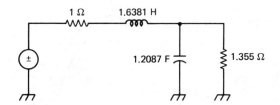

Figure 3.10 Normalized two-pole Chebyshev filter used as a prototype for a bandpass circuit.

Consider now some practical filter components. The choice is somewhat arbitrary although experience with such designs will provide the builder with intuition. Still, there is great flexibility. Assume that the resonators are to be constructed with inductors of 1 μH and an unloaded quality factor of $Q_u = 250$. Both are easily realized with powdered iron toroid cores at 10 MHz. Q_e is calculated as 48.98 using the methods of the previous section. The external resistor that would provide the required loading is $R_{pe} = Q_e \omega L = 3077.4 \Omega$.

A 1-μH inductor will resonate at 10 MHz with a nodal capacitance of 253.3 pF. Of the current flowing in this capacitance, some fraction must be coupled to the adjacent resonator. The fraction is just the coupling coefficient, K_{12}. Hence, using a top coupling capacitor between parallel resonators, the coupling capacitor is $K_{12} C_0$ = 7.2 pF. The capacitances used to tune each resonator must be reduced by this amount. The resulting circuit is shown in Fig. 3.11a. The passband response is shown as the solid curve in the transducer gain plot of Fig. 3.12.

A resistance of 3077 Ω is somewhat obscure and other resistances may be more convenient. The filter could be redesigned with smaller L values for somewhat

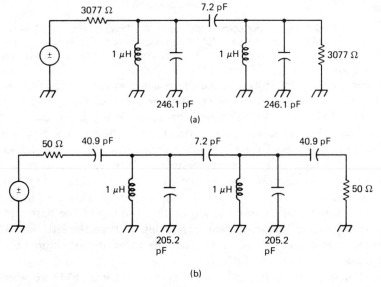

Figure 3.11 Chebyshev bandpass filter resulting from evaluation of the coupling and loading parameters of the low pass prototype of Fig. 3.10. The basic filter is shown at (a) while one with 50-Ω terminations is at (b).

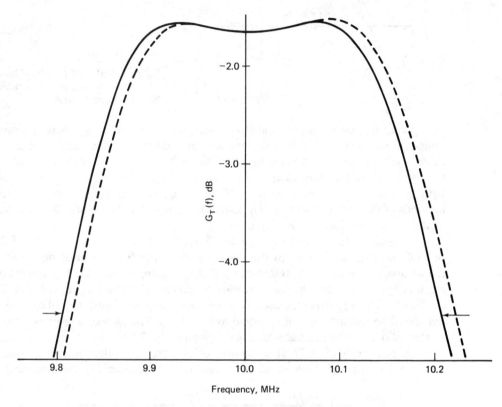

Figure 3.12 Frequency response of the filters of Fig. 3.11. The solid curve represents the case of 3077-Ω terminations at each end while the dashed curve shows the effect of the matching capacitors for a 50-Ω termination.

lower impedances. However, the components would become unreasonable for values as low as 50 Ω. A 50-Ω system could be matched using a series capacitor, shown in Fig. 3.11b. The 40.9-pF capacitors serve to transform the 50-Ω termination to 3077 Ω. The transducer gain of this filter is shown as a dotted line in Fig. 3.12. Note that the tuning capacitors are reduced in value by the end loading capacitors.

Although there are minor differences between the two responses of Fig. 3.12, both show an insertion loss of about 1.55 dB. The filter using capacitive matching at the ends is shifted slightly higher in frequency. This is of little consequence, for slight retuning will remove this problem. The top of the passband has a slight tilt with the higher frequency side slightly higher than the left. The 0.1-dB Chebyshev response is not strictly accurate. The differences are so minimal that they are rarely considered in a practical design.

The filters of Fig. 3.11 were equally terminated. There would have been no loss in the passband if there had been ideal components in the resonators. Such was not the case. The presence of loss, a finite inductor Q, has caused an insertion loss. Had the Q_u value been lower, the loss would have been even greater. Other than the loss though, the shape of the passband is unaltered. This is intuitively reasonable. The low pass prototypes used lossless components. The finite Q of the end sections resulted from the loading by the terminations and nothing more. The finite inductor Q_u was absorbed into the external loading to provide the proper loaded Q. This is not always true.

The design of third and higher order filters is similar to the process just described. The low pass prototype is used to derive k and q values. These are used with resonators of more or less arbitrary choice to determine a desired passband. There are $n-1$ coupling coefficients required, for the filter will contain n resonators. Only the end Q values must be considered.

A significant problem arises if the finite Q of the resonators is not considered. To illustrate this, a third order filter was designed using the methods outlined. No account was taken during the design for the unloaded Q. After design, the transducer gain of the filter was analyzed using the ladder method. The filter was designed for a center frequency of 10 MHz and a bandwidth of 300 kHz. Capacitive elements served as coupling between parallel resonators and for transformation to 50-Ω terminations. A family of curves was then plotted, as shown in Fig. 3.13. The topmost transfer function is that of the design, a 0.1-dB Chebyshev response with infinite unloaded Q. The next three curves are identical except that lower, finite Q_u values were assumed. The curve for $Q_u = 250$ is representative of good quality toroid inductors. The $Q_u = 100$ curve represent a case with slug-tuned inductors while the case with $Q_u = 50$ is typical of the poorest of miniature coils.

The differences are profound. The ideal resonator case shows the desired response, or something close to it. There is no loss in the passband. As the Q_u value decreases, insertion loss grows. In addition, the shape is changed, or distorted. The $Q_u = 250$ case resembles a Butterworth filter much more than it does the original 0.1-dB Chebyshev response. The lower Q_u examples take on a Gaussian-like shape.

The curves are plotted only for the first 20 dB of attenuation. As such, they do not show stopband details. As the input frequency is further removed from band center, the curves differ less and less except in insertion loss.

Consider the effect of a finite Q_u upon the poles of $H(s)$. This is done with a normalized low pass prototype. The three poles start for the lossless case on the ellipse typical of a Chebyshev filter. As loss is introduced in each element of the filter, and hence in each of the low pass elements, the poles move to the left in the s plane. Motion is in the direction of higher damping. The Q of the individual poles decreases. There is no change in the vertical positions though. Such movement is typical of a distorted filter.

The movement of the poles will be the same if the unloaded quality factor of each resonator is the same, the usual case in practice. It will be by an amount equaling the reciprocal of the normalized element q resulting from internal losses.

The way around the distortion problem is to predistort the responses. The critical denormalizing factor in the design of a bandpass filter was the filter Q, Q_f. The critical factor in design of a predistorted filter is q_o, the normalized Q. This is the unloaded Q_u of the resonators divided by Q_f. The values were $Q_f = 25$ and $Q_u =$

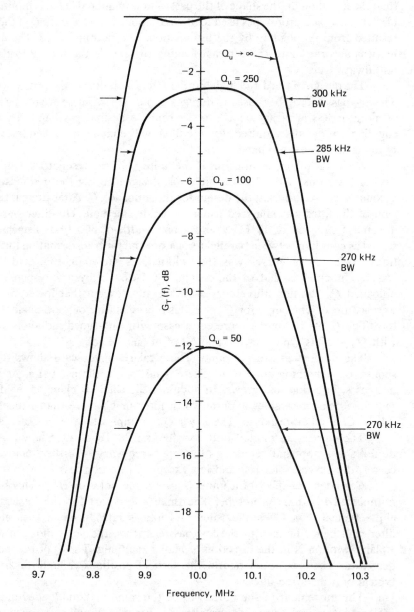

Figure 3.13 Frequency response of a third order Chebyshev bandpass filter. Several values are given for unloaded resonator Q. The high Q case was used in design. Finite Q increases insertion loss and distorts the filter shape. Stopband response is not severely altered by finite resonator Q.

250 for the two-pole filter designed earlier. Hence, the normalized Q was $q_0 = 10$.

Predistortion is straightforward, at least in concept. The presence of a finite Q_u will shift pole positions to the left by an amount dependent upon q_0. Hence, the process is to shift the poles of the ideal low pass prototype to the right by a corresponding amount such that the final result is the pole plot of the desired response.

There is a limit to the extent this procedure may be applied. The leftward

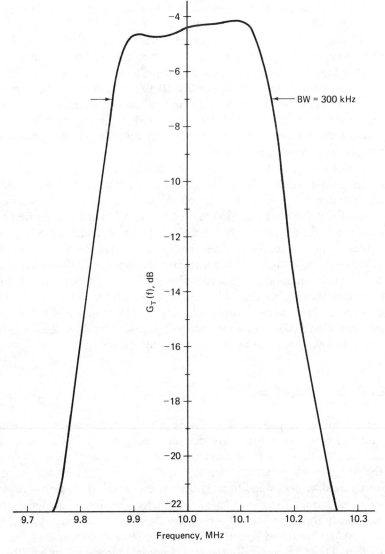

Figure 3.14 Frequency response of a third order Chebyshev filter designed using predistortion techniques. The "tilt" in the passband response may usually be eliminated during the tuning process.

shift in pole position from finite element loss is the same for each pole. The prototype poles must be shifted to the right by a constant amount. However, some of the poles are close to the vertical axis of the s plane. Excess movement to the right would carry the pole positions into the right half plane, a physically impossible condition for passive circuits (and an uncomfortable one for active ones!). There is a minimum value of q_0 for each type of filter shape for which predistortion can be applied. The value grows for filters of higher order.

Zverev has tabulated exhaustive tables of k and q values for predistorted filters of many different kinds and orders (2). These tables are indispensable for the rf filter designer. Figure 3.14 shows the response of a third order, 0.1-dB Chebyshev filter designed using the Zverev predistorted tables. The Q_u value assumed was 250 and the filter was designed to complement the responses of Fig. 3.13 where predistortion was ignored. Generally, the 0.1-dB Chebyshev response is preserved. There is a tilt in the passband shape from the utilization of capacitive coupling between resonators and into the 50-Ω terminations. Note that the insertion loss is higher than was true for the filter of Fig. 3.13 with $Q_u = 250$. This is the price that must be paid for preservation of filter shape.

We were able to design the two-pole Chebyshev filter of Fig. 3.11 without the use of predistorted methods for a fortuitous reason. The normalized k and q values are not dependent upon the q_0 for the second order Butterworth and Chebyshev filters. Only the filter insertion loss changes. As mentioned earlier, this is intuitively reasonable. There are many other filter types of great interest where the k and q values *do* depend upon q_0, even for filters of only the second order. This makes it much easier to design double tuned circuits of the Butterworth and Chebyshev types with simple equations. The less well-behaved filter types offer characteristics which are sometimes highly desirable though, especially so far as time domain behavior is concerned. This characteristic is of increasing importance in modern communications systems where pulse responses are of interest. The filter design procedures are the same though, using the tables presented by Zverev.

3.3 THE DOUBLE TUNED CIRCUIT

One of the more popular filter forms is that containing two resonators, the double tuned circuit. These filters are designed using the k and q values discussed in the previous section. As mentioned there, the Butterworth and Chebyshev filters have the desirable characteristic that the k_{12}, q_1, and q_2 values are not a function of the normalized q, q_0. In addition, q_1 and q_2 are equal. This leads to a relatively simple set of equations that form an algorithm for the design of double tuned circuits. It may easily form the basis of a program for a handheld calculator or home computer.

A double tuned circuit using parallel resonators was designed in the preceding section. These are generally the most popular. However, an equally useful form uses series tuned resonators. This is known as the mesh realization of the double tuned circuit.

Sec. 3.3 The Double Tuned Circuit

The mesh-filter design procedure is straightforward. The nodal capacitance is determined for a center frequency, bandwidth, and inductor with known Q_u. Normalized k and q values are known for a desired response. The k and q parameters are denormalized by use of $Q_f = f_0/\text{BW}$. Specifically, $Q = Q_f q$ and $K = k/Q_f$. Knowing the net end section loaded Q and the unloaded Q_u, the external Q, Q_e, is calculated. The required series resistance, R, to be placed in series with the resonator is chosen from Q_e and the known inductive reactance at the center frequency. This is shown in Fig. 3.15. More often than not R is small with respect to 50 Ω.

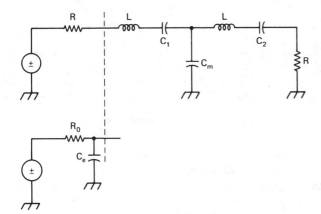

Figure 3.15 A mesh or series-resonator form of a second order bandpass filter. A desired characteristic impedance is often matched with shunt capacitors at the ends of the filter.

Some form of impedance transformation may be applied if a 50-Ω system is to be used as the termination for the filter. A simple one is a parallel capacitor which will transform the 50 Ω (or whatever) down to a series resistance R. The presence of this capacitor will add reactance in series with the resonator. This variation is also shown in Fig. 3.15.

The nodal capacitance was that required for resonance with the chosen inductor, L. Coupling between sections in the mesh-type filter is found by dividing the nodal capacitance by the denormalized coupling factor, K_{jk}. K_{jk} in a narrow bandwidth filter is a value much less than unity. Hence, the shunt coupling capacitor, C_m, is large with respect to the nodal capacitance.

The capacitors used to tune the filter, C_1 and C_2 of Fig. 3.15, must be larger than the nodal capacitance. Capacitive reactance was introduced into each loop by the coupling capacitor, C_m, and by the end capacitor, if it was used. The tuning capacitance required for C_1 of Fig. 3.15 is found by opening the second loop. The net capacitance in the first loop is then adjusted by C_1 such that the series equivalent is the nodal capacitance C_0. The same technique is applied to the second loop. The terminating resistances need not be identical.

Table 3.1 presents the equations for the double tuned circuit, either in series or parallel resonator form. The table contains three sections. The first set of equations is general. Parameters used for both resonator types are evaluated. The second set of equations describes the parallel resonator form while the mesh configuration is

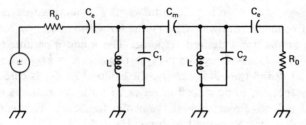

Parallel resonator form

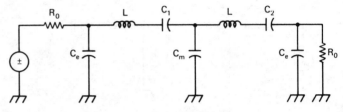

Series resonator form

Table 3.1

General Equations

$\omega = 2\pi f_0$ $\quad q_0 = Q_u/Q_f$

$Q_f = f_0/\text{BW}$ $\quad Q_e = \left(\dfrac{1}{qQ_f} - \dfrac{1}{Q_u}\right)^{-1}$

$C_0 = (\omega^2 L)^{-1}$ $\quad K_{12} = k_{12}/Q_f$

$\text{IL(dB)} = -20 \log\left(\dfrac{q_0}{q_0 - q}\right)$

Equations for Parallel Form

$C_m = K_{12} C_0$ $\quad C_e = \dfrac{1}{\omega}(R_{pe}R_0 - R_0^2)^{-1/2}$

$R_{pe} = Q_e \omega L$ $\quad C_1 = C_2 \approx C_0 - C_e - C_m$

Equations for Series Resonator Form

$C_m = C_0/K_{12}$

$R_s = \omega L/Q_e$

$C_e = \left(\dfrac{R_0 - R_s}{R_s \omega^2 R_0^2}\right)^{1/2}$

$C' = \dfrac{C_e^2 \omega^2 R_0^2 + 1}{C_e \omega^2 R_0^2}$

$C_1 = C_2 = \left(\dfrac{1}{C_0} - \dfrac{1}{C'} - \dfrac{1}{C_m}\right)^{-1}$

derived from the third. The equations presume that capacitors will be used for termination matching at the ends of either filter type. Also, the set of equations assumes that the filter is terminated in the same impedance at each end. Table 3.1 uses fundamental units, hertz, farads, and henrys. Schematics for the filters are included in the table.

Presented in the beginning part of the table is an equation for insertion loss. This equation is for a Butterworth filter. Similar equations exist for the Chebyshev responses and may be found in the Blinchikoff and Zverev text (3). Evaluation of the Butterworth insertion loss is a reasonable approximation to the Chebyshev value.

The k and q values of the desired filter type must be inserted into the equations of Table 3.1. They have been tabulated in Table 3.2 for the Butterworth and for the Chebyshev filters with ripple factors of 0.1, 0.25, 0.5, 0.75, 1, and 1.5 dB. Also shown in the table are the k and q values for third order filters. These apply only to the lossless resonator case, for in the Butterworth and Chebyshev filters of third and higher order, the k and q values depend upon the q_0. In addition, whenever lossy elements are encountered, k_{12} and k_{23} are not equal, nor are q_1 and q_3. The predistorted tables presented by Zverev should be used for more refined designs.

Table 3.2 k and q Values for Two- and Three-Pole Filters

Passband Ripple, dB	n	k	q
Butterworth	2	0.7071	1.414
0.1 dB	2	0.7107	1.638
0.25	2	0.7154	1.779
0.5	2	0.7225	1.9497
0.75	2	0.7290	2.091
1.0	2	0.7351	2.3167
1.5	2	0.7466	2.452
Butterworth	3	0.7071	1.000
0.1 dB	3	0.6617	1.4328
0.25	3	0.6530	1.6330
0.5	3	0.6474	1.8640
0.75	3	0.6450	2.0498
1.0	3	0.6439	2.2156
1.5	3	0.6437	2.5169

The equations of Table 3.1 may form the basis of a computer or calculator program for double tuned circuits. They may also be used for more complicated circuits. Assume, for example, that a third order filter is to be designed using predistorted tables where $k_{12} \neq k_{23}$ and $q_1 \neq q_2$. The program is first evaluated with q_1 and k_{12} and the resistance value to be used for termination at the input. This will establish C_e and C_1 as well as C_{12}, the coupling element between the first and second resonator. The procedure is then repeated using q_3 and k_{23}, yielding C_3, C_e at the output and C_{23}. Having already calculated the nodal capacitance, C_0, the tuning capacitor for the center resonator is evaluated from fundamentals.

The responses of the two forms of double tuned circuit are identical near the passband. However, the stopband characteristics can be considerably different. Figure 3.16 shows transducer gain evaluations for two different filters. Each is a double tuned circuit designed for a center frequency of 10 MHz with a 1-MHz bandwidth. The unloaded resonator Q was assumed to be 250.

The filter using parallel resonators has a response which drops sharply on the low frequency side but is less dramatic above the passband. The mesh, or series resonator form, is just the opposite. It has much better stopband attenuation characteristics above the passband.

This behavior is relatively obvious when the circuits of Table 3.1 are examined. Consider the parallel resonator example. The inductors tend to appear as very high impedances at high frequencies. Hence, their effect disappears, leaving only a network of capacitors. Their reactances all change inversely with frequency, but some are series elements while others are shunt ones. Hence, a constant stopband attenuation is predicted. The parallel resonator form degenerates into a high pass filter with a constant attenuation at frequencies well above the passband.

The behavior of the mesh form is opposite. C_1 and C_2 tend to become short circuits, leaving a low-pass–like structure at high frequencies. The inductive effects

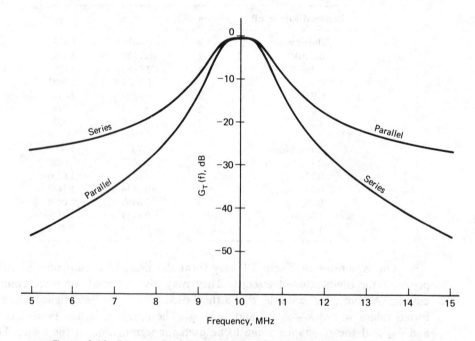

Figure 3.16 Frequency responses of two-pole Butterworth bandpass filters with a 10% bandwidth. One filter uses parallel resonators while the other uses the mesh-type resonators. The one using parallel tuned circuits degenerates into a high pass filter in the stopband while that using series resonators appears as a low pass in the stopband.

vanish at low frequencies leaving a network of capacitors, leading to a constant attenuation.

The lack of response symmetry is sometimes a problem. As filters become narrower though, the asymmetrical effects become less profound. A cascade of the two forms often yields excellent results in applications where a wide (10% relative bandwidth or so) is needed along with symmetry. The choice of the form to use is generally dictated by the intended application and the required stopband attenuation at various frequencies.

There is often a practical difference between the realizability of the two filter types. The coupling and end capacitors in a filter using parallel resonators are usually small. It is sometimes quite difficult to obtain precise components in small values. To the contrary, the coupling and loading elements in the mesh configuration are often large. Standard values are then available. Alternatively, the needed values can be fabricated from parallel combinations of standard units. The writer has found that the mesh configuration is sometimes easier to tune with simple instrumentation than is that with parallel resonators. This is especially true for wider bandwidth filters.

There is one characteristic of predistorted filters that should be emphasized which may not be obvious. The filters are rarely well matched.

Insertion loss in the passband vanishes when filters are built with ideal elements. If the termination impedances are identical, and there is no loss, a conjugate match must be present within the passband. The match, of course, disappears outside of the passband—it is the reflective (mismatch) nature of the elements that causes the filtering action. To the contrary, a filter constructed with elements of finite Q shows an input impedance, even within the passband, which rarely equals the driving impedance. A mismatch exists. Just because a filter is specified as being, for example, a 50-Ω filter, does not imply that the impedance looking into the filter is 50 Ω.

It is possible to design filters which are well matched in spite of lossy elements. This is generally out of the realm of those things which are easily done with a handheld calculator. A small computer (or a large one!) is needed. An optimization routine is used with impedance match being chosen as the desired characteristic rather than shape. Generally, filters which are physically symmetrical lend themselves well to such optimization.

3.4 EXPERIMENTAL METHODS, THE DISHAL TECHNIQUE

Our discussion of filters up to this point in the text has been theoretical. The emphasis has been on fundamental design concepts and their relation to the frequency domain pole positions. The examples have been practical to the extent that components and frequencies were from real filters and Q_u values were typical of those encountered in practice. Still, the filter components, especially with narrow bandpass circuits, tend to become impractical as higher frequencies are approached.

For example, a two-pole bandpass filter at 10 MHz with a 5% bandwidth built from parallel resonators would have a coupling capacitor of 8.96 pF. This assumes inductors of 1 μH. This is a completely practical component. If this same filter was scaled to 500 MHz, the inductor would decrease to 20 nH and the coupling capacitor would be 0.18 pF. The inductor is still practical although care would be required in construction. A section of transmission line would probably be used. The coupling capacitor, however, borders on the absurd. Even if such components could be purchased, they would not be precision parts. Moreover, the inductance of the capacitor leads would alter the effective reactance. The cynical microwave engineer would refer to this component as a "parallel plate inductor." Other methods are needed, not only for construction but for design and alignment.

The methods which allow the synthesis techniques presented to be extended and measured are the result of work by M. Dishal (4). The underlying concept is that the performance of a filter is completely controlled by the Q of the end sections and the coupling between resonators. Measurement of these parameters will ensure the desired filter characteristic.

The methods are presented here in terms of simple filters at 10 MHz. However, they are easily extended to virtually any frequency within the limits of available instrumentation. The methods are not limited to LC resonators, but may be applied equally to other forms. These include other electromagnetic resonators such as transmission line sections, helical resonators, and wave guides. Mechanical and acoustic resonators may also be used.

Figure 3.17 shows the filter we will use for a demonstration of the Dishal methods. The filter is centered at 10 MHz with a bandwidth of 500 kHz. The inductors are of 1-μH value with $Q_u = 250$. As mentioned, these are practical values with inductors wound on powdered iron toroid cores.

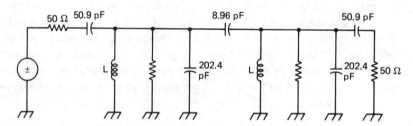

Figure 3.17 Double tuned circuit used to illustrate the Dishal methods for filter alignment.

The parallel resistors in Fig. 3.17 represent the core losses and at a given frequency are $R_p = Q_u \omega L$. The values for k and q are, respectively, 0.707 and 1.414 if this filter is to have a Butterworth response. Denormalization yields $K_{12} = 0.03536$ and $Q_1 = Q_2 = 28.28$. Experimentally, the goal is to measure the coupling coefficent and loading of the system of resonators.

Figure 3.18 shows the method used for measuring the loaded Q of the resonator. The filter was designed for a 50-Ω termination with an end matching capacitor of

Sec. 3.4 Experimental Methods, the Dishal Technique

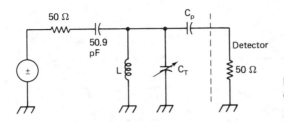

Figure 3.18 Measurement of loaded end section Q. The coupling capacitor, C_p, is very small.

50.9 pF. The tuning capacitor, C_t, is variable. This will most likely be required in practice anyway and will be different for this measurement than it is in the filter. The reason is that only a single resonator is being studied in Fig. 3.18 and the coupling capacitor is not present. C_p is a "probe" capacitor and is very small in value. This might consist, in practice, of nothing more than a wire from a coaxial connector placed in proximity to the resonator. The probe is used to couple energy into a sensitive detector. The detector, typically with a 50-Ω input impedance, could be a spectrum analyzer, an oscilloscope of suitable bandwidth, or a broadband amplifier driving a diode detector.

Analysis of the circuit of Fig. 3.18 will show that the only loading of significance on the resonator comes from the external generator and the unloaded Q losses of the coil.

The generator is first set to the desired 10-MHz center frequency. C_t is adjusted for a peak response in the detector. There will be considerable insertion loss. If there were not, it would indicate that the probe is loading the inductor excessively and that the value of C_p should be decreased. After tuning, the signal generator is swept and the frequencies where the response is decreased from the peak by exactly 3 dB is noted. The result may be read directly if a spectrum analyzer or calibrated receiver serves as the detector. A 3-dB attenuator may be alternately inserted and removed from the system to detect the difference if a broadband amplifier and diode detector are used. Ultimately, the careful use of a step attenuator is the best (most fundamental) method.

The bandwidth and center frequency may be used to calculate the loaded Q of the circuit, $Q_L = f_0/\text{BW}$. The value of the coupling capacitor to the 50-Ω load may be adjusted to produce the desired Q_L. Alternative methods of end loading may also be employed. Note that Q_u may also be measured with this system.

Curve A of Fig. 3.19 shows the results of probing and sweeping the circuit of Fig. 3.18. These results are calculated for purposes of illustration. However, experimental results are of identical form. The calculations, including those to follow, were done with the ladder method.

End section Q has been established. This may be done with each end of the filter as well as with individual resonators for determination of unloaded Q values. The coupling coefficient must now be evaluated. This is done with the system shown in Fig. 3.20. Two probe capacitors are used. One is driven by the signal source while the other drives the detector. The junction drives the first of the two resonators. Neither resonator is terminated in an external load.

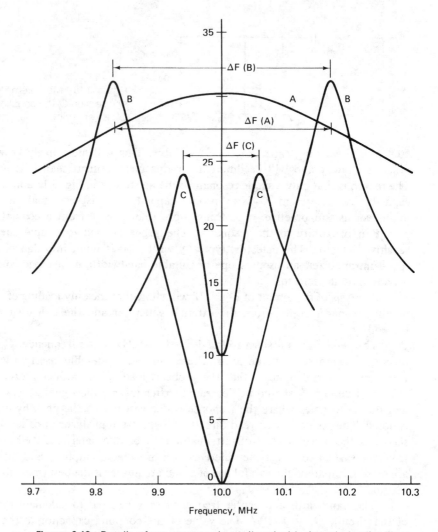

Figure 3.19 Results of measurements on the double tuned circuit with a probe coupled to a detector. Curve A shows the loaded Q of a single resonator, Fig. 3.18. Curve B shows the effect of coupling as measured with the method of Fig. 3.20. Curve C is an example of reduced coupling between resonators.

A switch, S_1, is shown across the second resonator. This is nothing more than a short piece of wire that is used to temporarily short the resonator. The second resonator may be severely mistuned to produce an identical result for higher frequency applications.

The generator is set to the filter center frequency and the first resonator is tuned for a peak by adjusting C_1 with S_1 closed. Then, S_1 is opened and the second resonator is tuned. The output seen in the detector will begin to decrease as the

Sec. 3.4 Experimental Methods, the Dishal Technique

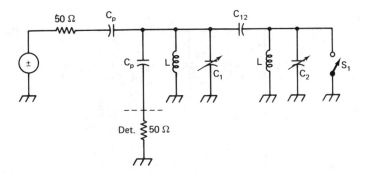

Figure 3.20 Use of the Dishal methods for measurement of coupling between resonators.

two frequencies begin to coincide. C_2 is adjusted until the output is at a minimum. Then, leaving the filter tuning in this condition, the generator is swept.

The detector response as the generator is swept is shown in curve B of Fig. 3.19. This curve corresponds to a coupling capacitor of 8.96 pF in the filter of Fig. 3.17. There are two distinct peaks in the response with a "dip" between which is greater than 30 dB below the peaks.

The frequency separation between peaks is recorded from the measurement. It is indicative of the coupling. Specifically, the separation is approximately equal to the coupling coefficient, K_{12}, multiplied by the center frequency. The separation is 0.36 MHz in this example. Noting the 10-MHz center frequency, the coupling coefficient is $\Delta f/f_o = 0.036$. The desired value, that used in the synthesis, was 0.03536.

Curve C of Fig. 3.19 shows a similar coupling curve where the capacitor between resonators has been reduced from 8.96 pF. to 3 pF. The separation between peaks has decreased as has the magnitude of the dip between peaks.

The simple relationship between peak separation, center frequency, and coupling coefficient is an approximation. Generally, it is close enough for simple filters though. More refined information is provided by Zverev (5).

The curves of Fig. 3.19 show an interesting feature that is not general. The separation between peaks during a coupling measurement is equal to the loaded bandwidth of the single loaded end resonators. This is a general characteristic of the two-pole Butterworth filter. However, it is not true for other configurations.

This method is not restricted to direct measurement of coupling coefficients and loaded end section Q values. It also is valuable for filter tuning. Consider the n-pole filter of Fig. 3.21. We assume that the coupling capacitors have all been properly adjusted or calculated ahead of time. Initially, all resonators are mistuned or shorted. Then, the first resonator is tuned for a peak response. Having achieved this, the second is tuned for a minimum detector output just as was done during the coupling measurement. Then, the third resonator is tuned for a peak response. The process is continued, producing alternate peaks and nulls with each stage of "tweeking." The resonators may have the final end loading in place during this procedure.

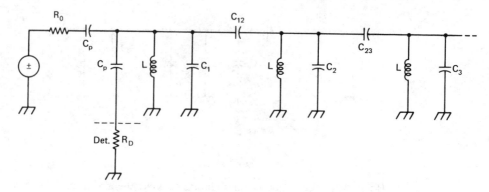

Figure 3.21 Filter alignment with the Dishal method. See the text.

The result of a swept measurement is complex but reasonable if there is no end loading during this method. There will be one peak in the swept response for each resonator.

It is important to note that this result comes from each resonator being tuned to the center frequency of the filter. Once tuning is completed, the end terminations may be applied with slight retuning of the end sections as needed. If the couplings and loadings are correct, the filter response will be that desired.

The implications of the Dishal method are truly profound, for they allow very complex filters to be designed, adjusted for proper coupling, and finally aligned. Consider as an example the design of a third order vhf filter. Because the filter must be physically small but have low insertion loss and a narrow bandwidth, high Q_u resonators are required. Helical resonators are a good choice in the vhf and low uhf frequency spectrum (6).

First, experimental resonators were built. Small probes were used to measure Q_u values of approximately 425 at a frequency of 110 MHz.

Next, several filters were built which contained but two resonators. The helical resonators consist of solenoidal coils inside a shielded enclosure with a small tuning capacitor at one end. From outward appearances they look like simple LC resonators. Actually, they are sections of a special helical transmission line. Holes, or apertures were cut in the wall separating the two resonators. The several filters constructed were identical except that the aperture size changed. The coupling coefficient was measured for each aperture diameter, producing a suitable curve. Preliminary experimental work was now finished.

The next step was the design of the filter. Because this filter had to have a reasonable step response, both the Butterworth and Chebyshev responses were avoided. Instead, a Gaussian prototype was chosen based upon the predistorted tables of Zverev. A filter was then designed using capacitor and inductor values that were completely unreasonable from a physical realizability viewpoint. This matters little to the computer or calculator, though. The response was optimized using computer methods. Then the resulting theoretical prototype was analyzed to determine end section Q values and coupling coefficients between resonators.

Having designed the *LC* prototype on paper, the *K* and *Q* values were then applied experimentally to the helical resonators. Aperture diameters were chosen for coupling. The end section loading was determined by placing a tap near the grounded end of the helices with a series capacitor for attachment to the 50-Ω coaxial connectors. The use of both a series capacitor and a tap provides a minimal sensitivity to tap position, aiding manufacturability. The Dishal method of alignment of Fig. 3.21 was used for initial adjustment. Final alignment was performed with a signal generator and detector operating at the center frequency of the filter.

The general procedure used in design and realization of this filter is much more significant than the actual details. It was possible with the Dishal method to transfer the results of lower frequency techniques into the vhf region. The prototype filter was physically impractical. However, through the concepts of coupling and end section loading, the prototype is experimentally "denormalized" to whatever frequency is desired. The method is universally applicable and certainly lends considerable physical meaning to the *k* and *q* ideas.

3.5 QUARTZ CRYSTAL FILTERS

One of the most popular forms of filter in the hf and lower vhf spectrum is that consisting of coupled quartz crystals. These elements behave as *LC* resonators with a very high unloaded *Q* and excellent frequency stability with variations in temperature. They allow the synthesis and construction of very narrow bandwidth filters.

Figure 3.22 shows a cross section view of a quartz crystal, a thin slab of single crystal quartz, usually in a disc shape. Quartz is a piezoelectric material. When an electric field is placed upon it, a physical displacement occurs. This is shown in the exploded view of Fig. 3.22b. In this case, the physical displacement is in a direction at right angles to the imposed electric field. This is a shear-mode displacement. The magnitude of the displacement, the distance shown as Δx, is proportional to the intensity of the driving electric field. The electric field is produced in the quartz crystal by placing a voltage between thin layers of metallization attached to the surface.

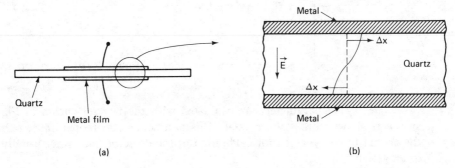

Figure 3.22 Cross section (a) and exploded (b) view of a quartz crystal resonator.

The piezoelectric effect is twofold. If some means is used to force a physical displacement, an electric field is produced within the material. This is detected as a voltage across the metallization plates. This phenomenon allows piezoelectric materials to serve as very sensitive transducers for the measurement of physical motion or displacement. Quartz is but one example of a piezoelectric material.

An interesting and useful characteristic is noted if a source of high frequency energy is used instead of a dc voltage. A frequency will be found where the current is maximum as the frequency of the ac source is swept. This frequency corresponds to one where the quartz crystal is mechanically resonant. One surface of the crystal will be moving to the right while the other is moving to the left, out of phase with the opposite surface. This motion is that depicted in Fig. 3.22b. The two surfaces will be mechanically oscillating back and forth at a radio frequency rate. They will produce an electric field which is in phase with that driving the crystal from the generator.

The motion will still occur at slightly different frequencies from the peak described. However, the mechanical boundary conditions will tend to make each surface move, in part, in the same direction. The mechanical mass of the crystal and its mounting will cause great damping of such motion, causing the losses to increase.

There will still be a current flowing through the crystal at frequencies far removed from the resonance point. This is a direct result of the capacitance formed by the parallel plates of metallization upon the quartz dielectric. This parallel capacitance, usually a few picofarads, may be measured at audio frequencies, far removed from the dominant resonances.

Figure 3.23 shows the traditional symbol for a quartz crystal and the equivalent circuit, a model. The parallel capacitance is shown as C_p. The resonance is shown as a series tuned circuit with a motional inductance, L_m, and a motional capacitance, C_m. A series resistance, R_s, is included to model losses. This loss can result from the mounting, from defects in the crystal or even from energy coupled to the gas (air, nitrogen, or argon) surrounding the quartz disc. The highest Q crystals are contained in a vacuum.

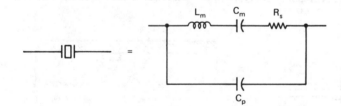

Figure 3.23 Schematic symbol and model for a quartz crystal.

The crystal parameters are evaluated with the test set shown in Fig. 3.24. A switch is shown across the crystal. This is a schematic representation that allows the crystal to be bypassed for calibration of the detector and is not usually included in an actual test set.

The power available to the 50-Ω detector is evaluated with the switch closed. The switch is then opened and the generator is slowly swept through the crystal

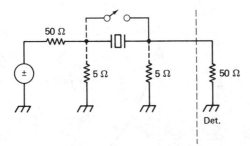

Figure 3.24 Setup for measurement of crystal parameters.

series resonant frequency. A pronounced peak is noted. The signal generator is set at this peak frequency and the detector response is recorded. It will be less than measured with the switch closed owing to the losses in R_s of the equivalent circuit, Fig. 3.23. The series resistance may be calculated from voltage divider action and the impedance of the source and the detector.

The next step is to evaluate the crystal Q. This may be done in a 50-Ω system. The Q_L is determined by sweeping through the response and noting the -3-dB points. The unloaded Q may be calculated from the relationship IL $= -20 \log (1 - Q_L/Q_u)$ where IL is the insertion loss in dB. Alternatively, a lower impedance system may be used. This is realized by shunting the 50-Ω source and detector with small resistors. Five-Ω units are shown in dotted lines in Fig. 3.24. The insertion loss with most crystals will be high enough that the Q measured in the 5-Ω system will be quite close to the unloaded value.

The motional inductance may be calculated directly from unloaded Q and the series resistance. The motional capacitance is then evaluated from knowledge of the resonant frequency. All crystal parameters are now known.

These measurements are best done with a spectrum analyzer and a tracking generator. However, this is by no means mandatory. They can also be done with simple equipment. An oscilloscope or other sensitive detector with a 50-Ω termination can replace the spectrum analyzer. Any stable 50-Ω signal source will suffice. A frequency counter is useful for precise frequency determination. One should be careful to limit the power that is dissipated in the series resistance, R_s. The maximum is around 0 dBm with typical quartz crystals, although lower values are preferred for initial measurements.

Figure 3.25 presents a plot of the response of a typical crystal. This data is calculated using the ladder method for a 5-MHz crystal. A typical series resistance of 30 Ω is assumed along with an unloaded Q value of 100,000. The motional parameters resulting are $L_m = 0.09548$ H and $C_m = 0.01061$ pF. A parallel capacitance of 7 pF is assumed. The horizontal axis of Fig. 3.25 is in kHz displacement with respect to the 5-MHz center frequency while the vertical axis is in dB. The insertion loss at resonance is just over 2 dB in a 50-Ω system.

A significant feature of Fig. 3.25 is the pronounced output dip just above the series resonance. This is the result of the parallel capacitance. The $L_m C_m$ combination appears as an inductance at frequencies above the series resonance. This produces a

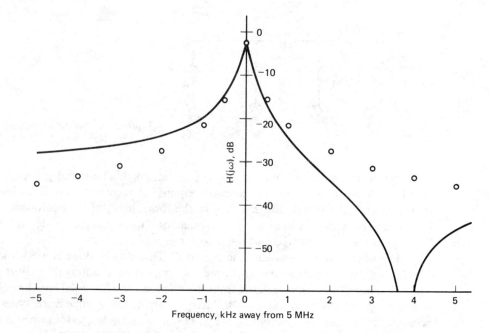

Figure 3.25 Frequency response of a 5-MHz quartz crystal driven from a 50-Ω generator and terminated in a 50-Ω load. The parallel capacitance is assumed to be 7 pF for the solid curve. The dots show the response when the parallel capacitance is allowed to vanish.

parallel resonance with C_p. A pole at the series resonant frequency is obtained as well as a zero at the parallel resonant frequency if the s plane transfer function is evaluated. For this reason, the frequency separation between resonances, 3.78 kHz in the example of Fig. 3.25, is known as the pole-zero spacing.

A series of circled dots is also shown in Fig. 3.25. These correspond to a crystal where the parallel capacitance is allowed to vanish. This response is symmetrical. If the parallel capacitance is allowed to become larger, the series resonance is unchanged, but the pole-zero spacing becomes smaller.

Examination of the transfer response, Fig. 3.25, shows a very narrow bandwidth with still a small insertion loss. The crystal would form a useful filter merely by insertion in a 50-Ω system. More often though it is desired to eliminate the effects of the parallel resonance brought about by C_p. This is realized by paralleling the crystal with a suitable inductor that will resonate with C_p. The inductance value may not be practical. The parallel resonance may alternatively be eliminated with the balanced circuit shown in Fig. 3.26. The source is modified by placing a transformer in cascade with the crystal. If desired, the impedance level may be altered with this transformer. Two secondary windings are employed. One drives the crystal while the other drives a variable capacitor. The voltage driving the added capacitor is 180 degrees out of phase with the signal driving the crystal. If the variable capacitor

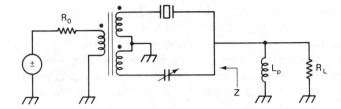

Figure 3.26 Crystal filter using a single crystal. The parallel capacitance of the crystal is "canceled" with the variable capacitor. The inductor tunes out the reactance appearing when looking back into the crystal.

is adjusted to have the same value as C_p, the effects of the zero in the transfer function are eliminated.

An inductor, L_p, is added to the circuit of Fig. 3.26. This is needed to tune out the effects of the parallel capacitances. If the impedance looking back into the crystal and balancing capacitor (shown in the figure as Z) is examined, a capacitance of $2C_p$ is observed. This will appear in the circuit as a shunt capacitive element which could alter circuit response. It is removed with the parallel inductor, L_p, which resonates with $2C_p$. The added inductor is not required in many applications.

Application of the balancing method of Fig. 3.26 has only a minor effect on the passband response of the single crystal filter. However, the effect is dramatic in the stopband. The attenuation at 4.5 and 5.5 MHz is only slightly more than 30 dB in the example response. The attenuation exceeds 75 dB if the effect of the parallel resonance is removed.

Coupled resonator filters may be built from crystals. Generally, the series resonance is stable and well defined while the parallel one is dependent upon C_p, a parameter that is subject to modification by external circuit details. Hence, the better filters are those that utilize the series resonant mesh-type circuits.

Assume for the present that the crystal may be modeled as a high Q series resonant circuit. The parallel resonance will be ignored. A double tuned circuit using the idealized crystals is shown in Fig. 3.27. The crystals are shown surrounded by dotted lines.

Design of this filter is identical with the mesh-type double tuned circuits described earlier, except that we have less freedom in choosing components. The motional components of the crystal are already determined. A bandwidth is chosen and Q_f is calculated. The normalized q_0 is then obtained from knowledge of Q_u of the crystals.

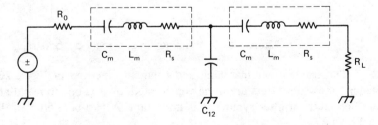

Figure 3.27 A mesh-type double tuned circuit using crystals as the resonant elements.

This may be used to evaluate the insertion loss for a desired filter polynomial. The suitable predistorted k and q values may be picked from tables.

The denormalized K_{12} and $Q_1 - Q_2$ values are obtained from Q_f and the normalized k and q information. External end section Q_e parameters are then evaluated from knowledge of Q_u. Terminating resistances are chosen. The total resistance, R_L and the crystal R_{loss}, must produce the desired end Q when compared with the inductive reactance.

The coupling capacitor, C_{12}, is calculated as the motional capacitance divided by the denormalized coupling coefficient, K_{12}. The center frequency of the resulting two-pole filter will be slightly higher than the series resonant point of the crystals. This is because the center frequency is the resonant frequency of the individual meshes with the adjacent meshes open circuited. Hence, the resonating capacitance will be, for the two-pole filter, the series combination of C_{12} and C_m. Owing to the extremely small value of the typical motional capacitances, the difference will be slight.

The design process just described has ignored the presence of the parallel capacitances of the crystal. This assumption is a very good one if the filter was to have a narrow bandwidth, specifically, much less than the pole-zero spacing of the crystal. The narrow band filter will be loosely coupled, resulting in a large value for C_{12}. The stopband attenuation is reasonable, for the large coupling capacitor forms a voltage divider with C_p to reduce out-of-passband response. The filter frequency is well separated from the parallel resonance.

The simple filters are no longer easily built as the filters become wider in bandwidth. A rule of thumb for building the simpler filters is that the bandwidth should be less than about 30% of the pole-zero spacing. Filters designed in this way are termed as lower sideband ladder types. A fifth order lower sideband ladder filter is shown in Fig. 3.28. The name for the filter type derives from the response shape which is asymmetrical. The shape, like that of a single crystal (Fig. 3.25), passes the lower sideband while offering high attenuation to frequencies above the passband. A ladder filter utilizing quartz crystals as parallel resonators is termed an upper sideband ladder.

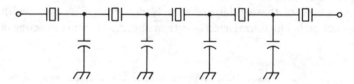

Figure 3.28 A fifth order lower sideband ladder filter.

The transformer matching and balancing method of Fig. 3.26 may be used if wider bandwidth filters are to be constructed, or if good stopband attenuation values are needed, or if filter symmetry is important. A two-pole filter of this type is shown in Fig. 3.29.

It is often desirable in filters using several crystals to use the balancing method at the end sections. This will help to preserve stopband attenuation and filter response

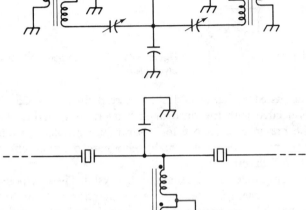

Figure 3.29 A double tuned circuit with the effects of parallel crystal capacitance removed by tuning.

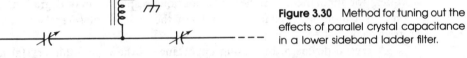

Figure 3.30 Method for tuning out the effects of parallel crystal capacitance in a lower sideband ladder filter.

symmetry. It is possible to utilize balancing at each coupling node in a wider bandwidth filter with many resonators. This is shown in Fig. 3.30.

The lower sideband ladder filter is probably the most popular form in communications equipment. It becomes increasingly difficult to design and construct as the bandwidth increases. Many other forms of crystal filter are available for such designs (7). One popular one is the half lattice. This is shown in Fig. 3.31. The series resonant frequencies in this filter, $Y1$ and $Y2$, are different. However, the balanced drive applied to the parallel capacitances tend to eliminate the effects of crystal C_p. A popular variation of this circuit is a four-pole filter, the cascade half-lattice shown in Fig. 3.32. The details of lattice filter design will not be presented here.

A simplified model was used for design. The real crystal has additional responses. One group is termed the overtones. If one examines the mechanical boundary conditions for a crystal to have stable oscillations, assuming a shear-mode vibration, the only constraint is that the surfaces must be moving in opposite directions. This will allow oscillation to occur not only at the fundamental frequency, but at overtones which are approximately equal to odd integers times the fundamental frequency.

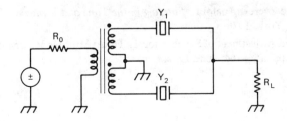

Figure 3.31 A half lattice crystal filter. The effects of parallel capacitance are removed by virtue of the crystals being identical except for a slight variation in center frequency.

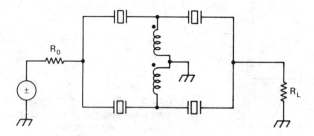

Figure 3.32 The cascade half lattice crystal filter.

Hence, the 5-MHz crystal discussed earlier would show a response at 15, 25, 35, ... MHz. Often, the Q_u associated with the third or fifth overtone is higher than that of the fundamental. The presence of an overtone is rarely a problem in filter design, for additional filtering in a system will usually eliminate overtone effects. This is realized with a simple LC circuit.

There are additional spurious modes in the usual quartz crystal. These responses are usually just above the series resonant frequency with a separation of from 20 to 200 kHz for fundamental modes. The spurious responses may degrade the filter response in a multiple crystal ladder filter if the "spurs" of several crystals coincide in frequency.

A crystal device of increasing importance is the monolithic crystal filter. This is shown schematically in Fig. 3.33 for a two-pole element. It is constructed by taking a quartz blank and placing two metallization layers on one of the surfaces. The layers are separated spatially such that each section will operate as a single resonator. However, the proximity of the two allow them to couple mechanically. The coupling coefficient determines the filter bandwidth along with the end terminations, which are controlled by the designer.

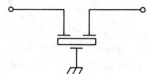

Figure 3.33 Schematic representation of a two-pole monolithic crystal filter.

REFERENCES

1. Zverev, Anatol I., *Handbook of Filter Synthesis*, John Wiley & Sons, New York, 1967.
2. See reference 1.
3. Blinchikoff, Herman J. and Zverev, Anatol I., *Filtering in the Time and Frequency Domains*, John Wiley & Sons, New York, 1976.
4. Dishal, M. "Alignment and Adjustment of Synchronously Tuned Multiple-Resonant-Circuit Filters," *Elec. Commun.*, pp. 154–164, June 1952.
5. See reference 1, Chap. 9.
6. See reference 5.
7. See reference 1, Chap. 8.

4 Transmission Lines

The previous two chapters presented information on passive components, capacitors and inductors. They were formed into complex networks which served primarily as filters, but also provided impedance transformations.

This chapter is devoted to transmission lines. They form a means of transferring energy generated at one place to another where the energy is to be used. The most familiar application of a transmission line is in the electrical power industry. It is a transmission line that conveys the electricity that probably provides the light used to read this text.

Transmission lines serve a similar role in the design of radio frequency circuits, that of transporting energy from one point to another. They have many other uses. They can act as impedance transforming elements just as a transformer or a series or shunt immittance does. Transmission lines serve as resonant circuits. As such, they may be combined as filters. They can, over a narrow bandwidth, serve as either inductors or capacitors (as can any resonant circuit). Perhaps of greatest significance though is that lines provide another means of viewing a circuit. It is this viewpoint that is commonly used in much of the design of rf circuits, especially at microwave frequencies. The virtues of this viewpoint are twofold. First, it is based upon measurements which are more easily performed than others. Second, it is based upon energy transport concepts, usually of greater fundamental importance than knowing voltage or current values.

4.1 TRANSMISSION LINE FUNDAMENTALS

A transmission line may be constructed in many different ways. Four of the common constructions are shown in Fig. 4.1. One is the parallel wire line, often used to convey energy to an antenna from a transmitter. The second, and one of the most common types in rf work, is the coaxial cable. A center conductor is surrounded by a dielectric material which, in turn, is surrounded by a shell of conductive material. The coaxial cable has the virtue of shielding. The outer shell may be connected to ground potential at one or both ends. The flow of rf current is restricted to a finite depth in a conductor because of the skin effect. The currents flow on the outer surface of the inner conductor and the inner surface of the outside conductor. The potential of the outside of the outer shell can be different than that on the inside.

The third transmission line, shown in Fig. 4.1c, is the strip-line. A conductor is placed a short distance above a ground plane. The strip-line can be shown to be equivalent to a parallel wire line using the concept of images of electromagnetic theory. A popular variation is the microstrip transmission line shown in Fig. 4.1d. This is usually formed with double-sided circuit board material or a suitable equivalent. One side of the circuit board is a continuous ground foil while the other is a thin strip. Microstrip line differs from the simpler strip-line only because the dielectric material surrounding the main conductor is segmented into two parts, usually with differing properties.

There are other transmission line types. Most, however, are variations of those shown. For the present, we will consider a transmission line to be a pair of parallel

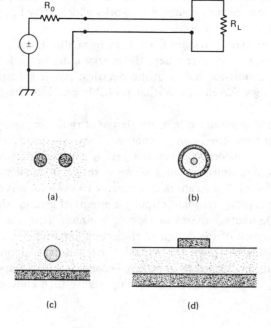

Figure 4.1 Basic transmission line types. Heavily shaded areas represent metal while lightly shaded regions are dielectric material. (a) shows the cross section of a parallel wire line while (b) is a coaxial cable. (c) shows a strip-line with a round conductor and (d) shows a related type, the microstrip line, usually formed on circuit board material.

conductors like that presented in Fig. 4.1a with the understanding that the concepts apply to any of the other forms.

The action–reaction viewpoint used in previous chapters has been associated with instantaneous response. For example, in Chap. 1 where low frequency models of active devices were presented, a voltage applied at the input to an amplifier resulted in an immediate output. At least, this was assumed in the model.

Delaying effects were noted when reactive networks were considered in Chap. 2. This is shown in the networks of Fig. 4.2. In the first, Fig. 4.2a, the switch is closed at $t = 0$. Current flows from the source immediately. The voltage at the output is initially zero. It begins to grow and eventually approaches the battery voltage at a later time. The current eventually decreases to zero.

The behavior of the inductive circuit, Fig. 4.2b, is similar. The switch is closed at $t = 0$. The voltage at the input to the inductor is immediately equal to that of the battery, but the initial current is zero. The output voltage is also found to be zero initially. However, like the capacitive circuit, the output voltage grows from zero at $t = 0$ to the battery value at a later time. Current continues to flow in the inductive circuit.

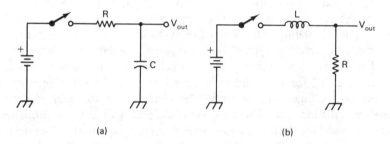

Figure 4.2 *RC* and *LR* networks showing a time delay between input and output response.

The transmission line shows a behavior that represents something of both types. It may be viewed as a resistor, but it also shows delayed effects like the networks containing capacitive and inductive elements. This is illustrated with the circuit of Fig. 4.3. A battery and a switch are again used. The switch is closed at $t = 0$ applying a signal to the input of a very long (strictly, an infinite) transmission line. Assume that we are able to observe the voltage and the current at several points along the line, all as a function of time.

Voltage is immediately present at the input to the line, x_0, when the switch is closed at $t = 0$. A current is also observed at that point. Both remain constant at that value for all later time. It is significant that the current does not change with time. This is equivalent to no phase difference between the input voltage and the current, indicating that the infinite line presents a resistive input impedance at $x = 0$.

No response is seen initially at $t = 0$ if the voltage and current at x_1 are observed. However, the battery voltage appears after a short delay at that point. Similarly, a

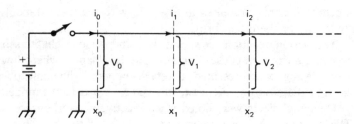

Figure 4.3 The beginning of an infinitely long transmission line excited by a battery and switch. Various positions on the line are shown for reference.

current then flows. The current and voltage remain constant for later times beyond the first appearance of a signal at x_1, which we will call t_1. Again a resistive impedance is inferred from the lack of variation in current with times after t_1. Similar results are obtained for other points, x_n, further down the line.

Evidently, the closing of the switch "launches" a wave front that propagates in a positive x direction. The velocity of propagation is calculated as $(x_n - x_{n-1})/(t_n - t_{n-1})$ and is found to be the velocity of light in the material surrounding the wires of the transmission line. The velocity for a real dielectric will be less than that of light in vacuum. The ratio of the velocity in the line to that in free space is called velocity factor. A typical value for many coaxial cables is 0.66.

The behavior for other points is unchanged if the line was finite in length, terminated at x_3, for example, with a resistor across the line equal to the ratio of the observed voltages and currents. The resistance is the characteristic impedance of the transmission line. The previous discussion will now be formalized. We will describe a model for the transmission line which will explain the observed phenomenon.

Examine a short length of line, Δx. The line segment will show some inductance, $L\Delta x$ where is L is the inductance per unit length. Similarly, the line has capacitance. The short length has a capacitance $C\Delta x$ where C is the capacitance per unit length. This model is shown in Fig. 4.4. Note that this schematic has the appearance of a low pass filter. However, it does not show a low pass characteristic. The reason is that the inductors shown in Fig. 4.4 become vanishingly small as the length Δx is allowed to vanish. The incremental capacitors also becomes small. This forces the low pass cutoff frequency to increase without bound.

There will be loss elements in a real transmission line. These would be modeled

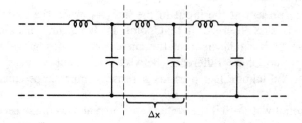

Figure 4.4 Section of a transmission line emphasizing the presence of both inductance and capacitance. Resistive loss elements, although present, are ignored in this analysis.

by a series resistance with each incremental inductor and by a parallel conductance with each incremental capacitor. We will ignore loss for this discussion.

The series inductor will lead to a voltage drop across the short segment of transmission line

$$\Delta V = -L\Delta x \frac{\partial I}{\partial t} \qquad (4.1\text{-}1)$$

A capacitive element will allow shunt current to flow. This must decrease the current flowing out of a segment over that entering. The current decrease is

$$\Delta I = -C\Delta x \frac{\partial V}{\partial t} \qquad (4.1\text{-}2)$$

The minus signs in the two equations indicate that both V and I decrease through the segment. Partial derivitives are used, for we wish to describe the time behavior of the voltage, $V(x,t)$, for any value of x.

If each of these equations is divided by the length increment and the interval is allowed to become vanishingly small, we obtain the partial differential equations

$$\frac{\partial V}{\partial x} = -L\frac{\partial I}{\partial t} \qquad (4.1\text{-}3)$$

$$\frac{\partial I}{\partial x} = -C\frac{\partial V}{\partial t} \qquad (4.1\text{-}4)$$

Equation 4.1-3 is now differentiated partially with respect to distance

$$\frac{\partial^2 V}{\partial x^2} = -L\frac{\partial^2 I}{\partial x \partial t} \qquad (4.1\text{-}5)$$

Similarly, Eq. 4.1-4 is differentiated with respect to time to obtain

$$\frac{\partial^2 I}{\partial t \partial x} = -C\frac{\partial^2 V}{\partial t^2} \qquad (4.1\text{-}6)$$

The dependent functions, V and I, are sufficiently well-behaved that we may reverse the order of differentiation in one. This is done to obtain an equation describing the voltage on the transmission line as a function of time and of distance

$$\frac{\partial^2 V}{\partial x^2} = LC\frac{\partial^2 V}{\partial t^2} \qquad (4.1\text{-}7)$$

A similar equation describing $I(x,t)$ results if Eq. 4.1-3 is differentiated with respect to time and Eq. 4.1-4 is differentiated with respect to distance

$$\frac{\partial^2 I}{\partial x^2} = LC \frac{\partial^2 I}{\partial t^2} \qquad (4.1\text{-}8)$$

Equations 4.1-7 and 4.1-8 are special cases of a general type found throughout all of physics, the wave equation. It is a second order partial differential equation relating time and spatial behavior of a given variable. The general form of the wave equation is

$$\frac{\partial^2 A}{\partial x^2} = \frac{1}{v^2} \frac{\partial^2 A}{\partial t^2} \qquad (4.1\text{-}9)$$

where v is the velocity of propagation of the wave.

Numerous methods may be used to solve the wave equation, depending upon boundary conditions. One of the most general is to assume a solution exists in the form of $A(y)$ where $y = t - x/v$. Direct substitution then shows that this is indeed a solution. Similarly, $A(y)$ with $y = t + x/v$ is a solution.

Consider $A(t - x/v)$. A will be constant if $y = t - x/v$ is a constant

$$K = t - x/v \qquad (4.1\text{-}10)$$

Multiplying by v and solving for x, we obtain

$$x = tv - K' \qquad (4.1\text{-}11)$$

where K' is a new constant, $K' = vK$. This equation states that a constant value of A exists for positions and times related by Eq. 4.1-11. x is an increasing function of time; hence, $A(t - x/v)$ describes a positive moving wave. Similarly, $A(t + x/v)$ describes a wave moving in the negative x direction.

A complete solution of the voltage wave equation, Eq. 4.1-7, is the sum of positive and negative moving voltage waves

$$V(x,t) = V(t - x/v) + V(t + x/v) \qquad (4.1\text{-}12)$$

We might infer from the previous discussion describing a step function traveling down a transmission line that the current in a line is found by dividing the voltage at the point by a characteristic impedance. This is not generally true. The example, Fig. 4.3, described only a positive going wave.

The voltage and current in a transmission line are related by considering the general voltage solution, Eq. 4.1-12, with Eq. 4.1-3,

$$-L\frac{\partial I}{\partial t} = \frac{\partial}{\partial x}[V(t-x/v) + V(t+x/v)]$$

$$= \frac{-1}{v}\frac{\partial}{\partial y_+}[V(t-x/v)] + \frac{1}{v}\frac{\partial}{\partial y_-}[V(t+x/v)] \quad (4.1\text{-}13)$$

where $y_+ = t - x/v$ and $y_- = t + x/v$. This equation is integrated partially with respect to time to yield

$$LI = \frac{1}{v}[V(t-x/v) - V(t+x/v)] + f(x) \quad (4.1\text{-}14)$$

$f(x)$ is a constant of integration with respect to time that we will ignore in this discussion. The integration removes the differentiations with respect to y_+ and y_-. Solving for the current

$$I = \frac{1}{Lv}[V(t-x/v) - V(t+x/v)] = \frac{V_+}{Z_0} - \frac{V_-}{Z_0} \quad (4.1\text{-}15)$$

V_+ and V_- are positive and negative traveling voltage waves. Z_0 is the characteristic impedance of the transmission line, $Z_0 = Lv$. But, by comparing Eq. 4.1-7 with the general wave equation, Eq. 4.1-9, $v = 1/\sqrt{LC}$, producing

$$Z_0 = \sqrt{\frac{L}{C}} \quad (4.1\text{-}16)$$

The current in a line, Eq. 4.1-15, is the difference between two traveling current waves. The negative sign in Eq. 4.1-15 is of considerable significance.

4.2 THE VOLTAGE REFLECTION COEFFICIENT AND STANDING WAVES

The current seen at a given point, x, when looking into a transmission line is not merely the voltage on the line divided by the characteristic impedance. The impedance of interest is shown in Fig. 4.5. It will be the ratio of the voltage at x to the current at the same position

$$Z_{\text{in}} = Z_0 \left(\frac{V_+ + V_-}{V_+ - V_-}\right) \quad (4.2\text{-}1)$$

This may be normalized by dividing by the characteristic impedance, Z_0

$$z = \frac{Z_{\text{in}}}{Z_0} = \frac{V_+ + V_-}{V_+ - V_-} \quad (4.2\text{-}2)$$

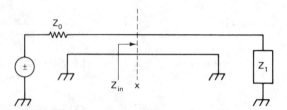

Figure 4.5 Section of line showing an input impedance at an arbitrary point.

The use of a normalized impedance, always Z_0, is very common and useful in transmission line work.

Equation 4.2-2 may be manipulated to obtain the ratio of the negative traveling wave to the incident positive traveling wave

$$\Gamma = \frac{V_-}{V_+} = \frac{z-1}{z+1} \qquad (4.2\text{-}3)$$

This ratio is defined as the voltage reflection coefficient and is the basis of much of transmission line analysis.

Before progressing with our discussion of transmission lines, the concept of a reflection coefficient will be examined in a more familiar context. Figure 4.6 shows a battery with a terminal voltage of 2V, a source resistance of 1-Ω and an arbitrary termination, R. The output voltage across the load, V_L is, from voltage divider action, $2R/(1 + R)$. We find that the maximum *power* is delivered when $R = 1$, that is, when a match exists to the source. The power delivered is less than the maximum for other values of R.

Rather than analyzing this familiar circuit, Fig. 4.6, with voltage divider action, we will use the voltage reflection coefficient. For this example, Γ is given by Eq. 4.2-3 as $\Gamma = (R - 1)/(R + 1)$. Note that for a matched load of 1 Ω the reflection coefficient is zero. The voltage for other terminations will be the sum of the incident value of 1 and the reflected voltage wave defined by Γ.

Assume for a first example, that $R = 0.5\ \Omega$. Then, $\Gamma = (0.5 - 1)/(0.5 + 1) = -0.333$. Hence, the forward wave is 1 and the reverse one is -0.333 leaving a net sum of $V_{\text{out}} = 0.667$. This is, of course, just that predicted from voltage divider analysis.

We obtain $\Gamma = +0.333$ if $R = 2$, leading to a net voltage of 1.333. The current may be calculated from Eq. 4.1-15. Noting that Z_0 is 1, $I = V_+/Z_0 - V_-/Z_0 = 1 - 0.333 = 0.667$, again consistent with a more direct analysis.

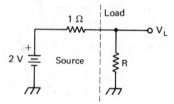

Figure 4.6 A simple 1-Ω generator with an arbitrary termination, R. The output voltage is calculated on the basis of reflected voltage waves.

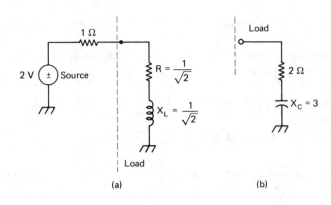

Figure 4.7 A 1-Ω generator driving complex impedances with (a) inductive and (b) capacitive reactance. Load voltages are inferred from the voltage reflection coefficient concepts.

The simplified method is easily extended to an ac source. There is no difference from the dc analysis for resistive loads. Consider a complex termination of $Z = 1/\sqrt{2} + j(1/\sqrt{2})$. This complex impedance has a magnitude of 1 and an angle of 45 degrees when evaluated in polar coordinates. This circuit is shown in Fig. 4.7a.

The voltage reflection coefficient is evaluated from Eq. 4.2-3 as $\Gamma = 0 + j0.4142$ or, in polar form, 0.4142 at 90°. The total voltage is the sum of the incident and reflected waves, $V = 1 + \Gamma = 1 + 0 + j0.4142 = 1.082$ at 22.5 degrees. Similarly, $I = 1.082$ at -22.5 degrees.

Consider as a final example, the network of Fig. 4.7b where a capacitive load of $Z = 2 - j3$ is used. The reflection coefficient is evaluated as $\Gamma = 0.7454$ at -26.6 degrees, or $\Gamma = 0.667 - j0.333$. Analysis then gives $V_L = 0.333 - j0.333$ and $I_L = 0.333 + j0.333$.

These examples were simple ones, essentially trivial. Moreover, they had nothing to do with transmission lines. However, they still work. The reflection coefficient is just a different way to represent a complex impedance. A detailed examination of the equations will reveal that the reflection coefficient has a magnitude of unity or less for any complex impedance with a positive real part. Hence, Γ is a mapping of the right half of a complex impedance plane into a circle of unity radius. The real utility of the mapping will become apparent presently.

Reflection coefficient can also be expressed in terms of admittance. If $Y_0 = 1/Z_0$, we have $\Gamma = (z - 1)/(z + 1) = (1 - y)/(1 + y)$ where $y = Y/Y_0$.

Returning to transmission lines, the case will now be considered of a sinusoidal generator at a frequency $\omega = 2\pi f$ at a position of $x = 0$. The intensity of the source will be $2e^{j\omega t}$ where it is understood that the real part of this is a cosine wave. A source impedance Z_0 is assumed, the same as the characteristic impedance of the line. The positive going wave is then $V_+(x,t) = e^{j\omega(t-x/v)}$. This satisfies the wave equation and is consistent with the stated source strength.

Assume that, at a position further down the line, x_L, there is a load with an impedance different from Z_0. Because $z \neq 1$, $\Gamma_L \neq 0$ and there will be a reflected wave. Γ_L is the reflection coefficient evaluated looking directly into the load. The total voltage at any x and t is the sum of the incident and reflected components

$$V(x,t) = V_+(t - x/v) + V_-(t + x/v) = V_+ e^{j\omega(t-x/v)} + V_- e^{j\omega(t+x/v)} \quad (4.2\text{-}4)$$

where V_+ and V_- are the incident and reflected magnitudes, respectively. The time dependence may be factored out to yield

$$V(x,t) = e^{j\omega t}(V_+ e^{-j\omega x/v} + V_- e^{j\omega x/v}) \quad (4.2\text{-}5)$$

Using earlier results (Eq. 4.1-15), the current is similarly calculated

$$I(x,t) = \frac{e^{j\omega t}}{Z_0}\left(V_+ e^{-j\omega x/v} - V_- e^{j\omega x/v}\right) \quad (4.2\text{-}6)$$

The factor ω/v may be replaced by β which is called the phase constant.

The impedance at any point on the line may now be evaluated as the ratio of the voltage to the current

$$Z(x) = Z_0 \left(\frac{V_+ e^{-j\beta x} + V_- e^{j\beta x}}{V_+ e^{-j\beta x} - V_- e^{j\beta x}} \right) \quad (4.2\text{-}7)$$

The time dependence cancels as it should to define an impedance.

Following the earlier derivations, the reflection coefficient is evaluated

$$\Gamma(x) = \frac{V_- e^{j\beta x}}{V_+ e^{-j\beta x}} = \Gamma e^{j2\beta x} \quad (4.2\text{-}8)$$

where Γ is complex and is the reflection coefficient evaluated at $x = 0$.

$V_+ = 1$ in this analysis and the generator was at $x = 0$ with a load at $x = x_L$. In a more general case, we place an arbitrary load with Γ_L at $x = 0$ and consider the reflection coefficient at an arbitrary point x

$$\Gamma(x) = \Gamma_L e^{-j2\beta x} \quad (4.2\text{-}9)$$

where

$$\Gamma_L = |\Gamma_L| e^{j\theta_L}$$

This is the single most significant equation we have derived about transmission lines. It says that an arbitrary load, Z_L, with a corresponding reflection coefficient Γ_L, yields Γ with a constant magnitude along an attached transmission line. Only the angle of the reflection coefficient changes along the line. Moreover, the interaction of the incident and the reflected waves cause the angle of the reflection coefficient to change at a rate double that of the phase of the incident wave. This one equation justifies our choice of Γ as a viable means for representing complex impedances.

For reasons that will become apparent, we shall refer to Eq. 4.2-9 as the "Smith-chart equation."

We will investigate the phenomenon of standing waves before we explore further use of the reflection coefficient. Consider the case of a very long Z_0 terminated line. A generator at $x = 0$ of matched source impedance drives the line. Because of the matched load at the far end, there is no reflected wave.

Assume the generator has a frequency f. The wavelength is given by $\lambda = v/f$. This means that, at one specific time, there is a spatial voltage pattern on the line with a difference λ between positive voltage peaks. However, at a slightly later time the same pattern will have propagated to the right. A real ac voltmeter placed on the line would show a steady state reading, one that was independent of position on the line. The voltmeter could be a simple diode detector attached to a dc voltmeter or it could be a power meter. Using the relationship between wavelength, velocity, and frequency, and assuming that the open circuit voltage of the generator was 2V, the instantaneous voltage on the line would be

$$V(x,t) = e^{j\omega t} e^{-j2\pi x/\lambda} \qquad (4.2\text{-}10)$$

The time and spatial factors have been factored. The argument for the position dependence suggests that line length may be conveniently measured in fractions of a wavelength. Alternatively, a line length or position on a line may be specified in terms of degrees where 360 degrees corresponds to one wavelength. The velocity in all cases is that in the transmission line; hence, the velocity factor must be taken into account. Both normalizations are used frequently and interchangeably.

Consider now a finite line length which is terminated in a load other than a perfect match. Reflected waves occur leading to a finite Γ at the load. Applying earlier equations, the voltage along the line is given as

$$V(\theta,t) = e^{j\omega t}[\, e^{-j\theta} + \Gamma e^{j\theta}] \qquad (4.2\text{-}11)$$

The position dependence is a line length in degrees or radians.

Assume for illustration that the termination is an open circuit. Hence, $\Gamma = 1$ at 0 degrees. We see a maximum at the load if the voltage is plotted as a function of position. However, when we have moved 90 degrees (one quarter wavelength) from the load, the voltage goes to zero. Another 90 degree movement along the line brings us back to a voltage maximum. The ratio of the maximum voltage to the minimum voltage is called the voltage standing wave ratio, or vswr. These standing waves occur with a periodicity of a half wavelength. The instantaneous voltage is, of course, modulated by the time dependence. The vswr is infinite for the example cited of an open circuit termination, for the voltage vanishes at critical points. The voltage will not vanish at the minima when a nonzero Γ less than unity is used as the termination.

Although we will not develop the details, the standing waves have several interesting and sometimes useful properties. First, the vswr is related to the reflection coefficient by

$$\text{vswr} = \frac{1+|\Gamma|}{1-|\Gamma|} \qquad (4.2\text{-}12)$$

Note that vswr is a scalar quantity—it provides only information about the magnitude of the reflection coefficient.

The position of a voltage maximum or minimum can be of great significance in measurements though. The angle of the voltage at a load is that corresponding to the line length between the load and the first maximum or minimum. A maxima occurs at points where the impedance is purely real.

The observations outlined describe the operation of a traditional instrument for microwave measurements, the slotted line. This apparatus is a length of transmission line with a narrow slot in it which serves as an opening for a small probe. The probe can be a short piece of wire, something that measures the electric field and, hence, the voltage on the line. Alternatively, it can be a loop, an element sensitive to the magnetic field and, hence, the line current. The probe is movable. An accurate scale is attached to the apparatus enabling the user to determine the relative position of the voltage or current maxima.

There was a time in the not too distant past when the slotted line was the workhorse of all microwave laboratories. Today, however, the slotted line often sits on a shelf in the corner, gathering dust, having been replaced by a more modern, accurate, and more easily used network analyzer. The slotted line is still used for the measurement of complex impedance, propagation velocity, and frequency at the top extremes of the microwave region where viable network analyzers have not yet become available.

Because of the historical importance of the slotted line, vswr is a parameter that is still used with surprising frequency. It is a measure of the magnitude of a match, or mismatch that is universally understood. In this text we will, however, describe impedance matches in terms of the magnitude of Γ. A related parameter is the return loss, formally given as

$$\text{Return loss} = -20 \log |\Gamma| \qquad (4.2\text{-}13)$$

The reason for the naming of the term will become apparent when measurements are discussed later.

All of the examples presented have used sources which were matched to the characteristic impedance of the transmission line. If this were not done, and the far end of the line was terminated in a mismatched load leading to reflections, problems could arise. Specifically, a launched incident wave impinges upon a poor load, leading to a reflected wave. That component travels in the $-x$ direction, reaching the mismatched source. It is then partially re-reflected. If the source is matched, the only reflection is that at the load. Re-reflection is common and often desired. An example would be the output of a radio transmitter, rarely a matched source of rf. The transmission line transfers energy to the antenna, the eventual termination.

4.3 THE SMITH CHART AND ITS APPLICATIONS

Equation 4.2-9 is an expression for the reflection coefficient on a transmission line as a function of line length and load, Γ_L. This relationship may be restated in terms of line length expressed either in fractions of a wave length or in degrees

$$\Gamma(x) = \Gamma_L \exp(-2\beta x)$$
$$\Gamma(x) = \Gamma_L \exp(-j4\pi x/\lambda) \qquad (4.3\text{-}1)$$
$$\Gamma(\theta) = \Gamma_L \exp(-j2\theta)$$

The vital property of these equations is that a lossless transmission line alters only the angle of a reflection coefficient, not its magnitude.

This feature can be used to great advantage in analysis and synthesis of circuits using transmission lines. One could plot the reflection coefficient of a complex impedance as a point on a polar graph. The impedances that result from connection to a transmission line of length θ corresponds to rotating that point through an angle of 2θ on the polar plot. A calculation would transform the impedance to a Γ value, the rotation is done graphically, and the final result is then transformed back into impedance coordinates.

P. H. Smith had a similar, but much more elegant approach to the same problem in 1939 (1). He used this graphical method for performing the analysis. However, instead of labeling the graphs with the polar representations of reflection coefficient, he drew lines on the graph representing contours of constant resistance and reactance. Hence, the chart is labeled directly in terms of impedance. This chart is shown in Fig. 4.8 and, quite appropriately, is known as the Smith chart. In spite of great strides in recent years in calculating ability, both in the form of sophisticated handheld programmable calculators and in high speed digital computers, the Smith chart continues to find considerable application in all phases of rf design.

The standard Smith chart is normalized to a characteristic impedance of 1 Ω. All resistance and reactance lines are labeled with respect to this normalized Z_0. The center of the chart, a single point, represents a normalized impedance of $1 + j0$. Other points on the horizontal line drawn through the center represent purely real impedances. Those to the right of the center are for resistance values greater than 1 while the lower resistances are to the left. Points above the center line represent complex impedances which are inductive with a positive imaginary part. Capacitive impedances are below the center line.

All of the lines on the Smith chart are circles. The lines of constant resistance are complete circles with their centers on the horizontal line. All lines of constant reactance are segments of circles with centers on a vertical line normal to the center

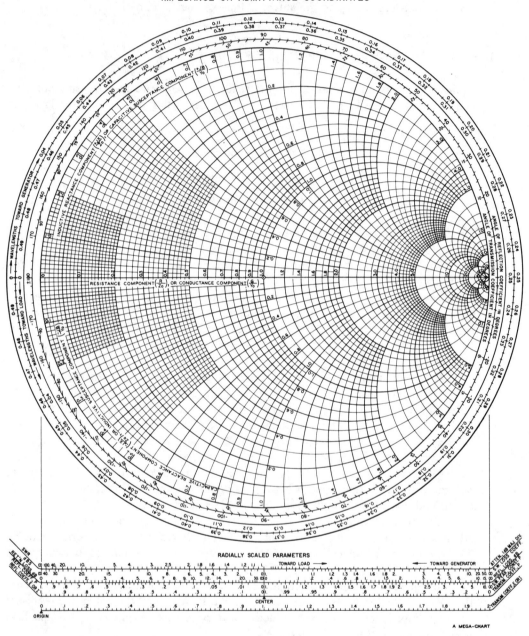

Figure 4.8 Smith chart, reprinted by permission of P. H. Smith, renewal copyright, 1976.

Sec. 4.3 The Smith Chart and Its Applications

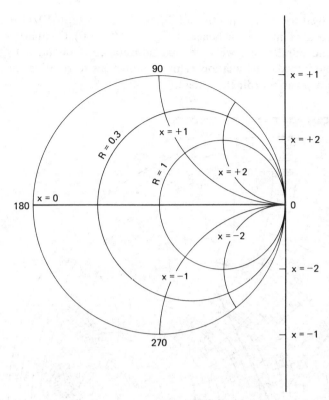

Figure 4.9 Construction of Smith chart. All constant resistance circles are centered on horizontal line and intersect the vertical line. The centers are located to the right of the center of the chart by ½ [1 + (r − 1)/(r + 1)]. All circles of constant reactance are centered on the vertical line. The centers of some value circles are shown. All circles are tangent to the horizontal line and have a radius of $|x|^{-1}$. The normalized resistance is r while x is the normalized reactance.

horizontal line and positioned tangent to the high impedance outer perimeter of the chart.

The method used for Smith chart construction is shown in Fig. 4.9. The pertinent equations are also contained in that figure. These relationships are especially useful when writing computer programs with a graphical output.

The Smith chart has numerous features with as many different applications. Many of the features will be given in some of the applications to be described below. The reader is urged to obtain a pad of Smith chart forms and to work through the examples as well as to solve problems of his or her own choice. Smith charts are available at reasonable cost from Analog Instruments Company, P.O. Box 808, New Providence, N.J. 07974.

Example 1. Evaluation of Admittance

Examination of a standard Smith chart reveals that lines of constant "resistance" are actually labeled both in terms of resistance and conductance. Similarly, the constant reactance lines are also suitable for evaluation of susceptance. It is not immediately obvious that this dual nature should exist.

The reflection coefficient, Γ, is related to normalized impedance by $\Gamma = (z - $

$1)/(z + 1)$. This may be solved for z as a function of Γ, yielding $z = (1 + \Gamma)/(1 - \Gamma)$. Admittance is the reciprocal of impedance; hence, $y = (1 - \Gamma)/(1 + \Gamma)$. Comparing equations for impedance and admittance, we note that admittance is obtained by substituting $-\Gamma$ for Γ in the impedance equation. This is equivalent to rotating the Γ vector by 180 degrees in a polar coordinate system.

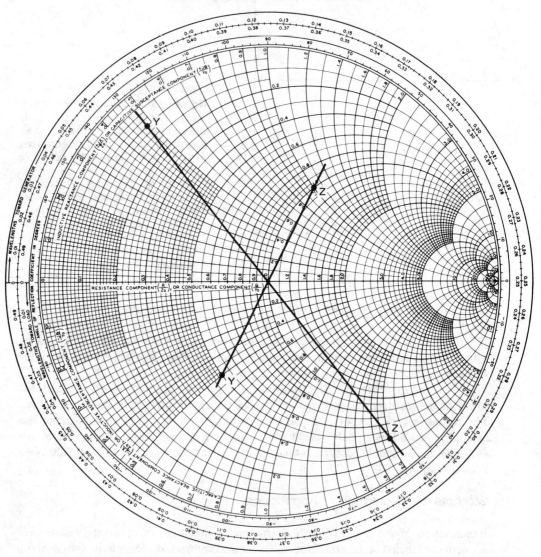

Figure 4.10 Use of the Smith chart to convert an impedance to an equivalent admittance.

The method for transforming from impedance to admittance on the Smith chart is clear from this analysis. A straightedge is used to draw a line from a known impedance point through the center of the chart. A compass is then used to mark a distance from the center on the line that is equivalent to that of the separation of the impedance point from the center.

This is shown in Fig. 4.10 with two examples. First, an impedance (normalized) of $1 + j1$ is marked. The construction is performed, showing that the corresponding admittance value is $y = 0.5 - j0.5$. If $Z_0 = 50$, then $Y = 0.01 - j0.01$. Similarly, an impedance of $0.4 - j2.0$ transforms to $y = 0.96 + j0.48$.

Admittance to impedance (and back) transformations are direct and easily performed with a handheld calculator, especially one with built-in polar to rectangular conversion. It is still important to be able to do the conversion on a Smith chart. The utility will become apparent with further examples.

It is often useful to read impedance and admittance simultaneously. This may be performed with two sheets of Smith chart paper. The normal one is positioned to read impedance as shown. The other is rotated by 180 degrees with the centers coincident and used to read a corresponding admittance. This operation is best performed if the charts are printed on nearly transparent paper.

Some versions of the Smith chart are printed in two colors. The standard impedance coordinates are printed in red, while lines of constant conductance and susceptance are superimposed in green. This type of chart is especially useful and is also available from Analog Instruments. The examples shown use the standard one color Smith chart.

Example 2. Evaluation of Impedance along a Transmission Line

For our second application, assume that a piece of 50-Ω transmission line is connected to a termination consisting of a 100-Ω resistor in series with a 100-Ω inductor. The normalized impedance is $2 + j2$. We wish to find the impedance seen at various points along this line when looking back toward the load. This is illustrated in Fig. 4.11.

The impedance, $2 + j2$, is marked on the plot as point A. A circle centered on the center of the chart is drawn through z. A line is also drawn from the center of the chart to the outer edge to allow measurement of angle, or line length. The line intersects the outside perimeter of the chart at a point marked 0.208 wavelength. We are moving "toward the generator," in a clockwise direction because we are traveling away from the load. This is consistent with the minus sign in the exponent of Eqs. 4.3-1.

The impedance is $4.3 + j0$, resistive and greater than unity in magnitude where the circle first crosses the horizontal line. The vswr is read at this point, 4.3:1. This position is marked B on Fig. 4.11. The line position found by drawing a line from the center to the perimeter is 0.25 wavelength. Hence, this impedance is realized with a movement along the line of $0.25 - 0.208 = 0.042$ wavelength, or 15.1 degrees.

Point C on the circle is at an impedance of $z = 1 - j1.6$. This occurs for a

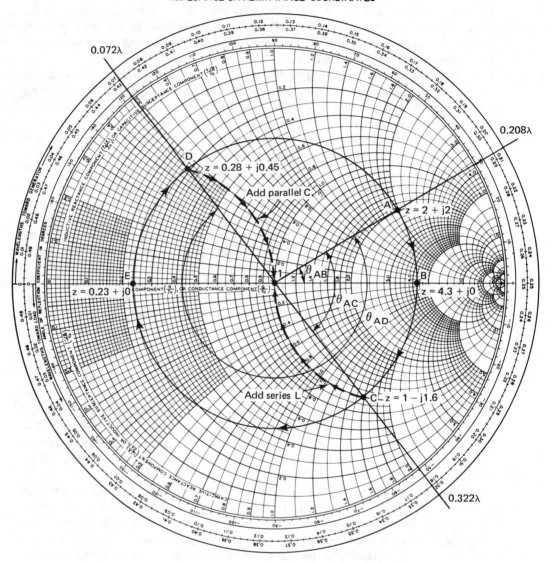

Figure 4.11 A normalized impedance of $z = 2 + j2$ is placed at the end of a line. This is represented by point A on the chart. The other points (B, C, D, and E) show the impedance as the position along the line varies. Point C intersects a line of constant unity resistance. Addition of a series reactance, an inductor, will match that impedance to the characteristic impedance. Point D intersects a line of constant unity conductance and may be matched with a shunt capacitance.

position on the perimeter labeled 0.322 wavelength. The rotation from the original point, then, is $0.322 - 0.208 = 0.114$ wavelength. This corresponds to 41 degrees of transmission line.

Point C is significant for it is a point where the real part of the impedance is unity. The reactance is capacitive because the point is in the lower half of the chart. This impedance could be matched to Z_0 by inserting a series inductor with a normalized reactance of 1.6 Ω. The denormalized value is $50(1.6) = 80$ Ω. The insertion of a reactive element corresponds to motion along a line of constant resistance. This motion is shown as dotted line C-1 on the chart of Fig. 4.11.

The impedance continues to change if the rotation is continued from point C (we have elected not to do the matching with the series inductor mentioned above). At the point marked D, the impedance is now in the upper inductive part of the chart with $z = 0.28 + j0.45$. The point on the perimeter shows a rotation from the original impedance of 0.364 wavelength, or 131 degrees of line length.

Point D is also significant so far as impedance matching is concerned. The admittance associated with point D is $1 - j1.6$. The vital factor is the conductance of unity. The impedance at point D may be matched to Z_0 by the addition of a capacitive susceptance of 1.6 S. The addition of parallel capacitance is equivalent to motion along a line of constant conductance and is shown as a dotted line, D-1, in the figure. Observe that points C and D are diametrically opposed in the diagram. This can be most useful in designing matching networks. The addition of a parallel capacitance is usually preferred over a series inductor owing to the reduced parasitic reactance of the undesired type.

Other useful information is also obtained from Fig. 4.11. When the compass has been adjusted to draw the circle, it may then be used with the scales at the bottom of the standard Smith chart (see Fig. 4.8). The vswr is read from one of these scales. In addition, the corresponding voltage reflection coefficient, 0.625, and return loss, 4 dB, are all read from these scales.

Another standard result, the quarter wavelength transformer, is well illustrated by the chart, Fig. 4.11. Points B and E are both on the zero reactance line, but are separated by a 90 degree line segment. The two impedances are reciprocals of each other. Two real impedances, R_a and R_b, may be matched with a quarter wavelength of transmission line with a characteristic impedance such that $Z_0^2 = R_a R_b$. Note also that the impedances repeat every 180 degrees if the line is continued without inductive or capacitive matching elements.

Example 3. Open and Short Circuited Transmission Lines

The previous example considered an initial impedance with both a real and a complex part, $z = 2 + j2$. It was located roughly halfway between the center of the chart and the outer perimeter. Of special interest in many applications are motions along a line which produce a purely reactive effect.

Consider a short circuit, a zero of impedance. This point, $z = 0 + j0$, is on the left extreme of the horizontal line. The impedance will continue to have a zero real part, but will become increasingly inductive if we move along a transmission line away from this termination (clockwise). This is shown in Fig. 4.12 with point A as the short circuit termination.

A series inductor was required in the previous example to match point C of Fig. 4.11 to Z_0. The required normalized reactance was 1.6 Ω. This reactance would be produced by a line of the same Z_0 as used before with a length of 0.161 wavelength. This element, a shorted stub, is placed in series with the original transmission line at point C of Fig. 4.11.

The alternative solution to achieving a match in Example 2 was the use of a shunt capacitor at point D of that line. The required normalized capacitive susceptance was 1.6 S. This could be achieved with a shorted stub of length to produce an admittance of j1.6, or a reactance of 0.63. The length required would be 0.41 wavelength.

An equally interesting termination is the open circuit, $y = 0 + j0$. This point is shown as B in Fig. 4.12. A clockwise rotation on the chart to the desired values of reactance will also produce the required open stubs to serve as either inductors or capacitors.

An especially important application of shorted and open stubs is as a resonator, a substitute for an LC tuned circuit. The impedance is an open circuit if a short circuit termination is used with a line of 90 degrees. This will occur for one frequency. At a slightly lower frequency, the line, while of the same physical length, is less than 90 electrical degrees long. Hence, it will have an input impedance that is inductive. Conversely, at slightly higher frequencies, the line will be capacitive. This is the same behavior displayed by an ideal LC combination.

The analogy is not exact. If the line is a quarter wavelength long at f, it will be three quarters wavelength long at $3f$. This length would correspond to one and

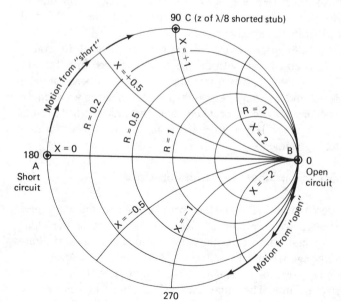

Figure 4.12 Effect of moving along a transmission line from either an open or short circuit.

a half trips around the Smith chart, again resulting in an open circuit input immittance. Similar resonances occur for frequencies nf where n is an odd integer. These resonances are often called reentrant modes.

A transmission line resonator need not be exactly 90 degrees in length. Consider a shorted eighth wavelength line. The input impedance is point C on Fig. 4.12, $z = 0 + j1$ or $y = 0 - j1$. The admittance of this condition is located at point D. The addition of a parallel capacitor moves the admittance back toward point A, resulting in the impedance at point B, again an open circuit. A length of about 45 degrees is common for resonators used in filter applications.

Example 4. Evaluation of LC Networks

Previous examples have dealt with transmission lines. This is, of course, where the Smith chart is put to best advantage. However, the chart is so powerful that the rf engineer becomes accustomed to thinking in terms of position on the Smith chart whenever complex networks are encountered. The chart is just as applicable to networks containing discrete inductors and capacitors as it is with transmission lines.

Shown in Fig. 4.13 is the impedance transformation effected by an L-network. The starting impedance is $2 + j1$. It is desired to match this to a Z_0 source. The initial impedance, point A, is transformed to the equivalent admittance at B, $y = 0.4 - j0.2$. A parallel inductor is then added. The admittance at B is inductive. Addition of further inductance moves the point along a line of constant conductance to point C, further into the inductive portion of the admittance plane. The admittance at point C is $y = 0.4 - j0.49$. This value was chosen such that, when transformed to an impedance, the resulting value (point D) has a resistive component of unity. The impedance at D is $1 + j1.22$. This is now matched to $z = 1 + j0$ with movement along a line of constant resistance, accomplished with addition of series capacitance.

Intuition might lead one to assume that if the network of Fig. 4.13 were analyzed from the opposite direction, one would merely backtrack over the original path. This is not true, and is also not true when transmission lines are used instead of lumped components.

Figure 4.14 shows the analysis of the same network starting at the center of the chart. The network consists of a series capacitor with a reactance of 1.22 Ω, and a shunt inductor with a reactance of 0.29 Ω. The capacitor is added starting with the characteristic impedance, moving the result to point A on the chart. A parallel element must next be added. Hence, the impedance is converted to an admittance, point B. The parallel inductor is added, moving the resultant admittance to that at C and the admittance is converted back to an impedance, point D. The final result is $z = 2 - j1$. This is the complex conjugate of the original impedance of Fig. 4.13. This example illustrates not only the directional nature of impedances, but the relationship between two impedances that are "matched."

The Smith chart equations are all straightforward and easily programmed on a calculator. In spite of this, the Smith chart is not obsolete. A programmable calculator is ideal for analysis, but provides little intuition about possible variations and is less easily used for circuit synthesis. An excellent method for design is to use both. A calculator is programmed with a Smith chart simulating program. Then, a problem is attacked with the calculator sitting beside a pad of charts. The initial analysis is done with the chart. However, the calculator is used to obtain more accuracy. This allows a degree of "slop" in the chart constructions that would otherwise require considerable care. Freehand sketching will often suffice.

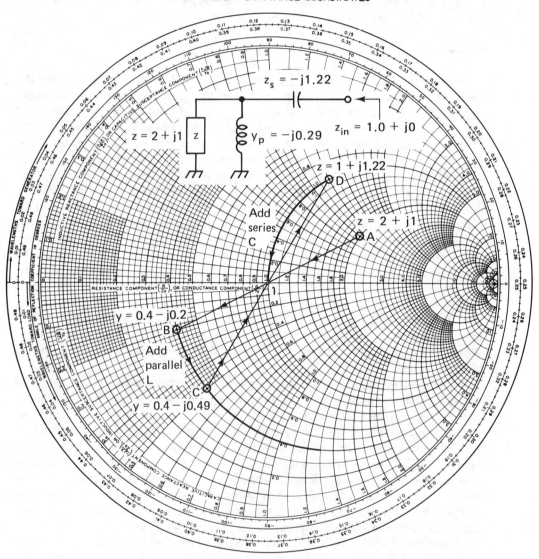

Figure 4.13 Analysis of a lumped element matching network by the Smith chart.

IMPEDANCE OR ADMITTANCE COORDINATES

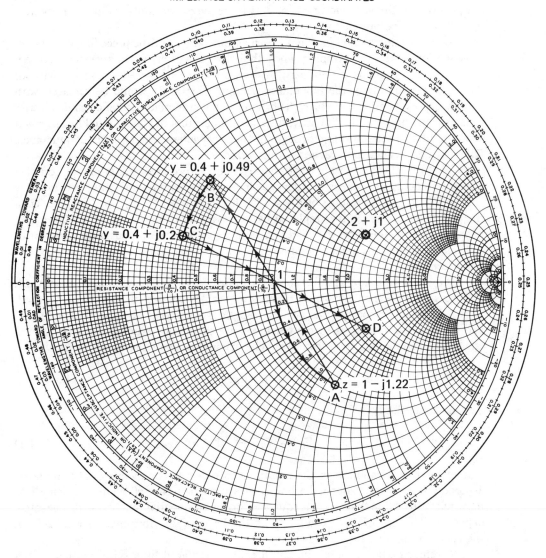

Figure 4.14 Analysis of the same matching circuit as used in Fig. 4.13 except that the analysis begins at the generator end. Note that the resulting impedance, point D, is the complex conjugate of the starting value for the previous example, Fig. 4.13.

4.4 PRACTICAL TRANSMISSION LINES

The basic concepts of the transmission line and the Smith chart were presented in the previous sections. Some practical details, required for transmission line applications, will now be presented.

Lossless transmission lines have been assumed in all of the preceding discussion. This was done at the beginning when the wave equation was derived and solved. Lossless lines, of course, do not exist. If the wave equation were to be solved without the lossless assumption, the inductance per unit length, L, would be replaced with a term representing the voltage drop across the inductive reactance and a series resistance per unit length. Similarly, the change in current in a differential line segment would result from that flowing through the shunt capacitance and a shunt conductance. Different conclusions are reached if this is done.

The first effect of loss is that characteristic impedance, Z_0, is now a complex quantity. The second effect is a change in the exponential functions. We found a change in the spatial voltage of a wave of the form $e^{-j\beta x}$ for the lossless line. With the presence of losses, this is replaced by expressions of the form $e^{-\alpha x - j\beta x}$. This may be compacted to the form $e^{-\gamma x}$ where γ is $\alpha + j\beta$. γ is the propagation constant of the line, a complex number composed of α, an attenuation constant, and β, the phase constant.

An approximation may be applied to most lossy lines encountered in rf work so long as loss is not excessive. First, the complex nature of the line is ignored. This is justifiable since the loss per wavelength is assumed to be low. The inductive and capacitive components in the equations are dominant. Second, the fundamental equations are modified by a loss term. The original basis of the Smith chart was

$$\Gamma(\theta) = \Gamma_L e^{-j2\theta} \tag{4.3-1}$$

This is replaced by

$$\Gamma(\theta) = \Gamma_L e^{-j2\theta}\, 10^{-A\theta/10} \tag{4.4-1}$$

where A is the attenuation of the line in dB per degree. The value of A is replaced accordingly if position along the line is measured in terms of a physical length or in wavelengths. Note that $10^{-A\theta}$ is equivalent to an exponential function with a negative exponent. The use of 10 instead of e as the basis of the formulation results from the common use of dB in loss measurements.

Circular motion around the Smith chart is replaced by a spiral motion with a continuously decreasing radius, a decreasing reflection coefficient magnitude. The approximation is valid so long as the magnitude of the reflection coefficient does not change too much over a half wavelength rotation. This simplifies calculations and allows us to continue using the Smith chart for analysis and synthesis.

Consider a transmission line with a loss of 3 dB per 100 ft length, a common attenuation value at 100 MHz for coaxial cables. The cable is to be used at 100

MHz, has a Z_0 of 50 Ω, and a velocity factor of 0.66. One wavelength corresponds to a length of 198 cm or 6.496 ft.

Assume a severely mismatched termination, $Z = 20 + j100$ Ω. $\Gamma_L = 0.855$ at 51.7 degrees and the return loss is -1.36 dB.

First attach a 10-ft section of line. If the line were lossless, the input impedance seen looking into the section of line would be $Z = 83.5 + j210.67$. The magnitude of the reflection coefficient and, hence, the vswr and return loss, would remain unchanged.

The results are different with loss. The input impedance becomes $Z_{in} = 106 + j184.6$. This is obtained by noting that the loss in the 10-ft cable is 0.3 dB. Motion around the Smith chart is by the corresponding distance of 1.539 wavelength. The magnitude of the reflection coefficient is multiplied by the factor $10^{-(0.03)}$, leaving a result of $\Gamma = 0.7982$. The angle of reflection coefficient is unchanged. Conversion of the modified Γ value to impedance yields the given impedance value.

The match to 50 Ω has improved with the decrease in the magnitude of Γ. The return loss is now -1.96 dB with a corresponding vswr of 8.91. This is still not a good match, but better than with a lossless cable.

Consider an extreme case, a 1000-ft spool of the same cable. The far end of it is again misterminated in $Z = 20 + j100$. However, the impedance seen at the input is now $49.99 + j0.085$ Ω owing to the 30-dB loss of the long cable. The return loss is -61.4 dB, corresponding to $\Gamma = 0.0009$, an excellent match.

The cable loss normally is measured in a matched system. The input return loss is twice the cable loss, accounting for the forward and the reflected wave, plus the original return loss of the load of -1.4 dB. This illustrates the rationale behind the term "return loss."

The last example also suggests an application. If a long, relatively lossy line is terminated in virtually any impedance, it functions well as a known termination for testing purposes.

The approximate loss equation (Eq. 4.4-1) is suitable for most rf applications. It is not well suited for situations where the loss in the line is large in one wavelength. An example would be a twisted wire transmission line operating at audio frequencies such as that used for telephone cables. More refined analysis methods must be used for such work.

The Smith chart may be used directly for analysis of systems with lossy lines. The simplest method is implicit in the meaning of return loss. An analysis is first performed with no loss assumed. Then, the loss of the line corresponding to the length used is calculated. The return loss, as determined on a scale below the standard chart, is then increased in magnitude by twice the loss of the line. The corresponding radial change in Γ is noted and the position of the chart is changed accordingly.

The same evaluations may be performed with a calculator using a program that simulates the Smith chart. This is one of those situations where interactive use of a calculator and a Smith chart is advantageous. Careful structuring of the program will often allow data to be entered to bypass intermediate calculations. For example, the preceding examples with lossy lines were performed with a calculator. The loss was entered in dB per 100 ft, a typical specification offered by cable manufacturers. Other input forms would be better if, for example, the user was doing microwave work with microstrip-type lines.

Assume that a 100-MHz resonator was to be built using an eighth wavelength of 50-Ω cable of the previous examples. The input impedance will be $0.28 + j49.99$ with a short circuit termination. Transforming to an admittance, we see that the inductor formed from the stub is equivalent to a 50-Ω inductive reactance in parallel with a resistance of 8914 Ω. The resonator Q_u would be 178.3. This is not especially good at 100 MHz. The Q of the same resonator would increase to 535 if the cable was replaced by a type with a loss of only 1 dB per 100 ft. This is a more respectable value, but still less than spectacular. Coaxial cable is often used for resonator elements, assuming that it will be low in loss. While sometimes valid, calculations should be performed.

Of the many forms of transmission line, three are most common in rf work. These are the coaxial cable, the parallel wire line, and microstrip. The characteristic impedance of coaxial cable is related to the dimensions by

$$Z_0 = \frac{138}{(\epsilon)^{1/2}} \log(D/d) \qquad (4.4\text{-}2)$$

where ϵ is the relative dielectric constant with air or vacuum being represented by $\epsilon = 1$. D is the inner diameter of the outside conductor while d is the outside diameter of the inner wire. The units chosen are of no significance, for they appear only as a ratio. Coaxial cables are commercially available in a wide variety of sizes and types of construction.

Coaxial resonators are often used at frequencies of 1 GHz and higher. It has been shown that the lowest loss (highest resonator Q_u) is obtained with a characteristic impedance of around 70 Ω (2). The resonators are usually fabricated by milling a piece of solid metal. A common length is 1/8 wavelength and the complete structure is plated with silver to improve conductivity. Q_u values of 4000 are not uncommon in a 2-GHz resonator.

The dimensions of a parallel wire transmission line are related to the characteristic impedance by

$$Z_0 = 120 \cosh^{-1}(D/d) \qquad (4.4\text{-}3)$$

where D is the center-to-center separation of the two wires and d is the outside diameter of each wire. It is assumed that the dielectric material surrounding the wires is air or vacuum. Parallel wire line is used primarily in the hf and vhf spectrum and has the virtue that loss is quite low.

Figure 4.15 shows a cross section view of a microstrip-type transmission line. This line, although complicated in analysis, is extremely popular for modern rf design. The reason is that such lines, in short lengths, are easily fabricated in integrated forms with other circuit elements. Circuits are often constructed at vhf and low

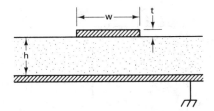

Figure 4.15 Cross section view of a microstrip transmission line.

uhf using microstrip methods on double-sided circuit boards. Ceramic substrates are often used at microwave frequencies. Metallization is applied using a variety of thin- or thick-film methods. The dielectric constant of most circuit board materials is around 2.5, although it may be much higher for ceramic substrates, in excess of 10. The complexity of the equations describing microstrip results from the mixture of two dielectrics, air and the substrate. An effective dielectric constant is used.

For W/h less than unity, where W is the line width and h is the dielectric thickness, both in centimeters, the characteristic impedance is given by

$$Z_0 = \frac{60}{(\epsilon_{\text{eff}})^{1/2}} \ln[8h/W + W/(4h)] \tag{4.4-4a}$$

where the effective dielectric constant is

$$\epsilon_{\text{eff}} = \frac{\epsilon_r + 1}{2} + \frac{\epsilon_r - 1}{2}[(1 + 12h/W)^{-1/2} + 0.04(1 - W/h)^2] \tag{4.4-4b}$$

If W/h is greater than unity

$$Z_0 = \frac{120\pi/(\epsilon_{\text{eff}})^{1/2}}{W/h + 1.393 + 0.677 \ln(W/h + 1.444)} \tag{4.4-4c}$$

and the corresponding effective dielectric constant is

$$\epsilon_{\text{eff}} = \frac{\epsilon_r + 1}{2} + \frac{\epsilon_r - 1}{2}(1 + 12h/W)^{-1/2} \tag{4.4-4d}$$

These equations presume zero thickness for the line. If more refined analysis is required, the reader is referred to the summary paper by Bahl and Trivedi (3). ϵ_r is the substrate dielectric constant.

Although rarely used in traditional applications, a type of great importance in resonators is the helical transmission line. A wire is formed into a solenoidal coil and is placed on the inside of an appropriate enclosure. The result is a transmission line. One may intuitively view the structure by assuming that a voltage wave launched into one end propagates along the wire of the solenoid. The axial velocity of propagation is then much less than for a traditional coaxial structure. As such, it is called a

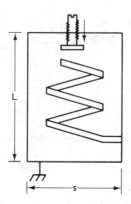

Figure 4.16 Cutaway view of a helical resonator.

slow-wave structure. Other applications for slow-wave structures are in traveling wave tubes and deflection plates for high speed cathode-ray tubes.

Like any transmission line, the helical type may be used to form a resonator. If one end is attached to ground and the other end is left open, the device will be resonant at the frequency where the line is a quarter wavelength long. The length may be much shorter than a similar traditional coaxial resonator because of the reduced axial propagation velocity. A helical resonator is shown in Fig. 4.16.

The quarter wavelength helical resonator may be constructed with either a circular or square cross section enclosure. The square ones are probably more common owing to the ease of fabrication. If the length of the solenoid is approximately equal to the dimension of one side of the square enclosure, S in Fig. 4.16, and the length to side ratio is about 1.5, the unloaded Q_u value that may be obtained is approximately given by

$$Q_u = 60S(f_0)^{1/2} \tag{4.4-5}$$

where S is given in inches and f_0 is the operating frequency in MHz. The proper number of turns is given by

$$N = \frac{1600}{f_0 S} \tag{4.4-6}$$

with a helix diameter of about $\tfrac{2}{3} S$. Additional information on helical resonator design is given in the excellent treatment in Chap. 9 of Zverev (4).

Some workers have described the helical resonator as being merely a high Q inductor in a shielded enclosure. This is not accurate, although the distinction between the two is sometimes subtle. Air core inductors also display a $\sqrt{f}$ dependence in Q_u. However, the helical resonator has one very dramatic characteristic that is not found with a lumped LC circuit—it has reentrant modes. Hence, a 100-MHz helical resonator will also be resonant at about 300 MHz. Higher order resonances also occur.

A helical resonator is usually built slightly shorter than a quarter wavelength.

Then, resonance is produced at the proper frequency by the addition of a small capacitance at the open end of the helix. This may be nothing more than a screw moving through the wall. High quality variable capacitors are often used at lower frequencies. The resonator may be probed for measurement by placing a coaxial connector in the side wall. The Dishal methods outlined in Chap. 3 should be followed.

Any of the standard methods may be used for coupling between resonators. However, some are less than practical. A suitable method is aperture coupling. This was described in an example in Chap. 3.

Loading the end resonator in a filter is realized with a capacitive probe, a loop near the grounded end of the helix, or by connection to a tap on the helix. The tap method is usually the most convenient. However, it can present a practical problem—the position of the tap is often very close to the grounded end of the helix, perhaps by a small fraction of one turn. It is useful to insert a series capacitor in the line from a coaxial connector to the tap. The capacitor transforms the usual 50-Ω termination to a higher impedance level and allows the tap position to move up the helix to a more easily controlled position. The capacitor may be adjusted if variable loading is required.

Materials used for helical resonators are often critical. The helix wire should be as large as practical to decrease loss and provide mechanical strength. The helix is usually self-supported in resonators for the uhf region. A form is used at lower frequencies to hold the larger number of turns required. It is often useful to silver plate both the wire and the interior of the enclosure. In one instance with a three-section 510-MHz helical filter, it was found that plating the wire increased the Q_u value by over 20%, even when the enclosure had been plated. The wire surface becomes less critical at lower frequencies.

Some workers have built helical filters as low in frequency as a few MHz. The upper limit is probably around 1 GHz. It is usually more practical to use traditional transmission line sections above that frequency.

There are many other filter types that utilize transmission lines. Narrow bandwidth filters constructed with eighth to quarter wave coaxial cavities are built with many of the same considerations mentioned for helical resonators. The theoretical tools used to extend our earlier filter work to include microstrip methods are Richards' transformation and the Kuroda identities (5).

4.5 IMPEDANCE TRANSFORMING NETWORKS

A primary function of the transmission line is for impedance transformation. One example studied was the quarter wave transformer. Another was the L-network. The L-network was presented as an illustration of the Smith chart being used with lumped elements (Figs. 4.13 and 4.14).

There are numerous networks like the L. While they may be analyzed and synthesized with the Smith chart, they are also easily designed with more conventional methods. This section will present an overview of these transformations.

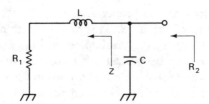

Figure 4.17 Analysis of the L-network, used for impedance matching.

The first network considered is the L-network shown in Fig. 4.17, so named because its shape represents the capital letter "L." Assume two resistive impedances, R_1 and R_2, are to be matched with R_2 being the larger of the two. Consider the impedance shown, $Z = R_1 + jX_L$. The corresponding admittance is $1/Z$. Upon multiplication by the complex conjugate of the denominator, $Y = (R_1 - jX_L)/(R_1^2 + X_L^2)$. Setting the real part of the admittance to $1/R_2$, the desired input immittance, the general equations emerge

$$X_s = (R_1 R_2 - R_1^2)^{1/2}$$
$$X_p = (R_1^2 + X_s^2)/X_s$$
(4.5-1)

The equations are presented in terms of X_s and X_p, series and parallel reactances. No specification has been made that one is inductive and the other capacitive. It is easily verified that Eqs. 4.5-1 still describe the transformation if Fig. 4.17 is redrawn into a form with a series capacitor and shunt inductor. One form is inherently a low pass filter while the other has a high pass transfer function. The method of design is based upon the desired impedances at a single frequency rather than upon the desired frequency domain transfer function. The network topology is, nonetheless, identical to a low or high pass filter of second order.

The network is designed for a single termination. The resistance at one end, R_1, is specified. The combination of series and parallel reactances produce an input resistance, R_2, at the other end. Conversely, a resistance R_1 will be seen when looking into the series arm if the end of the network with the shunt element is terminated in R_2.

The equations for the L-network were derived by looking into the series arm. Alternatively, the end with the parallel reactance may be terminated with R_2 and the admittance evaluated. The desired series and parallel immittances are calculated

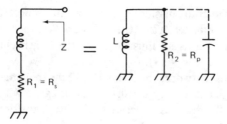

Figure 4.18 Viewing part of an L-network as a lossy inductor, allowing a Q to be evaluated.

for a required input, R_1. The equations appear different than those shown as Eqs. 4.5-1, but are identical to the extent that they predict the same results.

The derivation of Eqs. 4.5-1 is reiterated in Fig. 4.18 with a different emphasis. The original resistance, R_1, and the series reactance are presented. They are converted to an equivalent admittance such that the resistive portion is a parallel resistance, R_2. The parallel susceptance of the equivalent is resonated with an appropriate element, yielding the equivalent of a resonator. The loaded Q is calculated with the usual relationship, $Q = R_{par}/X_L$

$$Q = \frac{R_2}{X_p} = \frac{R_s X_s}{R_1^2 + X_s^2} = \left(\frac{R_2}{R_1} - 1\right)^{1/2} \qquad (4.5\text{-}2)$$

The network Q increases with the ratio of the transformed resistances.

The Q derived in Eq. 4.5-2 is exactly the same as was seen in Chap. 2 for a similar network. The method of evaluation is important. Q is properly defined and related to the energy stored in a resonator for a second order network, one with two reactive elements of opposite type. Q is not well defined for networks with three or more reactive elements. Still, Q is a frequently used parameter in the design equations for more complex networks. The meaning of Q is different when applied to such networks. It is the ratio of a resistance to a reactance when looking into one end of the network at one frequency. Network reduction methods are always used. The user should not deduce the bandwidth of the network by the Q used for design. This is illustrated in a later example.

L-networks are often used for transformation between impedances with reactive elements at one or both ports. Equations and design charts have been presented in other works. They will not be presented in this text in closed form. A reactive impedance is resolved into real and imaginary parts in either an impedance or admittance form. An L-network is designed to transform the real portion of the immittance. The reactive element is then lumped into the calculated reactances. A second method is to perform the network synthesis with the aid of a Smith chart.

A popular transformation network is the pi-type shown in Fig. 4.19. It is designed for a single termination. That is, if the left side is terminated in R_1, a resistance of R_2 is seen when looking into the right side. Conversely, a termination at the right of R_2 will yield a resistive input at the left of R_1. The design equations for a pi-network are

$$R_1 \geq R_2$$
$$X_{C1} = R_1/Q$$
$$X_{C2} = R_2 \left[\frac{R_1/R_2}{Q^2 + 1 - R_1/R_2}\right]^{1/2} \qquad (4.5\text{-}3)$$
$$X_L = \frac{QR_1 + R_1 R_2/X_{C2}}{Q^2 + 1}$$

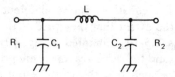

Figure 4.19 Schematic for the pi-network, used for impedance matching.

Q is an arbitrary parameter chosen by the designer. There is a lower bound on the Q. It must be high enough that $(Q^2 + 1)$ is greater than (R_1/R_2). X_{C2} vanishes when these parameters are equal and the pi-network degenerates into the L-network.

The presence of a third element in the pi-network provides an additional degree of freedom. Network Q is not rigidly coupled to the impedance transformation as it was with the L-network. For example, a symmetrical pi-network can be designed with $R_1 = R_2$, but with high Q values. This allows the builder to choose the network on the basis of other criterion, usually the frequency response and practicality of components.

The frequency response of a pi-network will depend upon whether single or double termination is employed. Figure 4.20 shows the response curves for a pi-network that was designed for a Q of 10 with $R_1 = R_2 = 50$ Ω. One curve is for double termination while the other uses current source drive. The latter application is common in the output of power amplifiers which are not matched for maximum gain, but instead for maximum power output (6).

Pi-networks are often used at the input of high impedance amplifiers to provide a voltage transformation. Although driven from a known source impedance, the output termination is so high that it may be assumed to be an open circuit. Figure 4.21 shows the response curves for three such networks. They were designed for 10 MHz and a Q of 10. The resistances seen looking back into the network were 50, 200, and 500 Ω while the source impedance is 50 Ω. The impedance seen looking into the network from the source will be purely reactive. The voltage at the output will be the square root of the transformation ratio times the open circuit generator voltage. While the network provides voltage gain, power gain is not defined owing to the open circuited output termination.

We see from the preceding figures that the frequency response of the pi-network is that of a peaked low pass filter. A bandpass-like response may be added with slight modification, shown in Fig. 4.22. A pi-network was designed for a Q of 10 and a frequency of 10 MHz. The inductor has a reactance value of 9.902 Ω. This was replaced with a larger inductor. The excess reactance was then cancelled with a series capacitor, leaving the net value at 10 MHz unchanged. Clearly, the bandwidth of the modified network is considerably narrower than the original network. However, the network reduction methods for evaluation of Q yield an identical value. The energy stored in the two networks is very different.

The pi-network design equations (Eqs. 4.5-3) are presented in standard form for a low pass type design. This is the more common. The dual form with shunt inductors and a series capacitor is just as viable from a transformation viewpoint. The equations are used as shown to calculate reactances. The substitution is made directly, maintaining equal reactance values in each position.

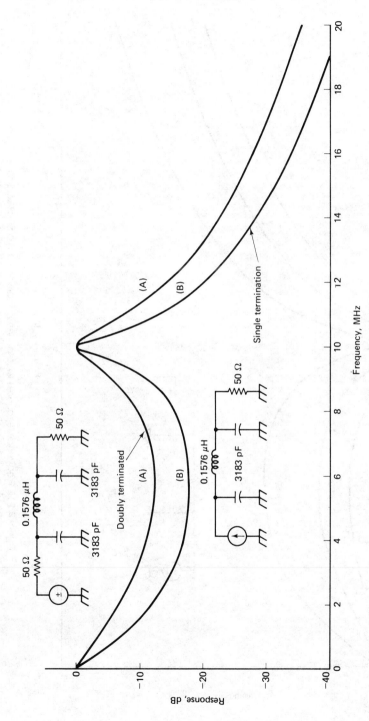

Figure 4.20 Frequency response of the pi-network shown in the figure, for double termination and drive from a current generator.

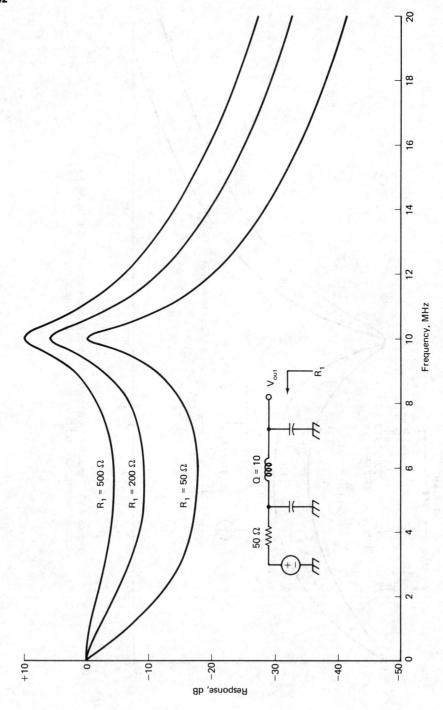

Figure 4.21 Frequency responses for a pi-network driven from a 50-Ω source, but without termination. This illustrates the voltage transforming properties of the circuit.

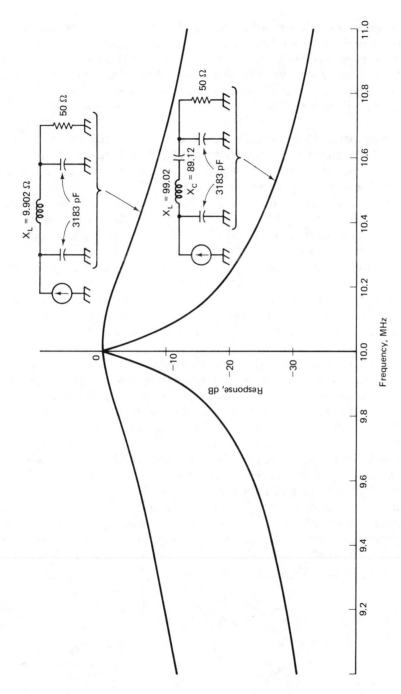

Figure 4.22 Frequency responses for two different pi-networks, shown with the curves. The networks have the same Q when viewed from one end, but have significantly differing selectivity. Care must be utilized when using the Q parameter for matching network design.

The pi-network displays a useful property, that of impedance inversion. Consider a symmetrical pi designed for 50-Ω terminations and a Q of 1. The value seen at one end is the same that would be seen if the network was replaced with a quarter wavelength of transmission line. This holds not only for any real, but for reactive terminations. Hence, if the 50-Ω pi-network is terminated in $Z = 10 + j10$, the impedance seen at the other end is $125 - j125$, exactly as predicted on a Smith chart.

The symmetrical pi-network has the easily remembered design parameters $X_C = X_L = Z_0$. Matching a 50-Ω source to a 300-Ω load is realized with a quarter wave line with impedance 122.5 Ω. A symmetrical pi-network designed with this impedance will also perform the transformation.

Not all pi-networks show this inversion property. For example, the symmetrical network with a Q of 10 will have an inductor of 9.901 Ω and 5-Ω capacitors. It will not appear as a quarter wavelength of line at the design frequency because the X_L and X_C values are not equal. It will, of course, appear as a quarter wave line at some frequency owing to the symmetry. The impedance will not be 50 Ω. The capacitors would be 3183 pF and the inductor would be 0.1576 μH if the network had been designed for 10 MHz. Calculation indicates that the network looks like a quarter wavelength line at 7.1 MHz. Moreover, the characteristic impedance at 7.1 MHz would be about 7 Ω. Careful analysis should be performed on the network chosen if impedance inversion is required.

Impedance inversion networks have wide application. One converts a series tuned circuit to a parallel resonator. For example, a quartz crystal appears as a high Q_u series resonant circuit. A parallel resonant circuit appears if this resonator terminates a quarter wavelength line, or equivalent network. The finite losses in the network or line will certainly degrade the Q of the equivalent parallel resonator over that obtainable with the original crystal.

A continuation of the impedance inversion pi-network is the half wave filter. This is a cascaded pair of symmetrical pi-networks, each with $Q = 1$. The filter behaves as a half wavelength of transmission line would.

While the pi and L are probably the most common impedance transforming networks, many others are also possible. Figure 4.23 shows a T-network using two capacitors and one inductor. This network is especially useful for matching to relatively low impedances from 50 Ω with practical components and low Q. The component values are given by the following set of equations.

$$R_2 > R$$
$$B = R_1(Q^2 + 1)$$
$$A = \left(\frac{B}{R_2} - 1\right)^{1/2} \qquad (4.5\text{-}4)$$
$$X_L = QR_1$$
$$X_{C1} = B/(Q - A)$$
$$X_{C2} = AR_2$$

Sec. 4.5 Impedance Transforming Networks

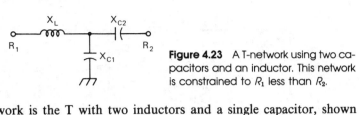

Figure 4.23 A T-network using two capacitors and an inductor. This network is constrained to R_1 less than R_2.

A similar network is the T with two inductors and a single capacitor, shown in Fig. 4.24. The design equations are

$$A = R_1(Q^2 + 1)$$

$$B = \left(\frac{A}{R_2} - 1\right)^{1/2}$$

$$X_{L1} = R_1 Q \qquad (4.5\text{-}5)$$

$$X_{L2} = R_2 B$$

$$X_C = A/(Q + B)$$

where A and B are intermediate parameters for both T-networks.

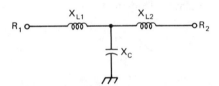

Figure 4.24 An *LCL*-type T-network.

One version of this network is of special interest, that with $X_{L1} = X_{L2} = X_C = Z_0$. This is the dual of the symmetrical pi-network with a Q of unity and has the same impedance inversion properties.

There are many variations of the networks discussed. Most are just cascaded versions of those presented. The more complex networks with added components are justified for preserving a better impedance transformation over a wide frequency range. The additional components will provide better filtering. Wide bandwidth transformations are best designed on a Smith chart or with sophisticated computer programs (7).

A combination of filters of special interest in communications is the diplexer. This is a combination configured so that the input impedance is constant over a wide range of frequencies. However, only some parts of the input spectrum are transferred through to a later stage; others are terminated in one leg of the filter, or perhaps in another load of interest.

A simple diplexer is shown in Fig. 4.25. This circuit is easily analyzed and is quite practical in many applications. Looking into the network at the arrow, the admittance of the leg containing the inductor is

$$Y_L = \frac{1}{R + sL} \qquad (4.5\text{-}6)$$

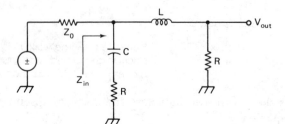

Figure 4.25 A simple diplexer network. Z_{in} is constant for all frequencies.

while the admittance into the capacitive portion of the circuit is

$$Y_C = \frac{sC}{1 + sCR} \qquad (4.5\text{-}7)$$

The two admittances are added and the result is inverted to produce

$$Z_{in} = \frac{R + sL + sCR^2 + s^2 RLC}{1 + 2sCR + s^2 LC} \qquad (4.5\text{-}8)$$

If this is set equal to R, manipulation leads to the condition

$$R^2 = \frac{L}{C} \qquad (4.5\text{-}9)$$

the constraint placed upon L and C to ensure a constant input impedance.

Energy incident at the input will split evenly between the two legs of the circuit when the magnitudes of the two currents are equal. This leads to a definition of the crossover frequency, $\omega_{c.o.}^2 = 1/LC$. Simultaneous solution of this relationship and Eq. 4.5-9 leads to expressions for the inductor and capacitor, $L = R/\omega_{c.o.}$ and $C = 1/(\omega_{c.o.} R)$.

Assume that the output of a circuit is to be applied to a 10-MHz amplifier with a 50-Ω input impedance. However, the circuit driving the amplifier must have a 50-Ω load, not only at 10 MHz, but at very much higher frequencies. A 30-MHz cutoff frequency is chosen, yielding $L = 0.265$ μH and $C = 106$ pF. This circuit presents a 50-Ω input impedance at all frequencies and will split the input power evenly at 30 MHz. The attenuation at 10 MHz is only 0.45 dB.

Diplexers take on many different forms. Two other possible combinations are shown in Fig. 4.26. That in Fig. 4.26a is a combination of a low pass and high pass filter, each with a third order response.

The circuit in Fig. 4.26b is the combination of a bandpass and bandstop filter. A series resonant circuit at the center frequency connects the source directly to the load. The parallel resistance is connected to ground through a parallel resonant circuit, a high impedance at the desired frequency. Minimal attenuation is offered to the input. The series leg "opens up" at other frequencies while the parallel resonator

Sec. 4.5 Impedance Transforming Networks

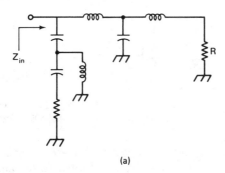

(a)

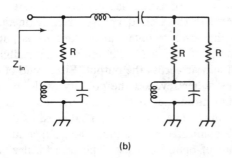

(b)

Figure 4.26 More complicated diplexer networks. The system at (a) uses a combination of three-pole low pass and high pass filters to maintain a constant Z_{in}. The network of (b) is a combination of bandpass and bandstop filters. If doubly terminated, this network will present a reasonable match at both the input and output for all frequencies.

becomes a short circuit, diverting input energy into the extra resistor. This circuit may be extended as shown to provide diplexing action at both the input and output ports.

Transformer action was discussed previously. The operation was for a "conventional" transformer. The discussion will now be extended to broadband transmission line transformers.

A narrow band transmission line transformer was presented in connection with the Smith chart. It is restricted to the spectrum where the line length is close to a quarter wavelength. Typically, bandwidths are limited to about 20% or less.

Consider a modified form of transmission line transformer. The normal transmission line is surrounded by a high permeability magnetic material and the ends of the line are connected to show transformer-like properties. The bandwidth is then significantly extended. The first transformers of this type were described in a classic paper by Ruthroff (8).

Figure 4.27 shows an example of the simplest type of transmission line transformer. A section of line, usually a twisted wire pair, is wound on a ferrite core. The most common core geometry is a toroid, although other types are also used. The input to the structure is driven from a source with a Z_0 characteristic impedance. Looking at the circuit only from a transmission line viewpoint, the current in the two legs must be equal, but of opposite direction. Neglecting transmission line considerations, the same result comes from transformer action—the magnetic core ensures that current entering a dotted terminal of a transformer causes an equal current to

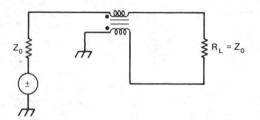

Figure 4.27 A wideband "balun" transmission line transformer.

leave the second dotted winding. The two viewpoints are in agreement. The load resistance in Fig. 4.27 is reflected through the transformer to the input (dotted) end.

There are some peculiarities to this structure. First, there is no dc isolation. The conventional transformers studied earlier had one dot associated with the primary while the other was connected to the secondary. No such distinction is made here. The other missing detail is the absolute potential of the two output terminals. It is not determined in the example of Fig. 4.27. The only thing established is that the input voltage difference will appear across the output. Either end of the load resistor may be grounded. The transformer reverses the polarity of the output with respect to the input if the end coming from the "hot" end of the source is grounded.

Figure 4.28 shows the same transformer with a balanced load. The load consists of two equal resistors, both connected to ground. The voltage at each transformer output terminal is equal, but of opposite phase. The transformer is often termed a "balun" because of this balancing action. This would be a 1:1 balun for there is no impedance transformation. Although the "balun" terminology is commonly accepted, this writer would prefer to call this structure a "pseudobalun." The transformer itself does not force the output voltage to be balanced. Only the presence of a balanced load will allow the transformer to function as a balun. Conversely, other transformers will force an inherent balance.

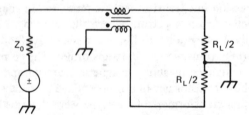

Figure 4.28 A balun transformer used to drive a balanced load.

Figure 4.29 shows another transmission line transformer. This is a balun. This results from the line length, a half wavelength. The phase shift of an incident wave is then 180 degrees. The characteristic impedance of the halfwave section is not of great importance, for it produces a single revolution on the Smith chart, repeating the terminating impedance. The two output voltages from the transformer are equal in magnitude and out of phase by 180 degrees.

The transformer of Fig. 4.29 has impedance transforming properties. Assume that the load is balanced with $R_{L1} = R_{L2}$. R_{L2} is transformed by the half wave line

Sec. 4.5 Impedance Transforming Networks

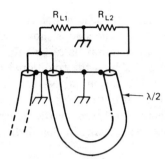

Figure 4.29 A narrow bandwidth balun transformer using a half wavelength of transmission line.

to a similar value at the junction with the main cable where it is paralleled with R_{L1}. The current delivered from the main line must be double that required in each of the two resistors. The sum of the two load resistances will, hence, equal four times the input impedance seen at the input of the transformer. A matched condition will exist for a total series load resistance of 200 Ω if 50-Ω cable is used. This balun transformer is used to drive balanced antennas which are fed with coaxial cable. The topology is also used in a microstrip form to provide a balanced drive to balanced mixers.

Figure 4.30 shows the equivalent transformer using a broadband transmission line structure. The transformer is drawn in three different ways to emphasize the details. Figure 4.30a emphasizes the analogy with the half wavelength balun of Fig.

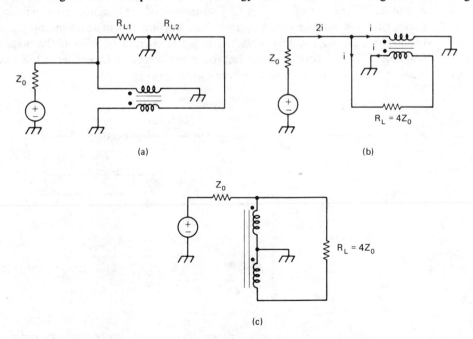

Figure 4.30 Three different schematic representations of a 4:1 impedance ratio balun transformer.

4.29. The transformer itself is applied as a polarity reversing device, serving the same function as the half wave line. The load is shown balanced. This is not required. The transformer forces the output terminals to be balanced with respect to ground.

The second version, Fig. 4.30b, is the traditional one for a 4:1 balun using a transmission line transformer. The current arrows are significant. One unit of current flows into the upper dot on the transformer. Hence, a similar current must be flowing out of the lower dot. But both are driven by the generator. The current delivered by the source is, hence, twice that flowing in the load, resulting in a 4:1 impedance transformation.

The transformer is drawn as a conventional auto transformer in Fig. 4.30c. The load appears across twice as many turns as does the input drive, again implying a 4:1 impedance transformation. All three configurations are used in schematics; they are all identical in operation. The three illustrate the need for a transmission line transformer to also have conventional transformer properties.

A "real" balun with a 1:1 impedance transformation is shown in Fig. 4.31. This transformer differs from the pseudobalun with the addition of a third winding, shown as B in the figure. Without this winding, the transformer is exactly the same as that of Fig. 4.27. However, the addition forces the output voltage to be balanced with respect to ground. This is most apparent in the schematic form of Fig. 4.31a. Winding B is called a magnetization winding. Allow the winding to be temporarily removed, leaving a pseudobalun construction. Current flows into the dot of winding A and out of the dot of winding C. Connection of the magnetization winding causes current to flow into its dot. However, load current out of the winding C dot would produce a cancelling current, leading to a net current of zero in the magnetization winding. This transformer shows improved performance, especially at low frequencies where transmission line behavior is not as dominant.

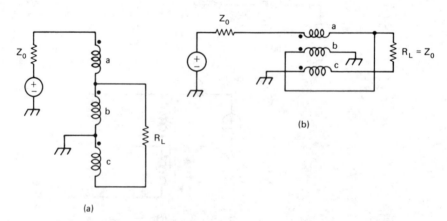

Figure 4.31 Two schematic representations of a 1:1 impedance ratio balun transformer. This is a real balun—the balance is forced by the transformer and does not depend upon a balanced load.

There are numerous other configurations and applications for the broadband transmission line transformer. The reader is referred to the literature (9).

Construction of transmission line transformers is not generally difficult. The length of the line should be somewhere around an eighth wavelength at the highest frequency of operation, although some transformers work well with a shorter line. The characteristic impedance of the line is critical. Low impedance lines are required to transform from 50 Ω down to lower impedances. Coaxial cables in miniature sizes are available with Z_0 as low as 25 Ω. A normal twisted pair of plastic covered wire with only a few twists per centimeter will have an impedance near 100 Ω. A tightly twisted pair of enamel-covered wires will have a characteristic impedance close to 50 Ω. Lower impedances are synthesized by winding two or more lines on a core. The windings are then parallelled at the ends. The use of wire with different insulation colors is a great advantage during the winding process.

Low frequency performance of transmission line transformers is almost completely dominated by conventional transformer action. Hence, the core material becomes critical. It should have high permeability to ensure adequate primary inductance. Care must be taken to select a core that is not excessively lossy. In addition, stray inductance in the interconnection of the windings should be avoided, for it can detract from the high frequency performance.

4.6 TRANSMISSION LINE MEASUREMENTS

Transmission line concepts have been introduced. The voltage and current in the line obey the usual physical laws and may be analyzed using traditional methods. However, we found that energy propagates along a transmission line in the form of waves. A method of analysis based upon these waves, the reflection coefficient, was introduced as an analytical tool. Further study led to the Smith chart.

The utility of the reflection coefficient concept comes from the interaction of two waves. One is an incident wave while the other is the result of a reflection. The ideal measurement would be one which separates the incident and reflected waves. The tools to do this are presented in this section.

Probably the most generally useful means for measuring reflection coefficients is the return loss bridge, shown schematically in Fig. 4.32. Much of its utility lies in its simplicity. Additionally it will be useful for other applications. The bridge consists of three resistors, a pseudobalun transformer with a R_0 characteristic impedance, a suitable enclosure and three coaxial connectors. All three resistors are identical with a value R_0, the characteristic impedance of the transmission line for which the bridge is designed.

The transformer serves two purposes. It allows a difference in voltage between points A and B to appear unbalanced at the detector port. The transformer also causes the detector impedance to appear across the bridge between points A and B. The bridge will have a loss of 6 dB when terminated in a load of R_0. There is then no voltage difference between the detector points, A and B. Hence, no power is

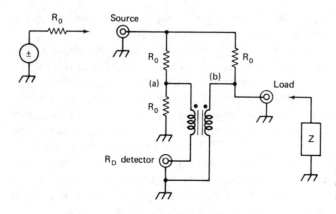

Figure 4.32 A return loss bridge, also known as a 6-dB hybrid combiner. This circuit is extremely useful for the measurement of complex impedances.

delivered to the detector. The power delivered by the generator is split into four equal resistors, one of them being the load, accounting for the 6-dB insertion loss. The impedance presented to the genertor is also R_0.

The bridge action is analyzed in Fig. 4.33. The circuit has been normalized to a characteristic impedance of 1 Ω. The open circuited generator voltage is set at 8 V. Arrows in the schematic show the assumed current direction. Three unknown voltages, corresponding to a three node circuit, are shown. E is the voltage applied to the bridge while V_1 and V_2 are the potentials at the detector port. At E, the nodal equation is

$$8 - E = E - V_1 + E - V_2 \qquad (4.6\text{-}1)$$

The nodal equation at the left detector point is

$$E - V_1 = V_1 + V_1 - V_2 \qquad (4.6\text{-}2)$$

while at the unknown load, the sum of the currents is

$$V_1 - V_2 + E - V_2 = V_2 y \qquad (4.6\text{-}3)$$

where y is the normalized, unknown admittance defined by the load.

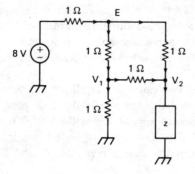

Figure 4.33 The circuit used for analysis of the return loss bridge where the characteristic resistance of the bridge is 1 Ω.

Simultaneous solution of Eqs. 4.6-1 through 4.6-3 yields the three unknown voltages

$$V_2 = \frac{4}{y+1}$$

$$V_1 = \frac{y+3}{y+1} \qquad (4.6\text{-}4)$$

$$E = \frac{3y+5}{y+1}$$

The voltage difference across the detector is calculated, and z, the normalized impedance, is factored into the result in place of y, yielding

$$V_2 - V_1 = \frac{1-y}{1+y} = \frac{z-1}{z+1} \qquad (4.6\text{-}5)$$

This is exactly the definition of the reflection coefficient. Hence, the voltage at the detector, a vector quantity, has a magnitude equal to the reflection coefficient magnitude and a phase angle with respect to the source phase equal to that of the reflection coefficient.

An 8-V source is an impractical restriction. A generator of arbitrary strength is attached and the load port is either open circuited or shorted. The power in the detector is noted. Then the unknown load is attached. The power at the detector will decrease. The magnitude of the decrease is the return loss in dB. Although not immediately obvious, the characteristic impedance of the source must be that of the bridge for accurate results. The same power will not be seen in the detector when going from an open to a short circuit at the load port if there is a generator mismatch.

The phase angle of the detector voltage is usually measured with a vector voltmeter such as the Hewlett-Packard HP-8407A. It may also be done with an oscilloscope triggered from the source generator, although accuracy is poor.

Other methods may be used for phase determination. The bridge is first used as described to measure the return loss (magnitude of Γ). Then, a sample of the generator is attenuated until it has the same amplitude as the signal at the detector port. The resulting two signals are algebraically added or subtracted in a resistive network. The result, still a scalar measurement, is combined with the value of attenuation used in the generator sampling line to infer the phase angle. Phasor analysis methods are used. The writer first learned of this method from Dr. R. D. Middlebrook (10).

Another application of the return loss bridge (RLB) is as a 6-dB hybrid combiner. This device is used to add two signals. The necessary characteristics are that system impedances be maintained and that the two sources be isolated. Operation of the

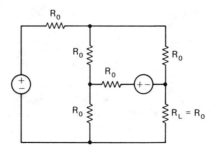

Figure 4.34 Use of a return loss bridge to combine signals from two sources while maintaining isolation.

RLB as a combiner is shown in Fig. 4.34 where the earlier circuit is drawn without the transformer and without any ground. The available energy from each generator appears at the load, but attenuated by 6 dB. The bridge action causes the energy from one generator to be cancelled at the other. A common application of the 6-dB hybrid combiner is for distortion measurements on amplifiers, mixers, or complete systems, discussed in Chap. 6.

The return loss bridge is never perfect. Some resistor values may be different than the bridge characteristic impedance. Stray capacitance and inductance may cause some arms of the bridge to differ from others. When a "perfect" Z_0 load is placed on the output, a finite reflection coefficient, or return loss will be measured. This is called the directivity of the bridge.

Return loss bridges are easily constructed, even in the home laboratory, with available components. They will show a directivity of 30 dB or more in the hf spectrum and at least 20 dB throughout the vhf spectrum if some care is used in construction (11).

RLB's using the same circuit as presented can be built that perform well into the microwave region. However, lead lengths, symmetry, and component quality all become critical (12).

The return loss bridge may be used for combining or splitting signals. It has the disadvantage that it dissipates power. Figure 4.35 shows a hybrid that overcomes this problem, the so-called zero-degree hybrid. The circuit is drawn in two forms to illustrate the circuits found in the literature. The form of Fig. 4.35a is used for analysis of the hybrid in a power splitter application. The assumed current directions are labeled.

Transformer current action forces $i_1 = i_2$. A nodal equation at the input shows $i_s = i_1 + i_2$. Finally, the voltage behavior of the transformer leads to $V_{in} - V_1 = V_2 - V_{in}$. These conditions are used to formulate a set of equations that may be solved for V_1, V_2, and V_{in} for specific component values. The results of such an analysis shows that R_B, the balancing resistor should have a value of $2R_1$ when R_1 and R_2 are equal. The input resistance is then $\frac{1}{2} R_1$. For example, if the two terminations are each 50 Ω, a 100-Ω unit is used for R_B and the input resistance is 25 Ω. The available power from the source is evenly split between the two loads, R_1 and R_2. If one load changes, the power delivered to the other remains constant. The effect of the change in a termination causes the excess power to be dissipated in the

Sec. 4.6 Transmission Line Measurements

balancing resistor. There is no loss in the circuit when equally terminated. The input resistance will, however, change as either output termination varies.

This circuit is also used for combining signals, as shown in Fig. 4.35c. The isolation between input ports is excellent, although driving impedances will be reflected through to the output. There is no phase shift caused by the circuit when used in either splitter or combiner applications.

Another hybrid is the so-called quadrature coupler. This type is not easily built to cover wide frequency ranges. The usual circuit at microwave frequencies uses

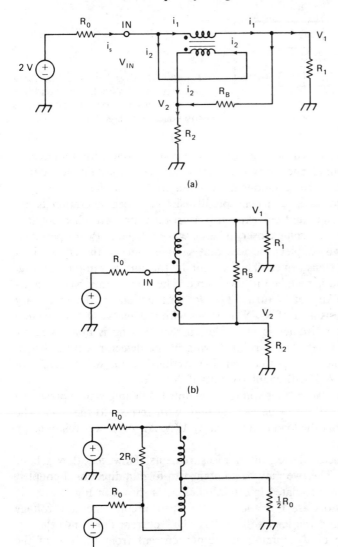

Figure 4.35 A zero degree hybrid. Analysis as a power splitter is done with the circuit at (a). (b) shows the same circuit with a different form used to depict the transformer. (c) shows the same circuit used to combine two signals.

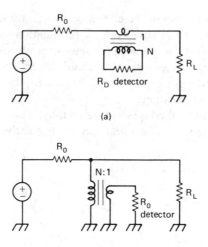

Figure 4.36 Couplers using transformers. That at (a) samples the current in the transmission line while (b) samples the voltage on the line.

microstrip techniques (13). The virtue of a quadrature coupler, one with a 90 degree phase difference between input and output, is that it allows a constant impedance to be presented at one port when variations are encountered at another.

The hybrids discussed have been conceptually simple. Their operation is explained on the basis of voltages and currents with no regard for wave phenomenon. Other bridge or coupler circuits come closer to being wave related in their operation.

Figure 4.36 shows two couplers, devices that sample some of the energy in a line. Each coupler uses a transformer. The current in the line is sampled at Fig. 4.36a. Assume that the load is matched to the source. The transformer has a single-turn primary and a secondary of N turns. The R_0 termination on the secondary has the effect of placing a resistance of R_0/N^2 in series with the line. If N is reasonably high, the equivalent series resistance is low. The current flowing is dominated by the load and not by the coupler. The current flowing in the detector will be that in the line divided by N. Hence, the power available at the detector will be $1/N^2$ of that flowing into the load. A 20-dB coupler results if $N = 10$.

Figure 4.36b is similar except that voltage is sampled. The impedance presented at the N-turn primary is $N^2 R_0$, a value usually high with respect to the load. The detector voltage is that across the load diminished by $1/N$. An N^2 power relationship still applies.

Both couplers are useful devices in rf measurements. However, they tell us nothing about impedances. The two may be combined to form a directional coupler, a device that will provide wave related information. This is shown in Fig. 4.37.

Assume that all resistors are 1 Ω and that the open circuit generator voltage is 2. The current flowing into the load will be 1 A. This current flows into the dot of the single-turn winding of T_1, forcing an output current from the dot of the N-turn winding of $1/N$ ampere. Because of symmetry, a current of $1/(2N)$ is forced into each of the detector resistors.

Sec. 4.6 Transmission Line Measurements

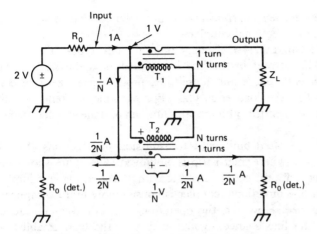

Figure 4.37 A directional coupler using two transformers.

Transformer T_2 samples the voltage across the load. A positive load voltage yields an output of $1/N$ volts at the secondary. A positive polarity appears at the dot on the single turn of T_2. This voltage causes a current to flow in the two terminating resistors, the detectors. The current is $1/(2N)$. These currents are indicated as arrows separated from the connecting wires in the schematic. The current resulting from the voltage sampling flows to the left, into the left termination, and out of the right detector. The total current into the left detector is $1/N$ while it is zero into the right detector. The coupling power ratio is again $1/N^2$.

Allow the role of the input and the output to be reversed with the output termination appearing at the "input" terminal and the generator at the "output." The load voltage is identical. However, the direction of the current is opposite. The net result is coupling of power in the ratio of $1/N^2$ to the right detector and no power to the left detector.

There will be a reflected wave if the load has a normalized value other than 1. The power at the right detector is representative of the reflected wave while that at the left detector is a sample of the incident wave. A reactive termination will lead to phase differences between the voltage and current in the load which is again coupled to the ports. The function of the directional coupler is exactly the same as the return loss bridge with the addition of a port that provides information about incident energy.

There is an asymmetry in the coupler. The output of T_1 is applied to one side of the secondary of T_2. The voltage sampling by T_2 is from one side of the primary of T_1. This is usually of little significance if N is large. The N-turn windings of each transformer should return to the center of the single-turn winding for best performance. This will improve the directivity of the coupler.

The characteristic impedance of the directional coupler will change if the impedance of the detectors is altered. Although directional couplers are usually used and explained in terms of wave phenomenon, they are really no different than the return loss bridge. Indeed, we note from Sec. 4.2 that the very concept of forward and reverse voltage waves was defined in terms of a specific source impedance. It is only the relationship of these concepts to transmission line behavior that adds relevance to the wave interpretation of directional couplers.

The coupler described in Fig. 4.37 is only one of many types. The basis is common to all though. The current in the line is sampled as is the load voltage. The two are compared in both phase and amplitude, producing a "zero reflected power" indication only when the load equals Z_0. For example, current sampling may be done with a small series resistor. Voltage sampling can be done with a capacitive voltage divider.

Directional couplers are often built with transmission line sections (14). An example is shown in Fig. 4.38 where microstrip is employed. The figure shows only the top pattern; the presence of a dielectric substrate with a ground plane is implicit. Analysis of a transmission line directional coupler is considerably more complicated than the other types described. However, the operation is intuitively reasonable. Current flowing through the line connecting the source to the load establishes a magnetic field. The close proximity of the coupled line will allow the field to induce currents. Similarly, capacitive coupling between lines will cause a sample of the load voltage to appear on the coupled line. This voltage will also lead to output current. The overall action is the same as was found with the transformer coupler of Fig. 4.37.

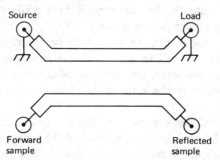

Figure 4.38 A transmission line directional coupler using microstrip. Only the top layer is shown—a dielectric layer and ground plane is assumed under the lines. Interaction of the magnetic fields of the line allow current sampling while electric field coupling allows the voltage on the line to be sampled. The operation is then identical to that of Fig. 4.38, although the bandwidth is restricted.

While our interest in the directional coupler has been for the measurement of complex impedances, with special emphasis on transmission lines, there are many other applications. It is often desired to sample energy from a line for the control of a system. Alternatively, the symmetry of a directional coupler suggests that energy may be injected onto a transmission line through a coupler. As such, they are often used in mixer or summing applications.

A wide variety of coupled transmission line type structures are especially useful at microwave frequencies. Figure 4.39 shows a microstrip-type bandpass filter. The input is a section of line that is short circuited. The input match would be poor if there were no other elements available, for all of the incident energy would be reflected by the short. However, the input line, l_1, is part of a directional coupler with l_2 forming the rest. The second line is not terminated at either end. l_2 is a half wavelength at the desired center frequency, thus forming a resonator. The third line is similar

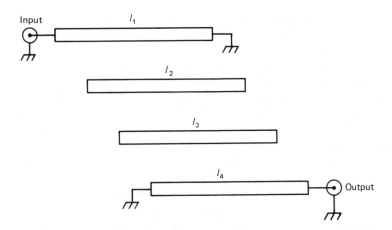

Figure 4.39 A double tuned circuit using sections of microstrip as directional couplers.

with the fourth line acting as a means of extracting energy from the filter. The response of this filter may be tailored to whatever filter response desired through the use of properly designed lines and spacings. All of the methods of Chap. 3 for bandpass filter design may be adapted to these structures. Measurements may be performed with a proper adaptation of the Dishal method.

REFERENCES

1. Smith, Phillip H., "Transmission Line Calculator," *Electronics,* **12,** January 1939.
2. Matthaei, George L., Young, Leo, and Jones, E. M. T., *Microwave Filters, Impedance-Matching Networks and Coupling Structures,* 2nd ed., ARTECH House, Dedham, Mass., 1980.
3. Bahl, I. J. and Trivedi, D. K., "A Designer's Guide to Microstrip Line," *Microwaves,* pp. 174–182, May 1977.
4. Zverev, Anatol I., *Handbook of Filter Synthesis,* Chap. 9, John Wiley & Sons, New York, 1967.
5. Wenzel, Robert J., "The Modern Network Theory Approach to Microwave Filter Design," *IEEE Trans. on Electromagnetic Compatability,* **EMC-10,** 2, p. 196, June 1968.
6. Hayward, Wes and DeMaw, Doug, *Solid-State Design for the Radio Amateur,* ARRL, Newington, Conn., 1977.
7. Hewlett-Packard Application Note 95, September 1968.
8. Ruthroff, C. L., "Some Broad-Band Transformers," *Proc. IRE,* **47,** pp. 1337–1342, August 1959.
9. Krauss, Herbert L., Bostian, Charles W., and Raab, Frederick H., *Solid State Radio Engineering,* John Wiley & Sons, New York, 1980.

10. Middlebrook, R. D., "Design-Oriented Circuit Analysis and Measurement Techniques," copyrighted notes for a short course presented by Dr. Middlebrook, 1978.
11. See reference 6, Chap. 7.
12. Dunwoodie, Duane E. and Lacy, Peter, "Why Tolerate Unnecessary Measurement Errors," Wiltron Technical Review, Wiltron Co. Palo Alto, Cal., October 1975.
13. Lange, J., "Interdigitated Stripline Quadrature Hybrid," *IEEE Trans. on Microwave Theory and Techniques,* **MTT-17,** pp. 1150–1151, December 1969. *Also see* Ho., Chen Y., "Design of Lumped Quadrature Couplers," *Microwave Journal,* pp. 67–70, September 1979.
14. See reference 2.

SUGGESTED ADDITIONAL READINGS

1. Skilling, Hugh Hildreth, *Electric Transmission Lines,* McGraw-Hill, New York, 1951.
2. Ramo, Simon, Whinnery, John R., and Van Duzer, Theodore, *Fields and Waves in Communication Electronics,* John Wiley & Sons, New York, 1965.
3. Harrington, Roger F., *Time Harmonic Electromagnetic Fields,* McGraw-Hill, New York, 1961.
4. Terman, Frederick Emmons, *Electronic and Radio Engineering,* 4th ed., McGraw-Hill, New York, 1955. Chap. 4 offers an excellent review of transmission lines.

5

Two-Port Networks

While some transistor amplifier circuits were presented in Chap. 1, all the examples considered were simple. The terminal immittances were scalar, containing no reactive terms. The gain was also modeled as frequency independent. These simplifications are often justified, for they will work. They are inadequate for many rf applications though.

The material presented in this chapter is confined to a small-signal description of networks. However, the methods are exceedingly powerful within that one constraint. They can be applied equally to passive and active circuits. The descriptions may result from modeling or from measurement of actual parameters of a transistor or whatever type of network is being studied.

5.1 TWO-PORT NETWORK FUNDAMENTALS

A two-port network is one with four terminals. The terminals are arranged into pairs, each being called a port. The general network schematic is shown in Fig. 5.1. The input port is characterized by input voltage and current, V_1 and I_1, while the output is described by V_2 and I_2. By convention, the currents are usually into the network.

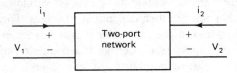

Figure 5.1 General configuration of a two-port network. Note the voltage polarities and direction of currents.

161

Many devices of interest have three terminals rather than four. Two-port methods are used with these by choosing one terminal to be common to both input and output ports. The two-port representations of the common emitter, common base, and common collector connections of the bipolar transistor are shown in Fig. 5.2. Similar configurations may be used with field effect transistors, vacuum tubes, integrated circuits, or passive networks.

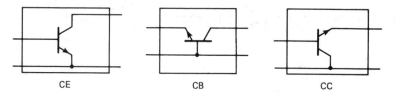

Figure 5.2 Two-port representations of the common emitter, common base, and common collector amplifiers.

The general concepts of two-port theory are applicable to devices with a larger number of terminals. The theory is expandable to N-ports (1). Alternatively, the bias on some terminals can be established with attention fixed only upon two ports of a multielement device. An example would be a dual-gate metal oxide silicon field effect transistor (MOSFET) in a common source configuration. The input port contains the source and gate-1 while the output port contains the source and drain leads. The fourth device terminal, gate-2, has a fixed potential. Signal currents at this terminal are ignored in the analysis. The common source, dual-gate MOSFET, is shown in Fig. 5.3.

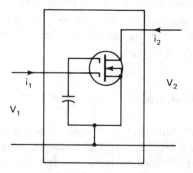

Figure 5.3 A dual gate MOSFET treated as a three terminal device in a two-port network.

There are four variables associated with any two-port network, two voltages and two currents. These are signal components. Any two variables may be picked as independent. The remaining pair are then dependent variables. These are expressed as an algebraic linear combination of the two independent quantities.

Assume that the two voltages are chosen as independent variables. The two currents are then expressed as linear combinations of the voltages, $I_1 = K_a V_1 + K_b V_2$ and $I_2 = K_c V_1 + K_d V_2$. The constants of proportionality, K_a through K_d, have the dimensions of admittance. The usual representation is

$$I_1 = y_{11}V_1 + y_{12}V_2 \qquad (5.1\text{-}1a)$$

$$I_2 = y_{21}V_1 + y_{22}V_2 \qquad (5.1\text{-}1b)$$

The independent and dependent variable sets are column vectors, leading to the equivalent matrix representation

$$\begin{pmatrix} I_1 \\ I_2 \end{pmatrix} = \begin{pmatrix} y_{11} & y_{12} \\ y_{21} & y_{22} \end{pmatrix} \begin{pmatrix} V_1 \\ V_2 \end{pmatrix} \qquad (5.1\text{-}2)$$

The y matrix for a two-port network uniquely describes that network. Consider the y parameters from an experimental viewpoint. The first y parameter, y_{11}, is the input admittance of the network with V_2 set to zero. Hence, it is termed the short circuit input admittance. y_{21} is the short circuit forward transadmittance. Similarly, if V_1 is set to zero, realized by short circuiting the input, y_{22} is the short circuit output admittance and y_{12} is the short circuit reverse transadmittance. A device is evaluated experimentally by short circuiting the input and output ports and measuring the responses accordingly.

The matrix subscripts are sometimes replaced by letters. The set y_{11}, y_{12}, y_{21}, and y_{22} are replaced by y_i, y_r, y_f, and y_o where the subscripts indicate respectively input, reverse, forward, and output. The subscripts are sometimes modified to indicate the connection of the device. For example, the short circuit forward transfer admittance of a common emitter amplifier would be y_{21e} or y_{fe}.

The y parameters are only one set of two-port parameters. The open circuited z or impedance parameters result if the two currents are treated as independent variables

$$\begin{pmatrix} V_1 \\ V_2 \end{pmatrix} = \begin{pmatrix} z_{11} & z_{12} \\ z_{21} & z_{22} \end{pmatrix} \begin{pmatrix} I_1 \\ I_2 \end{pmatrix} \qquad (5.1\text{-}3)$$

The parameter sets describe the same device; hence, they are related to each other. If Eq. 5.1-1a is multiplied by y_{22}, Eq. 5.1-1b is multiplied by y_{12} and the resulting equations are subtracted, the result is the input voltage as a function of the currents

$$V_1 = \frac{I_1 y_{22} - y_{12} I_2}{y_{11} y_{22} - y_{21} y_{12}} \qquad (5.1\text{-}4)$$

A similar procedure is used to find the output voltage as a function of the currents, leading to the general relationships

$$z_{11} = \frac{y_{22}}{\Delta y} \qquad z_{12} = \frac{-y_{12}}{\Delta y}$$
$$z_{21} = \frac{-y_{21}}{\Delta y} \qquad z_{22} = \frac{y_{11}}{\Delta y} \qquad (5.1\text{-}5)$$

where Δy is the determinant of the y matrix. The inverse transformations, yielding the y parameters when z parameters are known, are exactly the same as those in Eq. 5.1-5 except that y_{jk} and z_{jk} values are interchanged. The similarity is useful when writing transformation programs for a programmable calculator or computer.

Rudimentary methods were used to derive Eq. 5.1-5. The transformation is also obtained with a more formal application of matrix methods. The reader is urged to pursue this approach if possible when deriving two-port relationships (2, 3, 4).

The hybrid parameters are defined if the input current and output voltage are selected as independent variables

$$\begin{pmatrix} V_1 \\ I_2 \end{pmatrix} = \begin{pmatrix} h_{11} & h_{12} \\ h_{21} & h_{22} \end{pmatrix} \begin{pmatrix} I_1 \\ V_2 \end{pmatrix} \qquad (5.1\text{-}6)$$

The input term, h_{11}, is an impedance while h_{22} represents an output admittance. The forward term, h_{21}, is the ratio of the output to the input current, beta for a bipolar transistor. The reverse parameter, h_{12}, is a voltage ratio. The mixture of dimensions accounts for the "hybrid" name of the set.

Another popular set of parameters results if the output voltage and current are chosen as independent, yielding

$$\begin{pmatrix} V_1 \\ I_1 \end{pmatrix} = \begin{pmatrix} A & B \\ C & D \end{pmatrix} \begin{pmatrix} V_2 \\ -I_2 \end{pmatrix} \qquad (5.1\text{-}7)$$

These are termed the transmission parameters, the general circuit parameters or are merely called the "*ABCD* matrix." A different sign convention is used. Equation 5.1-7 contains a minus sign on I_2. If the sign of the output current is assumed to be outward, a positive sign appears in the corresponding equations.

The voltages and currents are sometimes renamed when using the *ABCD* matrix. The input variables are labeled with an s subscript, indicative of "sending," while the output V and I are subscripted with an r for "receiving."

Equation 5.1-7 gives the input or sending conditions as a function of those at the output. The roles may be reversed, leading to the reverse transmission parameters, sometimes called the $A'\ B'\ C'\ D'$ matrix.

All parameter sets are related. The y to z transformations were presented in Eq. 5.1-5. Transformations between y, z, h, and *ABCD* representations are summarized in Table 5.1. The choice of which parameter set to use is arbitrary. However, some are much more convenient than others when several two-port networks are to be combined. More will be said about the distinctions later.

Consider the y parameters for a transistor operating in the common emitter configuration, Fig. 5.4. The simple model used in Chap. 1 is assumed, a beta generator with an emitter resistor of $26/I_e$(mA, dc) where I_e is the bias current. Equations 5.1-1 are used for calculation. The output current as well as that at the input are independent of the output voltage. It makes no difference if the output is short circuited or not. I_1 and I_2 do not depend upon V_2. Hence, y_{12} and y_{22} must both be zero.

Sec. 5.1 Two-Port Network Fundamentals

Table 5.1 (Matrices in the same row in the table are equivalent)
$\Delta_x = x_{11}x_{22} - x_{12}x_{21}$

	[z]		[y]		[T]		[h]	
[z]	z_{11}	z_{12}	$\dfrac{y_{22}}{\Delta_y}$	$-\dfrac{y_{12}}{\Delta_y}$	$\dfrac{A}{C}$	$\dfrac{\Delta_T}{C}$	$\dfrac{\Delta_h}{h_{22}}$	$\dfrac{h_{12}}{h_{22}}$
	z_{21}	z_{22}	$-\dfrac{y_{21}}{\Delta_y}$	$\dfrac{y_{11}}{\Delta_y}$	$\dfrac{1}{C}$	$\dfrac{D}{C}$	$-\dfrac{h_{21}}{h_{22}}$	$\dfrac{1}{h_{22}}$
[y]	$\dfrac{z_{22}}{\Delta_z}$	$-\dfrac{z_{12}}{\Delta_z}$	y_{11}	y_{12}	$\dfrac{D}{B}$	$-\dfrac{\Delta_T}{B}$	$\dfrac{1}{h_{11}}$	$-\dfrac{h_{12}}{h_{11}}$
	$-\dfrac{z_{21}}{\Delta_z}$	$\dfrac{z_{11}}{\Delta_z}$	y_{21}	y_{22}	$-\dfrac{1}{B}$	$\dfrac{A}{B}$	$\dfrac{h_{21}}{h_{11}}$	$\dfrac{\Delta_h}{h_{11}}$
[T]	$\dfrac{z_{11}}{z_{21}}$	$\dfrac{\Delta_z}{z_{21}}$	$-\dfrac{y_{22}}{y_{21}}$	$-\dfrac{1}{y_{21}}$	A	B	$-\dfrac{\Delta_h}{h_{21}}$	$-\dfrac{h_{11}}{h_{21}}$
	$\dfrac{1}{z_{21}}$	$\dfrac{z_{22}}{z_{21}}$	$-\dfrac{\Delta_y}{y_{21}}$	$-\dfrac{y_{11}}{y_{21}}$	C	D	$-\dfrac{h_{22}}{h_{21}}$	$-\dfrac{1}{h_{21}}$
[h]	$\dfrac{\Delta_z}{z_{22}}$	$\dfrac{z_{12}}{z_{22}}$	$\dfrac{1}{y_{11}}$	$-\dfrac{y_{12}}{y_{11}}$	$\dfrac{B}{D}$	$\dfrac{\Delta_T}{D}$	h_{11}	h_{12}
	$-\dfrac{z_{21}}{z_{22}}$	$\dfrac{1}{z_{22}}$	$\dfrac{y_{21}}{y_{11}}$	$\dfrac{\Delta_y}{y_{11}}$	$-\dfrac{1}{D}$	$\dfrac{C}{D}$	h_{21}	h_{22}

From M. E. Van Valkenburg, *Network Analysis*, 3rd ed., Prentice-Hall, 1974, p. 337. Adapted with permission.

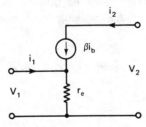

Figure 5.4 Simple transistor model, a beta generator with emitter resistance.

A nodal equation can be written and solved for I_1 as a function of V_1

$$\frac{V_1}{r_e} = I_1(\beta + 1)$$

$$I_1 = \frac{V_1}{r_e(\beta + 1)} \qquad (5.1\text{-}8)$$

Solving for $y_{11} = I_1/V_1$ and noting that $I_2 = \beta I_1$, the y matrix for the simple transistor model is

$$y_e = \begin{pmatrix} \dfrac{1}{(\beta+1)r_e} & 0 \\ \dfrac{\beta}{(\beta+1)r_e} & 0 \end{pmatrix} \tag{5.1-9}$$

where the e subscript indicates the common emitter configuration.

This example is somewhat trivial. The model is so simple that the y parameters are not really needed for any analysis. The y parameters are scalar quantities, having no imaginary parts. Later though, this model will be modified slightly to make better use of the resulting y parameters that have been derived.

This example is also unique with $y_{12} = 0$. Such an amplifier is said to be unilateral. As such, it offers perfect isolation—nothing that happens at the output will affect the input conditions. While unilateral amplifiers are not realistic in practice, it is nonetheless practical to sometimes make the assumption during analysis. This is especially true when the ratio y_{21}/y_{12} is large compared with unity.

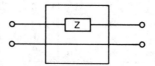

Figure 5.5 Two-port network representation of a series impedance.

Consider a passive network, Fig. 5.5, a series impedance represented by Z. The y matrix is derived by noting that output current is defined as going into the network. But the same current must be flowing in either end. Hence, $I_2 = -I_1$. From Ohm's Law, $V_2 = V_1 - I_1 Z$. Solving for I_1 and I_2

$$\begin{aligned} I_1 &= V_1/Z - V_2/Z \\ I_2 &= -V_1/Z + V_2/Z \end{aligned} \tag{5.1-10}$$

The resulting y matrix is then

$$y_{jk} = Y \begin{pmatrix} 1 & -1 \\ -1 & 1 \end{pmatrix} \tag{5.1-11}$$

where $Y = 1/Z$.

This two-port network has two common and useful characteristics. It is symmetrical in that it is identical when the input and output are exchanged. It is also passive; there are no controlled current generators contained within the network. In general, a passive network will have the following properties:

$$\begin{aligned} z_{12} &= z_{21} \\ y_{12} &= y_{21} \\ h_{12} &= -h_{21} \\ AD - BC &= 1 \end{aligned} \tag{5.1-12}$$

Sec. 5.1 Two-Port Network Fundamentals

A symmetrical network is one with the following characteristics:

$$z_{11} = z_{22}$$
$$y_{11} = y_{22}$$
$$A = D$$
$$\Delta h = 1$$
(5.1-13)

Not all networks will have a y matrix. Consider a shunt impedance. The input and output voltages must be equal. An equation can also be written that relates the terminal voltage to the total current flowing into the network. However, there is nothing to force the manner in which input and output currents split.

There is a practical answer to this formal difficulty. The y matrix exists for a series impedance followed by a shunt impedance. A very small series resistance is used in a practical analysis. The two element resistive network is shown in Fig. 5.6 and the y parameters are

$$y_{jk} = \begin{pmatrix} Y_s & -Y_s \\ -Y_s & Y_s + Y_p \end{pmatrix} \quad (5.1\text{-}14)$$

Although the y matrix for a shunt resistor could not be written, it is straightforward to write the associated z matrix. That matrix has a determinant which vanishes; hence, the z matrix is not transformable to admittance parameters according to Eqs. 5.1-5.

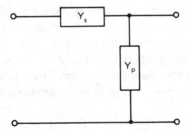

Figure 5.6 A two-port network with a series and shunt impedance.

For any set of y parameters for a given configuration, such as the common emitter bipolar transistor, it seems reasonable that the corresponding common base parameters should be calculable. Both parameter sets represent the same device with identical internal immittances and controlled generators. Only the means of application and extraction of signals has changed. These transformations are called exchanges, for the common terminal is exchanged with one of the others. The most fundamental method for performing this calculation for y parameters is to form the so-called three-port y matrix. The reader is referred to the literature for the details (5). The transformations derivable from these manipulations are summarized in the following equations.

To obtain ce from cb or cc,

$$y_{11e} = y_{11b} + y_{12b} + y_{21b} + y_{22b} = y_{11c}$$
$$y_{12e} = -(y_{12b} + y_{22b}) = -(y_{11c} + y_{12c})$$
$$y_{21e} = -(y_{21b} + y_{22b}) = -(y_{11c} + y_{21c})$$
$$y_{22e} = y_{22b} = y_{11c} + y_{12c} + y_{21c} + y_{22c}$$
$$y_{22e} = y_{22b} = y_{11c} + y_{12c} + y_{21c} + y_{22c}$$
(5.1-15)

To obtain cb from ce or cc

$$y_{11b} = y_{11e} + y_{12e} + y_{21e} + y_{22e} = y_{22c}$$
$$y_{12b} = -(y_{12e} + y_{22e}) = -(y_{21c} + y_{22c})$$
$$y_{21b} = -(y_{21e} + y_{22e}) = -(y_{12c} + y_{22c})$$
$$y_{22b} = y_{22b} = y_{11c} + y_{12c} + y_{21c} + y_{22c}$$
(5.1-16)

To obtain cc from ce or cb

$$y_{11c} = y_{11e} = y_{11b} + y_{12b} + y_{21b} + y_{22b}$$
$$y_{12c} = -(y_{11e} + y_{12e}) = -(y_{11b} + y_{21b})$$
$$y_{21c} = -(y_{11e} + y_{21e}) = -(y_{11b} + y_{12b})$$
$$y_{22c} = y_{11b} = y_{11e} + y_{12e} + y_{21e} + y_{22e}$$
(5.1-17)

Similar equations are derivable for z, h, or transmission parameters.

It is important to view the manipulations presented as rather general transformations or mappings. The y parameters relate terminal currents to terminal voltages. One can envision two isolated planes. One is a current plane; points in the plane correspond to points with coordinates I_1 and I_2. Similarly, the voltage plane has all possible points with coordinates V_1 and V_2. The y matrix transforms or maps points in the V plane into corresponding points in the I space. Much of the subject of linear algebra is devoted to mappings of this sort. As such, all of the power of matrix algebra may be invoked to manipulate the matrices. Numerous other transformations may be invented—some may even be of practical significance.

5.2 INTERCONNECTION OF TWO-PORT NETWORKS

The greatest virtue of viewing a transistor amplifier or a filter element as a two-port network is the generality. By working with the matrices, gain, input and output immittances, and other factors may be evaluated in terms of the network parameters. The general equations will apply to any device where the appropriate parameters are available.

To be of greatest value, we must have the ability to not only evaluate the characteristics of a network, but to assemble more complex networks from simple ones. The composite network, with corresponding two-port parameters, is then evaluated with the same ease that a simple one was. Hence, we must perform interconnection operations.

There are three basic interconnections of interest. One is the cascade where the output of 1 two-port is applied as an input excitation of a second. A second interconnection is paralleling networks. An example is a transistor amplifier with a parallel feedback resistor. The third interconnection is a series connection of two networks. An example is the addition of an emitter degeneration resistor in a transistor amplifier.

The interconnections are best described if the networks are viewed with respect to the defining two-port equations. This is shown in Fig. 5.7. The first part, Fig. 5.7a, shows the *y*-parameter representation. The input port is represented by an immittance, y_{11}, paralleled with a current generator controlled by the output voltage through the reverse parameter, y_{12}. The input admittance is just y_{11} when the output is shorted. With something other than a short circuit termination at the output, the input port current has an additional component beyond that predicted by y_{11}. The action at the output port is similar.

Figure 5.7b shows a *z*-parameter representation of the two-port network. Each port is represented by a voltage generator (current controlled) in series with an immittance. If one port is open circuited, the corresponding current vanishes. The voltage generator at the opposite port then vanishes, leaving a port immittance resulting from only the series elements. It is significant in both representations that no common

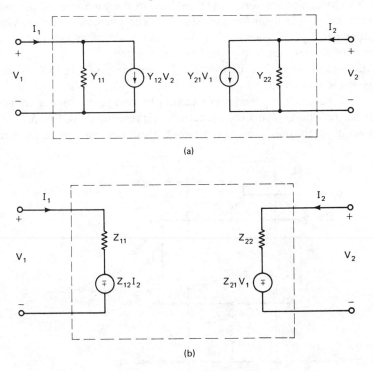

Figure 5.7 Interpretation of (a) *y* parameters and (b) *z* parameters of a two-port network.

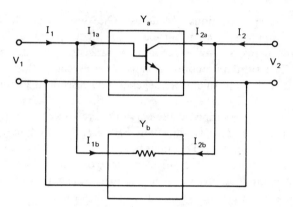

Figure 5.8 Parallel connection of two-port networks.

path exists between ports other than those described by the forward and reverse generators.

We can now consider some multiple connections of two-ports. Figure 5.8 shows a parallel connection. We see from the defining equations for admittance parameters that the total currents are $i_1 = i_{1a} + i_{1b}$ and $i_2 = i_{2a} + i_{2b}$. The input voltages are the driving sources, the independent variables in the equations. Hence, the y matrix of the paralleled networks is just the sum of the y matrices of the individual networks. This is a simple form for analysis. Very complicated equations would be involved if other parameters were used. It is usually convenient in practice to transform z, h, or $ABCD$ matrices into y parameters if networks are to be paralleled. An inverse transformation is then performed. The example of Fig. 5.8 shows a transistor and feedback resistance.

Figure 5.9 is a series connection of two-ports, here a transistor with emitter degeneration. Applying the z-parameter representation of Fig. 5.7b, we see that current flowing in the input of one network also flows in the series elements. It follows

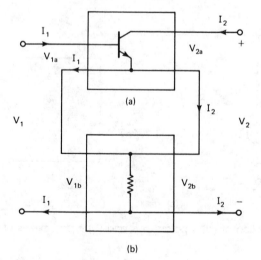

Figure 5.9 Series connection of two-port networks.

Sec. 5.2 Interconnection of Two-Port Networks

that the z parameters of series connected networks is just the sum of the z matrices of the individual networks. Again, the relationships would be much more complicated if a series connection was tried with y parameters.

The third interconnection, the cascading operation, is important, for multiple stage amplifiers must often be analyzed. Further, we must often analyze an amplifier in cascade with impedance matching networks. The cascade of two y-parameter networks is shown in Fig. 5.10.

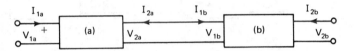

Figure 5.10 A cascade of two-port networks.

Examination of the networks provides two relationships

$$V_{2a} = V_{1b}$$
$$I_{2a} = -I_{1b} \tag{5.2-1}$$

Using these, along with the defining equations for y parameters, we obtain $V_{1b} = V_{2a}$ as a function only of the voltages at the outside of the composite two-port

$$V_{1b} = \frac{-y_{21a}V_{1a} - y_{12b}V_{2b}}{y_{11b} + y_{22a}} \tag{5.2-2}$$

Further substitution into the defining y matrix leads to

$$I_{1c} = \left[y_{11a} - \frac{y_{12a}y_{21a}}{y_{11b} + y_{22a}} \right] V_1 + \left[\frac{-y_{12a}y_{12b}}{y_{11b} + y_{22a}} \right] V_2 \tag{5.2-3}$$

and

$$I_{2c} = \left[\frac{-y_{21a}y_{21b}}{y_{11b} + y_{22a}} \right] V_1 + \left[y_{22b} - \frac{y_{21b}y_{12b}}{y_{11b} + y_{22a}} \right] V_2 \tag{5.2-4}$$

The four terms in brackets are the composite y parameters for the cascaded networks.

These equations may be used for computation. An alternative approach is to use the *ABCD* matrix format. The defining equations for the transmission parameters contained a minus sign with the "received" current, I_2. This was done to maintain a consistent form for the general two-port network. Using the alternative interpretation shown in Fig. 5.11a, the defining equations become

$$\begin{pmatrix} V_1 \\ I_1 \end{pmatrix} = \begin{pmatrix} A & B \\ C & D \end{pmatrix} \begin{pmatrix} V_2 \\ I_2' \end{pmatrix} \tag{5.2-5}$$

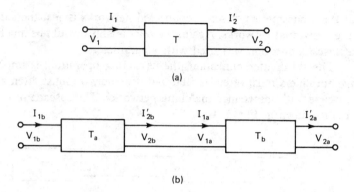

Figure 5.11 Redefinition of direction of output current in a two-port (a) to aid visualization of the *ABCD* matrix. A cascade of two-ports described by *ABCD* matrices is shown in (b).

The input is found with Eq. 5.2-5 if an output for a cascade is known, V_{2b} and I'_{2b}. The resulting input conditions equal the output of the first network. The input to the composite is found with a second application of Eq. 5.2-5. Hence, the composite of two cascaded T matrices is the matrix product of the individual ones

$$T_c = (T_b)(T_a) \qquad (5.2\text{-}6)$$

The order of multiplication is important, as is generally true for matrix products. This relationship applies to the standard form, Eq. 5.1-7.

The ease of evaluation of cascaded two-ports with the *ABCD* matrix is the justification for formulating such a network description. Many of the modern computer programs written for two-port network analysis use the *ABCD* matrix extensively. However, the programs are usually written such that the user is completely unaware of this. The program input and output are in other forms. It is possible to evaluate a cascade of two networks with but one form, the *y* parameters. Generally, any interconnection may be evaluated with any representation. The question of which form to use is largely a choice of the user (or the computer programmer) and has no bearing on the final results. The ultimate decision for calculation depends upon the efficiency of doing the assorted calculations. The best input and output forms are those most meaningful to the user.

5.3 THE HYBRID-PI TRANSISTOR MODEL

Only scalar transistor models have been discussed in detail. There has been no difference in performance with frequency. The phase of the immittances and the gains were all indicative of purely resistive behavior. This is, of course, completely unrealistic.

Sec. 5.3 The Hybrid-Pi Transistor Model

The usual bipolar transistor has a gain which decreases with increasing frequency. At low frequencies, in the audio spectrum, the characteristics will be the same as seen at dc. The current gain will be β_0. At a critical frequency, usually termed the F_β of the transistor, the current gain is decreased by 3 dB. Moreover, the input ceases to be resistive—it appears as a resistance in parallel with a capacitance of equal reactance at F_β. Current gain decreases at a linear rate with frequency well above F_β. That is, gain drops by half for a doubling in operating frequency. The point where $\beta = 1$ is called the F_t, or the gain-bandwidth product. The input is predominantly capacitive and the current gain shows a changing phase at frequencies well above the F_β.

Transistor behavior is modeled with a simple modification to the traditional scalar model we have used. Figure 5.12a shows the model we have used, a beta generator with an emitter resistor, $r_e = 26/I_e$. Figure 5.12b shows an equivalent model. While topology is different, this model will exhibit the same two-port parameters as that of Fig. 5.12a. The frequency dependent behavior is modeled with the addition of a capacitor at the input, shown in Fig. 5.12c. The value of the capacitor is chosen to have a reactance equal to r_e at F_β. Figure 5.12d is an equivalent representation. The model is identical with the scalar model except that the current gain is complex. The equation describing the variation of current gain with frequency is given with the model in the figure.

While many salient features of transistor behavior are well described by the models of Figs. 5.12c and d, two additional factors are often included. One is a feedback capacitor between the collector and the base. The value is predictable from the transistor physics. The modified model is shown in Fig. 5.12e.

Also shown in the figure is an external emitter resistor and inductor. Both are often used in a ce amplifier. They are easily included with the internal r_e when calculating the two-port parameters.

The other feature that may be included is r_b', the base spreading resistance. This appears in series with the input terminal of Fig. 5.12e. While not shown with the model, it is easily added with a cascading operation.

The hybrid-pi two-port parameters are easily calculated from first principals using nodal analysis. The ease of doing this with h parameters justifies the name of the model. Alternatively, the y parameters may be evaluated using a paralleling operation. The y matrix for a scalar model, Fig. 5.12a, was evaluated earlier and given in Eq. 5.1-9. While β was a scalar for that analysis, the equation still holds for a complex β. The y matrix for a parallel capacitor is written easily and added directly to the transistor matrix, yielding

$$y_e = \begin{pmatrix} \dfrac{1}{Z_e(\beta+1)} + j\omega C_{cb} & -j\omega C_{cb} \\ \dfrac{\beta}{Z_e(\beta+1)} - j\omega C_{cb} & j\omega C_{cb} \end{pmatrix} \quad (5.3\text{-}1)$$

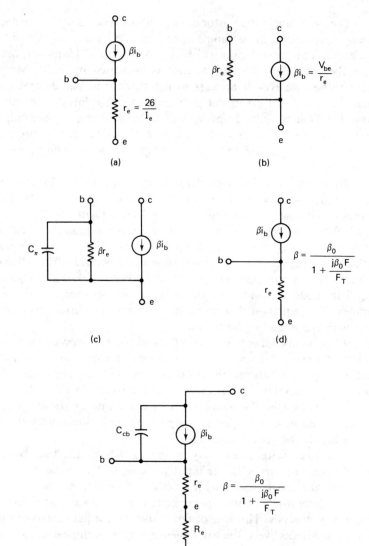

Figure 5.12 Evolution of the hybrid-pi model of the bipolar transistor. (a) and (b) show scalar models. A capacitor is added at (c) to simulate the frequency dependance of current gain. An equivalent network is shown at (d) where β is complex. A collector-base capacitor is added at (e) through a paralleling operation with y parameters. The emitter impedance includes external degeneration elements.

Sec. 5.3 The Hybrid-Pi Transistor Model

where β is complex

$$\beta = \frac{\beta_0}{1 + j\beta_0 F/F_t} \quad (5.3\text{-}2)$$

and the complex emitter impedance is given as

$$Z_e = r_e + R_e + j\omega L_e \quad (5.3\text{-}3)$$

This formulation is especially useful for calculator programs or for subroutines in a computer program. Equation 5.3-3 is simplified to $Z_e = r_e$ if the external components are not included. Although simple, this model is surprisingly accurate.

A simplification is suggested by Eq. 5.3-2. If the frequency is well below F_t, but still above F_β, the imaginary part of the denominator is much larger than unity. Hence $\beta \simeq -jF_t/F$.

Two-port parameters may be used to further refine a transistor model. Most transistors, even at very low frequencies, display some output admittance. This is modeled as a large resistor shunting the output. This is added to the hybrid-pi with a cascading operation. The usual output capacitance of a transistor may also be added. Lead inductances may be very significant, especially when modeling microwave devices. As a general rule of thumb, a bond wire to a transistor or a lead from a transistor case will have an inductance of about 1 nH per mm of lead length. Even 1 mm can materially change the two-port parameters.

The hybrid-pi model allows us to consider, at least on an intuitive level, one of the most important characteristics of two-port analysis, the problem of stability. When a capacitor is added to the input, the amplifier takes on integrator characteristics. The phase is shifted as well as having magnitudes of voltage and current altered. A similar action arises from the collector to base feedback capacitance. The result is an amplifier with the potential for oscillation.

The capacitive action is illustrated with a simple example. Consider the chain of amplifiers shown in Fig. 5.13. All of the amplifiers are assumed to be ideal op-amps. The first is a simple inverting amplifier. The gain is negative, indicating the usual 180 degree phase shift. The two following stages are integrators. Each has a frequency domain gain of $-Z_f/R$ where Z_f is the impedance of the capacitor. Hence, their voltage transfer function is $H(s) = -1/sRC$. Assume for simplicity that the two integrators are identical. One represents the integrating action of the hybrid-pi input capacitance. The other will simulate the feedback capacitor or an output capacitance.

Each integrator has a gain that decreases linearly with frequency, but has a fixed phase shift of 90 degrees. The total transfer function of the chain of amplifiers is, in the time domain

$$H(j\omega) = \frac{-R_f}{R_i}\left(\frac{j}{\omega RC}\right)^2 = \frac{R_f}{R_i}\frac{1}{\omega^2 R^2 C^2} \quad (5.3\text{-}4)$$

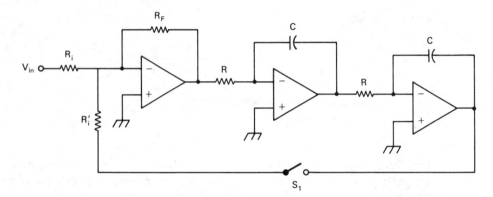

Figure 5.13 An idealized system with an inverting amplifier and two integrators. This chain of operational amplifiers illustrates possible instability in a two-port network.

The combination of the 180 degree phase shift of the inverting amplifier plus 90 degrees from each of the two integrators leads to an output that is in phase with the input. The amplifier will oscillate if the switch, S_1, is closed. It might be stable in practice. This, however, would occur only if real amplifiers were used, units with a finite gain.

The analogy with the transistor is very real. The reverse two-port parameters, y_{12}, are nonzero if the transistor is not unilateral. A feedback path does exist. The capacitances lead to excess phase shift. Hence, the amplifier has the potential for oscillation. Termination of either the input or the output in suitable loads, usually reactive to introduce further phase shift, will allow the oscillation to occur. No amplifier design is complete without a thorough analysis of stability at all frequencies where the transistor is capable of gain. Methods for this analysis are presented in the following section.

5.4 AMPLIFIER DESIGN WITH ADMITTANCE PARAMETERS

Two-port parameters have already been introduced. Some of the relationships between parameter sets and means of forming composite sets from individual matrices were presented. The subject of this section is to finally use the two-port parameters to design an amplifier. Some basic relations will be derived; others will be stated without proof, leaving the details to the reader. The literature on the subject is extensive and many references are listed at the end of the chapter.

The analysis will be limited to y parameter representations for a transistor (or whatever). However, similar equations exist for any of the parameter sets.

Consider the generalized amplifier shown in Fig. 5.14 with the active device represented by a y matrix. The amplifier is driven by a current source paralleled by a characteristic admittance, Y_s. The output is terminated in a load, Y_L. Although

Sec. 5.4 Amplifier Design with Admittance Parameters

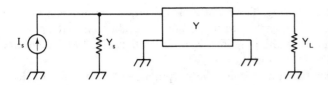

Figure 5.14 A two-port driven by a current source with characteristic admittance Y_s and terminated in a load, Y_L.

the source and load are shown as resistors, they may be and generally are complex impedances.

The first calculation evaluates the input admittance. From fundamentals, $Y_{in} = I_1/V_1$. However, from the defining equations for y parameters, I_1 is linearly related to both V_1 and V_2. The output voltage is related to the current by the load admittance, $V_2 = -I_2/Y_L$. Substitution in Eq. 5.1-1b yields

$$I_2 = y_{21} V_1 - \frac{y_{22} I_2}{Y_L} \tag{5.4-1}$$

Solving for I_2

$$I_2 = \frac{y_{21} Y_L V_1}{Y_L + y_{22}} \tag{5.4-2}$$

Eliminating V_2 from the equation for input current

$$I_1 = y_{11} V_1 - \frac{y_{12} I_2}{Y_L} \tag{5.4-3}$$

The result of Eq. 5.4-2 is now substituted, yielding

$$I_1 = y_{11} V_1 - \frac{y_{12} y_{21} V_1}{Y_L + y_{22}} \tag{5.4-4}$$

with the final result

$$Y_{in} = \frac{I_1}{V_1} = y_{11} - \frac{y_{12} y_{21}}{Y_L + y_{22}} \tag{5.4-5}$$

The output admittance is similarly evaluated. The output is driven from a current generator paralleled with an immittance, Y_L, while the input is terminated in Y_s. The result may be written directly by noting the symmetry of the equations

$$Y_{out} = y_{22} - \frac{y_{12} y_{21}}{Y_s + y_{11}} \tag{5.4-6}$$

These results are not surprising. The input and the output immittances depend not only upon the admittance associated with the port of evaluation, but upon the termination of the other port. If the amplifier is unilateral, $y_{12} = 0$ and complete isolation occurs, yielding $Y_{in} = y_{11}$ and $Y_{out} = y_{22}$.

Consider the amplifier transducer gain, the output power delivered to a load divided by the power available from the generator. The amplifier is driven from a current source. The input voltage will be that developed by the current source across the parallel combination of Y_s and Y_{in}

$$V_1 = \frac{I_s}{Y_s + Y_{in}} = \frac{I_s}{Y_s + y_{11} - \dfrac{y_{12}y_{21}}{y_L + y_{22}}} \tag{5.4-7}$$

or, with manipulation

$$V_1 = \frac{I_s(Y_L + y_{22})}{(Y_s + y_{11})(Y_L + y_{22}) - y_{12}y_{21}} \tag{5.4-8}$$

We found $V_2 = -I_2/Y_L$ from the evaluation of Y_{in}. Substitution in Eq. 5.4-2 yields

$$V_2 = \frac{-y_{21}V_1}{Y_L + y_{22}} \tag{5.4-9}$$

Substitution in Eq. 5.1-1b produces the output current as a function of only V_1

$$I_2 = \frac{I_s Y_L y_{21}}{(Y_s + y_{11})(Y_L + y_{22}) - y_{12}y_{21}} \tag{5.4-10}$$

The output voltage is then

$$V_2 = \frac{-I_2}{Y_L} = \frac{-I_2 y_{21}}{(Y_s + y_{11})(Y_L + y_{22}) - y_{12}y_{21}} \tag{5.4-11}$$

Part of the output termination is reactive. No power is delivered by reactive current flow. Hence, the output power is

$$P_{out} = |V_2|^2 G_L = \frac{I_s^2 G_L |y_{21}|^2}{|(Y_s + y_{11})(Y_L + y_{22}) - y_{12}y_{21}|^2} \tag{5.4-12}$$

where G_L is the real part of the load Y_L.

The driving element is a current generator with strength I_s. The source admittance, Y_s, will in general be complex. A conjugate match is required to obtain the maximum power from the source, the available power. This is achieved by shunting the source admittance with a susceptance equal in magnitude to the source susceptance and a resistance equal to the real part of the source admittance. The source current will then split evenly into the load and the source conductance. The voltage from the source, when matched, would be $I_s/2G_s$ where G_s is the source conductance, the real part of Y_s. The available power is then $P_a = I_s^2/4G_s$, leading to the desired transducer gain

$$G_t = \frac{4G_sG_L|y_{21}|^2}{|(Y_s + y_{11})(Y_L + y_{22}) - y_{12}y_{21}|^2} \tag{5.4-13}$$

Corresponding expressions may be derived for any of the parameter sets. They all have a similar form. The numerator contains the square of the forward term. The denominator has terms related to the terminations and a product of the forward and reverse parameters.

Other gains may also be evaluated. One of the most interesting is termed G_{max}. This is the gain that is obtained if both the input and the output of the two-port network are conjugately matched. Evaluation of G_{max} and the terminations required to achieve it is a subtle task. Consider an example. An amplifier is evaluated using the previously derived equations, yielding G_t, Y_{in}, and Y_{out}. The terminations used for the calculation were arbitrary. One might think that G_{max} could be achieved merely by changing the source and load immittances to the complex conjugates of the calculated input and output immittances. This is not completely true. It would be if the amplifier were unilateral, an unusual (and essentially trivial) situation. If the amplifier is not unilateral, a change in the output termination, as might be performed in an attempt at matching, changes the input admittance. Hence, the predicted source admittance is different than first envisioned.

Through proper manipulations, a set of equations may be derived which, when solved simultaneously, will yield the terminations required to achieve G_{max} (6). The resulting gain may then be evaluated by insertion of the calculated terminations into Eq. 5.4-13. The conditions for obtaining G_{max} are summarized in the following equations

$$G_s = \frac{1}{2g_{22}}\{[2g_{11}g_{22} - \text{Re}(y_{12}y_{21})]^2 - |y_{12}y_{21}|^2\}^{1/2} \tag{5.4-14}$$

$$B_s = -b_{11} + \frac{\text{Im}(y_{12}y_{21})}{2g_{22}} \tag{5.4-15}$$

$$G_L = \frac{1}{2g_{11}}\{[2g_{11}g_{22} - \text{Re}(y_{12}y_{21})]^2 - |y_{12}y_{21}|^2\}^{1/2} \tag{5.4-16}$$

$$B_L = -b_{22} + \frac{\text{Im}(y_{12}y_{21})}{2g_{11}} \tag{5.4-17}$$

where $Y_s = G_s + jB_s$ and $Y_L = G_L + jB_L$. In all of the above, g_{11} is the real part of y_{11} and b_{11} is the imaginary part. Also, $y_{22} = g_{22} + jb_{22}$.

Stability of two-port networks is analyzed through the use of negative resistances. To illustrate a negative resistance, a simple example will be presented. It is similar to a procedure that might be used to design an oscillator.

Envision a transistor described by a scalar model. The only exception is the presence of an output capacitance with $X_c = 1000 \, \Omega$. The beta will be 10, for it is operating at a frequency well above F_β. The assumed bias current will be 2.6 mA, yielding $r_e = 10 \, \Omega$. Neglecting the output capacitance for the present, the common emitter y matrix is

$$Y_e = \begin{pmatrix} \frac{1}{110} & 0 \\ \frac{10}{110} & 0 \end{pmatrix} \tag{5.4-18}$$

The transistor will be used in the circuit of Fig. 5.15. The device is operated in the common base configuration. Using transformations presented earlier, the common base y matrix is

$$Y_b = \begin{pmatrix} 0.1 & 0 \\ -0.091 & 0 \end{pmatrix} \tag{5.4-19}$$

The matrix is now modified to include the effects of a finite collector to emitter capacitance. The matrix for an assumed capacitor with a reactance of 1000 Ω is added to the common base matrix to yield

$$Y_{net} = \begin{pmatrix} 0.1 + j0.001 & -j0.001 \\ -0.091 - j0.001 & j0.001 \end{pmatrix} \tag{5.4-20}$$

The output admittance is now evaluated. This is done for a source admittance corresponding to a capacitor with a 100-Ω reactance, $Y_s = +j0.01$. Application of Eq. 5.4-6 yields $Y_{out} = -8.9 \times 10^{-5} + j9.9 \times 10^{-4}$. Of profound significance is the negative real part of this admittance. The equivalent circuit is a capacitor in parallel with a resistor of $-11.2 \, k\Omega$. If the goal of this exercise was to build an oscillator, a parallel resonant circuit could be connected to the output. If the Q of the resonator

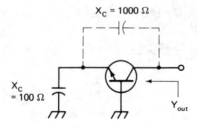

Figure 5.15 A transistor in a configuration to produce a negative output admittance.

was high enough that the parallel resistance was greater than 11.2 kΩ, the circuit would oscillate. The net parallel conductance would still be negative.

Returning to the problem of amplifier design where oscillations are not desired, the goal is now to evaluate a two-port network to see if it will be stable. This is done, in principle, by applying all possible terminations with positive real parts to each port. The opposite port is then investigated; the amplifier is unconditionally stable if and only if the port admittances have positive real parts.

The analysis described above would be unwieldy if performed literally. However, it may be used as a mathematical condition for derivation of a stability factor. The result is the Linvill stability parameter, C, named for its originator, John Linvill of Stanford University (7).

$$C = \frac{|y_{12}y_{21}|}{2g_{11}g_{22} - \mathrm{Re}(y_{12}y_{21})} \tag{5.4-21}$$

A two-port network is unconditionally stable if C is positive but less than unity. $C = 1$ represents a conditionally stable amplifier, a situation little better than an oscillator. If C is negative or greater than unity, the amplifier will probably oscillate if the wrong termination is used at one of the ports.

Another stability parameter of great utility is that attributed to Stern (8)

$$k = \frac{2(g_{11} + G_s)(g_{22} + G_L)}{|y_{12}y_{21}| + \mathrm{Re}(y_{12}y_{21})} \tag{5.4-22}$$

The Stern factor, k, differs from the Linvill factor in that it contains information about the source and load conductances. If k is greater than unity, stability is indicated. It is common to investigate transistor amplifiers at a specific frequency and known terminations to find both C and k greater than unity. This indicates that the amplifier is stable with the terminations chosen; however, if the terminations are removed, oscillations could still occur, and probably will! If C is between zero and unity, unconditional stability is ensured. Because it is more conservative, the Linvill factor is usually used for stability analysis. Some care is required when using computer programs to know which form of stability factor is used. This is complicated by the use of upper case K to indicate the reciprocal of the Linvill factor. Stability is ensured if K is greater than unity. This form is useful because it has utility in other calculations.

We now have the basic tools required for designing an amplifier. We can evaluate the stability, terminal immittances, and gain, and can calculate the terminations required to achieve the maximum possible gain. In addition, the reverse isolation of the amplifier is easily evaluated. By exchanging y_{21} and y_{12}, the gain in the reverse direction is obtained.

The first step is to choose a transistor or other suitable equivalent device. It should have an F_t greater than the desired operating frequency. (This is not strictly accurate. Although current gain decreases to less than unity above F_t, power gain does not always follow. Some transistors have a maximum frequency of operation somewhat higher than F_t. Microwave oscillators sometimes use such transistors.) When a transistor has been chosen, the y parameters should be obtained. They are usually available from the manufacturer's data sheets, or they may be measured. A hybrid-pi model may be used if the amplifier is to operate at relatively low frequencies where data is not available.

The next step in the procedure is the evaluation of stability. Some discussions of amplifier design suggest that the transistor should be discarded or rebiased to a different level if stability is not ensured with C positive and less than unity. While this is certainly a conservative viewpoint, it is sometimes restrictive. It is often acceptable to operate the transistor in a manner that provides more flexibility, while still ensuring *circuit* stability.

There are three common methods that may be used to stabilize a transistor that displays C greater than unity. The first is loading. If a relatively low value resistor is paralleled with the output port, the combination will often be unconditionally stable even if the transistor alone is not. A cascading operation with the matrices is used for analysis. As a general rule of thumb, stability is improved as the collector of a ce amplifier moves toward a short circuit termination.

The second method of stabilization is shunt feedback. In principle, it is possible to find a network that will "unilateralize" any two-port. However, this is difficult. Degeneration of the gain through the application of a parallel feedback resistor will often ensure low frequency stability. An inductor may be inserted in series with the feedback resistor to maintain high frequency stability.

A third method that has some subtle benefits is the application of emitter degeneration. The usual added emitter resistance has the bandwidth extension properties mentioned in Chap. 1. In addition, the decrease in low frequency gain aids stability. The use of emitter inductance also tends to improve stability, so long as the inductance is not excessive. One of the virtues of using an emitter inductor is that the input resistance of an amplifier is increased, often affording a direct match to 50 Ω in the uhf spectrum.

Clearly, combinations of the three methods may be used. Combinations are especially useful for stabilization of broadband amplifiers.

The usual discussions of amplifier design stress stability, as they should. However, the two-port parameters at the frequency of operation are often used for all analysis of a narrowband amplifier. Stability is investigated in detail only at the operating frequency. Herein lies the common error, often a fatal one. Even if an amplifier is to operate only over a narrow bandwidth, stability analysis should be performed over the entire spectrum where the transistor is active. Oscillation in an amplifier can be a severe problem, even if the oscillation is far removed from the operating frequency. Broadband stability is essential!

Once stability of a composite network containing the active transistor as well as feedback and loading networks is ensured, the rest of the design may progress. The transducer gain is evaluated in a convenient system impedance, usually 50 Ω for each termination. The performance may be adequate, and thus this step is the final one. However, care should be taken to assure that the 50-Ω (or whatever is used) terminations are preserved under application. If broadband analysis is performed with that assumption, it makes little sense to then use the amplifier to drive a filter, a circuit that may present a 50-Ω load at one frequency, but becomes highly reactive at others.

Matching networks are designed if additional gain is required. The impedances required to realize the maximum possible gain are calculated using Eq. 5.4-14 through 5.4-17. The gain that results from matching may be more than required. Feedback may be applied to reduce the gain. Alternatively, only partial matching might offer a good solution. Careful impedance mismatching is a useful tool to extend the bandwidth of wideband amplifiers. If there are many possible alternatives, the one used should be that which will best ensure amplifier stability. Impedance matching methods were discussed in Chap. 4.

Stability can be especially difficult in the design of multistage amplifiers. Impedance variations at, for example, the output of the composite cascade of two stages will reflect through the output transistor to present ill-defined impedances at the output of the first amplifier. This might have undesired effects on system gain, impedance characteristics, and stability. The most conservative design approach is the utilization of local feedback or loading at each stage to be certain of unconditional stability of each, again on a broadband basis.

Clearly, there is considerable flexibility left to the designer of high frequency amplifiers. Only some guidelines have been offered. The better designs will probably require a considerable amount of "number crunching." A programmable handheld calculator may be used for much of the work. The high speed digital computer is the preferred design tool though.

Programs are commercially available that are specifically aimed at microwave and rf design using two-port concepts. An especially attractive program is COMPACT, available from COMPACT Engineering, Inc., Palo Alto, Cal. This program offers considerable flexibility. Transmission lines, including coupled structures, are included within the program, allowing ease of design of microstrip networks and filters. The program is based upon S parameter input and output. Some nodal analysis is possible. Of greatest significance, though, is the optimization capability. Selected components are allowed to vary. The computer will then change them as required to produce the required response. This is the feature that allows the program to aid the designer in circuit synthesis, going beyond the more direct analysis. The largest problem with programs of this kind is that the designer must have considerable intuition about the circuit prior to optimization. The intuition is not provided by the program unless it is used with great care.

5.5 DESIGN EXAMPLES WITH ADMITTANCE PARAMETERS

Amplifier design using two-port methods is illustrated with several examples in this section. The transistor chosen is the Nippon Electric Co. type NE02135. The ce y parameters for this transistor are listed in Table 5.2. Bias conditions of 10 V between collector and emitter and $I_e = 10$ mA are used for all examples.

Table 5.2 Y-Parameters for the NEC type NE02135 bipolar transistor. Bias conditions are $V_{ce} = 10$ V and $I_e = 10$ mA. Admittances in S.

Freq. (GHz)	Y_{11}	Y_{12}	Y_{21}	Y_{22}
0.1	0.0038 + j0.0045	−0.00003 − j0.0004	0.261 − j0.04	−0.0001 + j0.0008
0.5	0.014 + j0.022	−0.00014 − j0.002	0.235 − j0.153	−0.001 + j0.005
1.0	0.039 + j0.031	−0.001 − j0.004	0.119 − j0.256	−0.001 + j0.013
2.0	0.072 − j0.01	−0.008 − j0.007	0.146 − j0.168	0.018 + j0.033
3.0	0.043 − j0.06	−0.014 + j0.003	0.166 + j0.04	0.056 + j0.014
4.0	0.01 − j0.04	−0.008 + j0.009	−0.045 + j0.075	0.052 − j0.015

The designs presented were done with a computer program. The program was written in **BASIC** for ease of use with a variety of computer types. None of the sophistication contained in COMPACT or similar programs was included. Instead, this one was aimed at simple analysis using the equations presented so far in this chapter. The program is based upon y parameters although a subroutine is included for data entry in the form of S parameters. Conversions from ce to cb and ce to cc are included, as are subroutines for conversion to z parameters. The latter is used for series combination operations of networks.

The first amplifier to be designed emphasizes simplicity. The goal is to build a simple amplifier that provides as much gain as possible at low frequencies with usable gain extending into the uhf region.

The first step in the design is a stability analysis of the transistor. The y parameters are evaluated at all frequencies with somewhat alarming results. The transistor is unconditionally stable with C less than unity at 2 GHz and above. C is greater than unity at lower frequencies indicating potential instability with some terminations. The Stern factor, k, was greater than unity in a 50-Ω system, indicating stability.

The amplifier is easily stabilized with an output loading resistor. The first value tried was 100 Ω. The result was $C = 0.97$ at 500 MHz and $C = 0.65$ at 1 GHz, but $C = 1.06$ at 100 MHz. While the amplifier with a 100-Ω collector resistor would be stable at uhf, it is not stable at low frequencies. This illustrates the necessity for investigating the stability of an amplifier at all frequencies, even if outside the range of interest.

The next value picked for collector loading was 68 Ω. This was low enough to ensure broadband stability. The circuit is shown in Fig. 5.16. The result of the analysis is shown in Table 5.3. Only data below 1 GHz is shown in the table. The transducer gain values were obtained in a 50-Ω system. Source and load impedances presented are those required to achieve $G_{\max}$.

The amplifier of Fig. 5.16 is a very practical one. It will provide useful gain at quite high frequencies and is unconditionally stable. Several such amplifiers may be cascaded, although the resulting gain will not merely be the sum of the 50-Ω transducer gains. This discrepancy comes from the impedance mismatch between stages.

For a second example, assume that the previous amplifier was to be used at

Sec. 5.5   Design Examples with Admittance Parameters   185

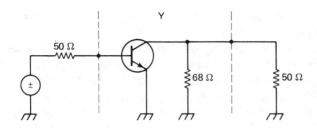

Figure 5.16 A common emitter amplifier using the NE02135 transistor. The 68-Ω collector resistor will ensure unconditional stability at all frequencies for all possible terminations. This amplifier will display useful gain to over 1 GHz.

Table 5.3

f, GHz	C	G_t, dB	G_{max}, dB	Z_s	Z_L
0.1	0.78	21.5	25.0	$40 + j110$	$33 + j43$
0.5	0.80	15.3	18.3	$9 + j21$	$22 + j30$
1.0	0.54	10.1	12.7	$10.6 + j6$	$28 + j19$

100 MHz. As much gain as possible is desired. The conditions for achieving this were presented in Table 5.3. The required source impedance is $40 + j110$ while a load of $33 + j43$ is needed. Examination shows that both are inductive and that the mismatch arises from reactive components. A reasonable match may be obtained by providing the proper reactive terminations with no change in the resistive components. Hence, inductors are added in series with each of the 50-Ω terminations of Fig. 5.16. The resulting amplifier is shown in Fig. 5.17. Analysis at 100 MHz shows $C = 0.78$, $G_t = 24.8$ dB, $G_{max} = 25.0$ dB. The terminations required to achieve G_{max} are $Z_s = 40 + j0$ and $Z_L = 33 + j0$. As expected, there is no change in C or in G_{max}. The difference between the 50-Ω transducer gain and G_{max}, 0.2 dB, is so slight that additional matching would probably be redundant. Parallel capacitors at each termination would provide the proper match. Each inductance would increase slightly. The greatest virtue of this refinement would be that the capacitors could be adjustable, allowing for transistor variations.

The inductors were added to the existing y matrix through cascading operations in the preceding analysis. Such modifications have the same effect as using reactive terminations on the amplifier of Fig. 5.16; hence, the Linvill stability parameter does not change.

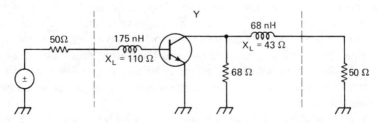

Figure 5.17 The amplifier of Fig. 5.16 with partial matching to increase gain at 100 MHz.

Assume that the simple amplifier of Fig. 5.16 is to be matched for maximum gain at 500 MHz. The required terminations are given in Table 5.3. The input source required is $9 + j21$ Ω. The natural matching network is an L-type. A 21-Ω series inductor is inserted in series with the base, yielding an impedance of $9 + j0$. An L-network is now designed to transform this impedance to 50 Ω. The L-network reactances are $(R_1 R_2 - R_2^2)^{1/2} = X_s$ and $X_p = R_1 R_2 / X_s$. R_1 is the larger of the two resistances to be transformed. Choosing the network with an inductive series element, we find $X_s = 24.7$ Ω while the parallel capacitive reactance is 42.5 Ω. The two inductors are combined to one with a reactance of 45.7 Ω.

Similarly, the output impedance of the amplifier, when driven from the source impedance required to achieve G_{max}, is $22 - j30$. The output has two possible equivalent circuits. One is a voltage generator in series with a 22-Ω resistor and a capacitor with 30-Ω reactance. The other, the parallel equivalent, is a current generator in parallel with a 63-Ω resistor and a capacitor with a reactance of 46 Ω. This resistance is so close to 50 Ω that no transformation is required for the real part. Only the reactive part of the termination must be tuned. This is realized with a shunt inductance of 46 Ω at the collector port. The final circuit is shown in Fig. 5.18. The 50-Ω transducer gain of this composite circuit should be quite close to the 18.3-dB value of G_{max} predicted from Table 5.3. A series output capacitor would produce a "perfect" match.

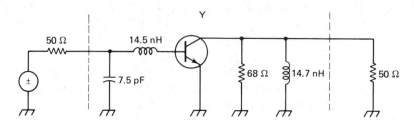

Figure 5.18 The amplifier of Fig. 5.16 with matching to increase the gain at 500 MHz. A slight improvement could be made with capacitive matching at the output.

Upon examination of the output network of Fig. 5.18, one might reason that the low inductance value would appear as a short circuit at lower frequencies, making the 68-Ω loading resistor redundant. This could be dangerous. An inductive load can produce an input impedance with a negative real part if the original y parameters (before the addition of the cascaded resistor) are investigated with a variety of possible terminations. Similarly, an inductance as the source admittance may yield a negative real part output impedance. Neither condition is possible with the 68-Ω collector resistor. The resistance can, of course, be the result of using a low Q inductor at the output.

The inductors used in the matched 500-MHz amplifier are quite small and may be impractical. The preferable way to build amplifiers at this frequency (and higher) is with the microstrip-type transmission lines described in Chap 4. They

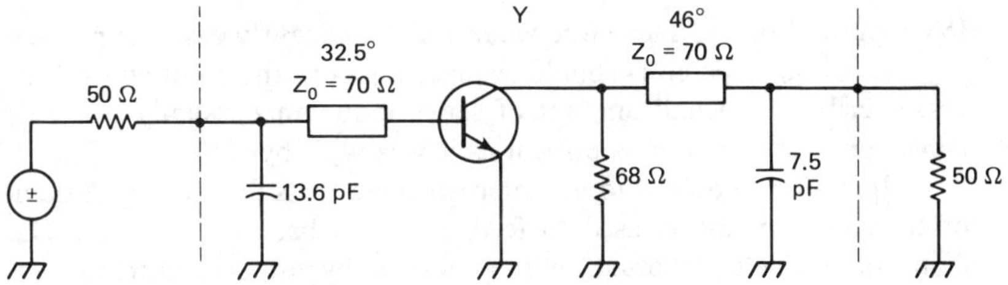

Sec. 5.5 Design Examples with Admittance Parameters 187

Figure 5.19 A 500-MHz amplifier using a microstrip transmission line. This configuration is more practical at 500 MHz and higher.

may be designed using a Smith chart or with a program, either using a calculator or a computer. Figure 5.19 shows an amplifier which is virtually identical to that of Fig. 5.18. However, microstrip methods replace the inductors.

The characteristic line impedance was chosen on a more or less random basis at 70 Ω. A line length was calculated which produced an admittance at the amplifier terminal with a real part of 0.02 S, the reciprocal of 50 Ω. The imaginary part of the admittance was evaluated to determine the required shunt element. If needed, the characteristic impedance of the microstrip is changed so that a shunt capacitor is used instead of an inductor.

Capacitive shunt elements are used in this design. They would probably be variable in practice. Alternatively, different characteristic impedances and line lengths could have been chosen to allow the use of shunt inductors. These are then synthesized with short circuited sections of microstrip. Also, open circuited stubs may be used as the capacitors in the design of Fig. 5.19.

The practical details of construction of uhf amplifiers cannot be neglected. Considerable care must be taken in the choice of components. A poor quality capacitor might have lead lengths that are long enough to show significant inductance. An inductor will, similarly, have some capacitive characteristics. If quality components are not available, the ones used should be evaluated with the instrumentation used for measurement of transistor parameters, with the results then used in the analysis.

The amplifiers described had no emitter degeneration. The slightest inductance in the emitter leads will change the results. The emitters of the NE02135 (there are two emitter leads) should be grounded as close to the case as possible.

Bias is easily delivered to the amplifiers designed, for the 68-Ω resistor provides a convenient path into the collector. The "grounded" end should be carefully bypassed. The leads on the bypass capacitor should be kept as small as possible. Recall that a wire lead on a component has an inductance of approximately 1 nH per mm. The bypass capacitor should probably be 500 to 1000 pF for a 500-MHz amplifier, or a broadband design that must work at uhf. Even smaller capacitors should be used for higher frequency designs. A parallel combination of several capacitors is often effective. Larger values are paralleled if the design must function well at low frequencies, or to maintain stability of a uhf design.

Base bias may be fed to the transistor through a resistor with a value of a few thousand ohms. This large value will significantly ease the problem of bypassing.

Blocking capacitors should be placed in both the input and output of the amplifiers described. A small amount of series inductance usually has no great effect, so capacitor quality is not as critical as it was with bypasses.

It is often practical to use microstrip methods for biasing. A quarter wavelength of transmission line is used to feed bias to a base or collector. The end removed from the device is "shorted" with a suitable bypass capacitor.

The amplifiers described have used a transistor biased at 10 V and 10 mA. A 12- to 15-V regulated power supply is usually available. A PNP transistor can serve as a convenient means of biasing the amplifier or, alternatively, an op-amp could be used.

As a final example, consider the design of a broadband amplifier. Assume that a gain of 13 dB is required out to a frequency of 500 MHz in a 50-Ω system. The gain should decrease beyond that frequency. The amplifier should have a similar gain at 100 MHz, but lower frequency operation is not important. The NE02135 at 10 V and 10 mA will again be used.

On the basis of previous work, a good candidate for this design would be one using a combination of emitter degeneration and shunt feedback. The topology used as a starting point is shown in Fig. 5.20a. The transistor y parameters are converted to z parameters, the z matrix for the emitter degeneration is added directly and the result is converted back to a y matrix. Some inductance, 1 nH, is included to account for the less than ideal nature of the emitter resistor. The y matrix of the 500-Ω feedback resistor is then added. The composite y parameters for the amplifier are analyzed with the results shown in Table 5.4.

Examination of the data is immediately encouraging. The 50-Ω transducer gain is high enough, even at 500 MHz, without matching. Moreover, the 500-MHz transducer gain is close to the value for G_{max}, indicating that no severe mismatch is present. However, there is a stability problem at 2 GHz. Recall that the same transistor was unconditionally stable at 2 GHz with a well-grounded emitter. Insertion of degeneration in a common arm is something that must always be investigated for its effect upon stability.

The stability problem is solved by cascading the existing amplifier with a shunt 500-Ω output resistor. This is shown in Fig. 5.20b. Because the only stability problem was at high frequency, it would also have been possible to stabilize the amplifier with a "damping" network at the collector, the series combination of a capacitor and a resistor. The results obtained with the amplifier are shown in Table 5.4, part B. The amplifier is now unconditionally stable. While the 50-Ω transducer gain at 500 MHz is below the desired 13 dB, the value for G_{max} is still high enough.

Table 5.4 does not show the terminations required for achieving G_{max}. They are $Z_s = 29 + j15$ and $Z_L = 69 + j30$, both at 500 MHz. The values are $Z_s = 20.6 + j5$ and $Z_L = 69 + j44$ at 1 GHz; they are not significant to this design.

The next step is to perform matching to increase the gain at the upper band edge, 500 MHz. The effect of the matching networks must be evaluated throughout the band to ensure that the low frequency gain is not severely degraded. The first

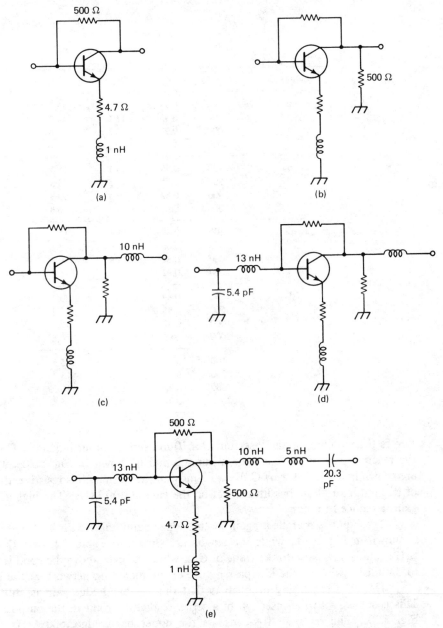

Figure 5.20 Evolution of a wideband amplifier using the NE02135 transistor. The initial configuration is the multiple feedback amplifier (a). Emitter inductance is included in the analysis. A collector resistor is added (b) to ensure broadband stability. A series inductor is added at the output (c) and an L-network at the input (d) to improve the impedance match at the high end of the band of interest, 500 MHz. A 500-MHz series tuned circuit is added at the output (e) to decrease the gain at lower frequency, leaving the 500-MHz gain unchanged.

Table 5.4

Configuration	f, MHz	C	$G_T(50\ \Omega)$	G_{max}
a	100	0.94	15.6 dB	15.9 dB
	500	0.93	13.4	14.2
	1000	0.92	10.0	11.3
	2000	1.02	5.7	—
b	100	0.89	15.1	15.3
	500	0.85	12.8	13.3
	1000	0.82	9.5	10.3
	2000	0.92	5.3	6.8
c	100	0.89	15.1	15.3
	500	0.85	12.6	13.3
	1000	0.82	8.8	10.3
	2000	0.92	2.1	6.8
d	100	0.89	15.1	15.3
	500	0.85	13.2	13.3
	1000	0.82	3.8	10.3
	2000	0.92	−17.5	6.8
e	100	0.89	13.9	15.3
	500	0.85	13.2	13.3
	1000	0.82	2.6	10.3
	2000	0.92	−20.0	6.8

step in the matching is the insertion of a 10-nH series output inductor. This removes the reactive part of the collector mismatch and is shown in Fig. 5.20c. The results are shown in Table 5.4, part C. Because most of the mismatch is resistive, the addition of this inductor alone has little effect on the gain at 500 MHz. The higher frequency gain is reduced though.

The input is matched, again at the upper band edge, with an L-network. This is shown in Fig. 5.20d while the results are shown in Table 5.4, part D. The 500-MHz gain is now close to the value of G_{max} while the gain above the band is beginning to diminish owing to the low pass nature of the matching network at the base.

The only remaining problem is that the low-band-edge gain is still too high. This is solved with the insertion of a series resonant circuit in the output, shown in Fig. 5.20e. The resonant frequency is the upper band edge, 500 MHz. The Q is kept low so no serious problems will be introduced at lower frequencies.

The final results are shown in Table 5.4, part E. Note that the low frequency gain is now only half a dB higher than that at the top edge. The gain above the band diminishes rapidly though.

This broadband amplifier illustrates a commonly used method, selective mismatching. It is also interesting to compare the results with those predicted for a simple feedback amplifier using nothing but a scalar model. The similarity is striking.

Only frequencies at the band edges were used in the analysis presented. This

Sec. 5.6 Scattering Parameters

is often a valid design approach. Analysis could be extended to frequencies between points where manufacturer's data was not available. The available two-port data is interpolated linearly with frequency between existing points. The more refined computer programs for two-port analysis have subroutines to perform the extrapolation.

5.6 SCATTERING PARAMETERS

The two-port network theory presented so far has dealt with four simple variables, voltages and currents at the ports. The variables are interrelated by appropriate matrices. The choice of which matrix is used depends upon which of the four variables are chosen to be independent.

There is no reason to limit the variables to simple ones. Linear combinations of the simple variables are just as valid. The more complicated variables chosen should be linearly independent and, ideally, should have some physical significance.

A transformation to other variables is certainly not new to us. Logarithmic transformations such as the dB or dBm are so common that we used them interchangeably with the more fundamental quantities without even mentioning that a transformation has occurred. The matrix elements are often admittances or impedances when working with voltages and currents as signals of interest. A new viewpoint can be of great utility in working with transmission lines. An impedance is replaced by a voltage reflection coefficient, $\Gamma = (Z - Z_0)/(Z + Z_0)$. This has especially attractive properties for evaluation of transmission line phenomena.

Scattering parameters are nothing more than a repeat of this viewpoint. Instead of considering voltages and currents to be the fundamental variables, we use four "voltage waves." They are interrelated through an appropriate matrix, the S parameters.

Figure 5.21 shows the traditional two-port network and an alternate one with voltage waves incident on and reflected from the ports. The voltage waves are defined with the letters a_1, b_1, a_2, and b_2. The a waves are considered to be incident waves and are the independent variables. The b waves are the results of reflection or "scatter-

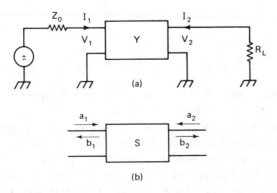

Figure 5.21 A two-port network viewed as being driven by voltages and currents (a) or voltage waves (b). The voltages and currents are related by y parameters while the voltage waves are related by scattering parameters.

ing," and are the dependent variables. The waves are related to voltages and currents by

$$a_1 = \frac{V_1 + Z_0 I_1}{2(Z_0)^{1/2}} \tag{5.6-1}$$

$$b_1 = \frac{V_1 - Z_0 I_1}{2(Z_0)^{1/2}} \tag{5.6-2}$$

$$a_2 = \frac{V_2 + I_2 Z_0}{2(Z_0)^{1/2}} \tag{5.6-3}$$

$$b_2 = \frac{V_2 - I_2 Z_0}{2(Z_0)^{1/2}} \tag{5.6-4}$$

The voltage waves are defined with respect to a characteristic impedance, Z_0. This was the case with the voltage waves used with transmission lines. The very concept ceased to have formal meaning without definition of a working impedance.

The scattered waves are related to the incident ones with a set of linear equations just as the port currents were related to the port voltages with y parameters. The relating equations are

$$\begin{aligned} b_1 &= S_{11} a_1 + S_{12} a_2 \\ b_2 &= S_{21} a_1 + S_{22} a_2 \end{aligned} \tag{5.6-5}$$

or, in matrix form

$$\begin{pmatrix} b_1 \\ b_2 \end{pmatrix} = \begin{pmatrix} S_{11} & S_{12} \\ S_{21} & S_{22} \end{pmatrix} \begin{pmatrix} a_1 \\ a_2 \end{pmatrix} \tag{5.6-6}$$

Consider the meaning of S_{11}. If the incident wave at the output, a_2, is set to zero, Eqs. 5.6-5 reduce to $b_1 = S_{11} a_1$ and $b_2 = S_{21} a_1$. S_{11} is the ratio of the input port reflected wave to the incident one. This reduces, using the defining equations for a_1 and b_1, to

$$S_{11} = \frac{b_1}{a_1} = \frac{V_1 - I_1 Z_0}{V_1 + I_1 Z_0} = \frac{Z - Z_0}{Z + Z_0} \tag{5.6-7}$$

where V_1/I_1 is the input impedance, Z. The form is that of a reflection coefficient.

Similarly, S_{21} is the ratio b_2/a_1. It is the voltage wave emanating from the output as a result of an incident wave at the network input. If the output of the network is terminated in Z_0, the impedance used in the definition of the S parameters, b_2, now a wave incident upon the load, will have no reflection. Hence, a_2 is zero. The condition for evaluation of S parameters is termination in the proper characteristic impedance just as short circuits were used for evaluation of y parameters. The signifi-

Sec. 5.6 Scattering Parameters

cance of S_{21} is that it represents a forward gain. Specifically, $|S_{21}|^2$ is the transducer, or insertion power gain when the network is placed in a Z_0 system.

The measurement implications are profound. We often found amplifiers that had a Linvill stability factor, C, that was greater than unity, indicating potential instability in the examples studied in the previous section. The transistor would most likely have oscillated if a short or open circuit termination had been used for direct measurement of y or z parameters. Conversely, k, the Stern stability factor, was always greater than unity in a 50-Ω system, at least for the examples studied. Hence, the transistors would not oscillate in such an environment. It is reasonable to expect from the linearity of the equations that y or z parameters could be inferred by performing measurements in an arbitrary system with proper algebraic manipulation. Scattering parameters are the formalism used for this manipulation.

The other two scattering parameters, S_{12} and S_{22}, have similar significance. S_{22} is the output reflection coefficient when looking back into the output port of the network with the input terminated in Z_0. S_{12} is the reverse gain. Specifically, $|S_{12}|^2$ is the insertion power gain in a Z_0 system when the output port is driven and the signal at the input port is detected. An amplifier with perfect isolation, one that is unilateral, will have $S_{12} = 0$.

The reflection coefficient nature of S parameters makes the Smith chart especially convenient in use and specification. Figure 5.22 shows an example, the S parameters for a NE02135 microwave transistor. The graphs represent the transistor S parameters with a 10-V collector bias and a 30-mA collector current. The tables below the figure present numeric values for the transistor under a variety of bias conditions. The left part of the figure is actually a Smith chart. S_{11} and S_{22} are plotted as a function of frequency on the chart with the frequencies being shown as dots on the curve. The right-hand chart is a polar representation showing the magnitude and angle of the gain parameters, S_{21} and S_{12}. The data presented in Fig. 5.22 are typical of the form used by most manufacturers for representing active devices. Indeed, the y parameter data presented earlier for the NE02135 (Table 5.2) were calculated from the data in Fig. 5.22.

The instrument used for S parameter measurement is the network analyzer. It is essentially a collection of directional couplers for the measurement of reflection coefficients and gains with a vector voltmeter for evaluation of the outputs in vector form. All modern network analyzers use swept frequency signal sources, allowing evaluation to occur over a wide spectrum with one measurement. A polar display is often used, allowing output data to appear in exactly the forms shown in the data sheet of Fig. 5.22.

All of the characteristics of a two-port network may be expressed in terms of the associated S parameters. The relationships will not be derived here. They are available in the literature (9).

The input reflection coefficient of a network is a function of both the network itself and the output termination. Defining the input by Γ_{in}

$$\Gamma_{in} = S_{11} + \frac{S_{12}S_{21}\Gamma_L}{1 - S_{22}\Gamma_L} \qquad (5.6\text{-}8)$$

NE021, LOW NOISE L/S-BAND NPN AMPLIFIER AND OSCILLATOR SERIES

NE02108 AND NE02135 COMMON EMITTER SCATTERING PARAMETERS

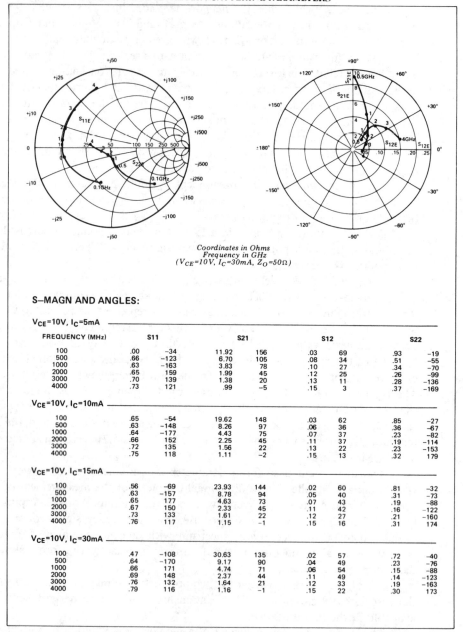

Coordinates in Ohms
Frequency in GHz
($V_{CE}=10V$, $I_C=30mA$, $Z_O=50\Omega$)

S—MAGN AND ANGLES:

$V_{CE}=10V$, $I_C=5mA$

FREQUENCY (MHz)	S11		S21		S12		S22	
100	.00	−34	11.92	156	.03	69	.93	−19
500	.66	−123	6.70	105	.08	34	.51	−55
1000	.63	−163	3.83	78	.10	27	.34	−70
2000	.65	159	1.99	45	.12	25	.26	−99
3000	.70	139	1.38	20	.13	11	.28	−136
4000	.73	121	.99	−5	.15	3	.37	−169

$V_{CE}=10V$, $I_C=10mA$

FREQUENCY (MHz)	S11		S21		S12		S22	
100	.65	−54	19.62	148	.03	62	.85	−27
500	.63	−148	8.26	97	.06	36	.36	−67
1000	.64	−177	4.43	75	.07	37	.23	−82
2000	.66	152	2.25	45	.11	37	.19	−114
3000	.72	135	1.56	22	.13	22	.23	−153
4000	.75	118	1.11	−2	.15	13	.32	179

$V_{CE}=10V$, $I_C=15mA$

FREQUENCY (MHz)	S11		S21		S12		S22	
100	.56	−69	23.93	144	.02	60	.81	−32
500	.63	−157	8.78	94	.05	40	.31	−73
1000	.65	177	4.63	73	.07	43	.19	−88
2000	.67	150	2.33	45	.11	42	.16	−122
3000	.73	133	1.61	22	.12	27	.21	−160
4000	.76	117	1.15	−1	.15	16	.31	174

$V_{CE}=10V$, $I_C=30mA$

FREQUENCY (MHz)	S11		S21		S12		S22	
100	.47	−108	30.63	135	.02	57	.72	−40
500	.64	−170	9.17	90	.04	49	.23	−76
1000	.66	171	4.74	71	.06	54	.15	−88
2000	.69	148	2.37	44	.11	49	.14	−123
3000	.76	132	1.64	21	.12	33	.19	−163
4000	.79	116	1.16	−1	.15	22	.30	173

Figure 5.22 Reprinted by permission of California Eastern Laboratories, Santa Clara, Calif.

where Γ_L is the reflection coefficient of the load. Similarly, the output reflection coefficient is given by

$$\Gamma_{\text{out}} = S_{22} + \frac{S_{12}S_{21}\Gamma_s}{1 - S_{11}\Gamma_s} \qquad (5.6\text{-}9)$$

where Γ_s is the reflection coefficient of the driving source.

An alternative representation is often found in the literature. The input reflection coefficient, Γ_{in}, is given as S'_{11}. The prime indicates that the network is terminated in an arbitrary impedance for the measurement. A similar form is used for Γ_{out}, being shown as S'_{22}.

The form of these equations is similar to those for other matrices. If the network is unilateral, with $S_{12} = 0$, the input state is defined completely by the input parameter, S_{11}. Isolation is perfect. A nonunilateral network provides less than perfect isolation.

If the network is terminated by a source and load represented respectively by Γ_s and Γ_L, the transducer gain is given by

$$G_t = \frac{|S_{21}|^2(1 - |\Gamma_s|^2)(1 - |\Gamma_L|^2)}{|(1 - S_{11}\Gamma_s)(1 - S_{22}\Gamma_L) - S_{12}S_{21}\Gamma_L\Gamma_s|^2} \qquad (5.6\text{-}10)$$

The similarity of this relationship to the corresponding one for y parameters, Eq. 5.4-13, is striking. The numerators contain a square of the forward parameter and termination related terms. The denominators contain terms that compare the terminations with the input and output parameters and a term that accounts for the lack of perfect isolation between ports.

If a network is terminated in Z_0, the impedance used in defining the network's S matrix, the source and load reflection coefficients are both zero. All energy impinging upon the termination is absorbed with none reflected. With such terminations Eq. 5.6-10 reduces to $G_T = |S_{21}|^2$.

Stability is of great concern with any active two-port network. Instability was indicated by an input or output immittance with a negative real part. When scattering parameters are used, an unstable condition is indicated by an input or output reflection coefficient greater than unity magnitude. This is easily demonstrated by evaluating a reflection coefficient $\Gamma = (Z - Z_0)/(Z + Z_0)$, with Z being any impedance with a negative real part. The reflected wave from such an immittance will be greater than the incident wave.

Only immittances with Γ less than unity magnitude were considered in the discussion of the Smith chart in Chap. 4. A reflection coefficient greater than unity is still viable. It will, however, lie outside the perimeter of the standard chart.

It is sometimes useful to plot the reciprocal of a reflection coefficient on the Smith chart. The magnitude will then lie within the unity circle. This is often done in the design of oscillators where accurate information is needed. It is not usually done for amplifier designs where, more often than not, knowledge that Γ is in the unstable region is information enough.

The usual scattering parameter indicator for stability is the Rollett factor, K, defined by

$$K = C^{-1} = \frac{1 + |\Delta S|^2 - |S_{11}|^2 - |S_{22}|^2}{2|S_{12}S_{21}|} \qquad (5.6\text{-}11)$$

where ΔS is the determinant of the scattering matrix, $\Delta S = S_{11}S_{22} - S_{12}S_{21}$. An upper case K is used for the Rollett factor to distinguish it from the lower case k signifying the Stern stability factor. A network is unconditionally stable if K is greater than $+1$. The K factor, like C, depicts the stability of the amplifier without regard to termination; that is, all possible terminations are considered. The reason for using a slightly different form, a reciprocal, is that the numeric value of K is useful in other calculations.

$|S_{21}|^2$ is the transducer gain in a Z_0 system while the transducer gain in an arbitrary set of terminations is given by Eq. 5.6-10. The maximum possible gain of the network occurs when both ports are conjugately matched. This condition leads to the gain

$$G_{\max} = \left| \frac{S_{21}}{S_{12}} [K - B_3(K^2 - 1)^{1/2}] \right| \qquad (5.6\text{-}12)$$

where K is the Rollett factor. The parameter B_3 is given by

$$B_3 = B_2/|B_2| \qquad (5.6\text{-}13)$$

where B_2 is defined by

$$B_2 = 1 + |S_{22}|^2 - |S_{11}|^2 - |\Delta S|^2 \qquad (5.6\text{-}14)$$

The form of Eq. 5.6-12 is slightly different than usually presented in the literature. B_3 is the signum function of B_2. The signum function of a variable x is $+1$ if x is positive, -1 if x is negative, and zero for $x = 0$. The form presented is useful for calculations with a calculator or computer.

The load immittance required to achieve $G_{\max}$ is given by

$$|\Gamma_{ML}| = \frac{B_2 - B_3(B_2^2 - 4|C_2|^2)^{1/2}}{2|C_2|} \qquad (5.6\text{-}15)$$

where C_2 is defined as

$$C_2 = S_{22} - (\Delta S)(S_{11}^*) \qquad (5.6\text{-}16)$$

The asterisk attached to S_{11} indicates the complex conjugate of S_{11}. The angle of the load reflection coefficient required to achieve $G_{\max}$ is

Sec. 5.6 Scattering Parameters

$$\theta_{ML} = -\theta_{C2} \qquad (5.6\text{-}17)$$

where θ_{C2} is the angle of C_2

The source reflection coefficient required to achieve G_{max} is given in the literature by equations similar to Eqs. 5.6-15 through 5.6-17. Alternatively, we note that G_{max} occurs when both ports are conjugately matched. The load required for this is given by the equations above. Hence

$$\Gamma_{MS} = \left\{ S_{11} + \frac{S_{12}S_{21}\Gamma_{ML}}{1 - (\Gamma_{ML}S_{22})} \right\}^* \qquad (5.6\text{-}18)$$

Consider an example that was investigated earlier with y parameters, the common emitter NE02135 operating at 500 MHz with bias of 10 V and 10 mA. The S parameters are given for this status in Fig. 5.22.

The first step, as always, is to investigate stability. Using only the 500-MHz data, the stability is evaluated by calculating the Rollett factor, K. The result is $K = 0.555$, indicating potential instability.

We know from earlier work that unconditional stability is assured by shunting the output port with a 68-Ω resistor. This cascading is best performed by converting to another parameter set when working with S parameters. The conversion was to y parameters, the matrix of the 68-Ω resistor was added using the cascading equations derived earlier, and the result was converted back to a scattering matrix.

The modified S parameters are $S_{11} = 0.61$ at -136 degrees, $S_{12} = 0.042$ at 41 degrees, $S_{21} = 5.80$ at 102 degrees, and $S_{22} = 0.25$ at -138 degrees. Repeating the stability calculation yields $K = 1.25$, indicating unconditional stability.

The stability has only been evaluated at one frequency. This is not a proper design method. The stability should be evaluated on a broadband basis before progressing. We will bypass this vital step for this specific example, for it was already done when the same amplifier was investigated with y parameters. We found that the 68-Ω resistor guarantees broadband stability.

The other parameters may be evaluated with the modified scattering matrix. The 50-Ω transducer gain, $|S_{21}|^2$, is 15.3 dB. Using Eq. 5.6-12, we find that when both ports are conjugately matched the gain is $G_{max} = 18.4$ dB. The matching conditions are found from Eqs. 5.6-15, 5.6-17, and 5.6-18. The results are $\Gamma_{ML} = 0.531$ at 110 degrees while $\Gamma_{MS} = 0.745$ at 131 degrees. Using $Z = Z_0(1 + \Gamma)/(1 - \Gamma)$, the corresponding impedances are found, $Z_{ML} = 22 + j30$ and $Z_{MS} = 9 + j22$.

A simplification is often used in the design of an amplifier. The amplifier is first investigated for stability. The circuit is modified as was done in the preceding example if instability is indicated. The parameters of the stable configuration are then calculated. Once stability in ensured, the assumption is then made that the amplifier is unilateral, setting S_{12} to zero. Analysis is performed with the simplified scattering matrix. The parameters other than S_{12} are left unchanged.

With Eq. 5.6-10 as a basis, the transducer gain of the unilateral amplifier is calculated

$$G_{Tu} = |S_{21}|^2 \left(\frac{1-|\Gamma_s|^2}{|1-S_{11}\Gamma_s|^2}\right)\left(\frac{1-|\Gamma_L|^2}{|1-S_{22}\Gamma_L|^2}\right) \qquad (5.6\text{-}19)$$

This equation is significant. It is the product of three separate factors. The first is the Z_0 transducer gain, $|S_{21}|^2$. The second term concerns only the matching conditions at the input while the third describes the output match.

The other virtue of the unilateral approximation is simplification. Note that most of the calculations required for evaluation of G_{Tu} are scalar in nature. This greatly simplifies calculator programs. Once a computer becomes available though, the unilateral approach to design is rarely used.

The corresponding unilateral value for the maximum matched gain, $G_{Tu\text{-max}}$, is evaluated by noting that a unilateral match is obtained with $\Gamma_s = S_{11}^*$ and $\Gamma_L = S_{22}^*$, yielding

$$G_{Tu\text{-max}} = |S_{21}|^2 \frac{1}{1-|S_{11}|^2} \frac{1}{1-|S_{22}|^2} \qquad (5.6\text{-}20)$$

$$G_{Tu\text{-max}} = G_{x\text{-max}} G_{s\text{-max}} G_{L\text{-max}}$$

Using Eq. 5.6-20, the maximum unilateral gain of the example amplifier is 17.6 dB. $G_{\max}$ of the original amplifier was 18.4 dB. The difference is slight.

A parameter of utility when working with the unilateral approximation is the so-called unilateral figure-of-merit, U

$$U = \frac{S_{11}S_{12}S_{21}S_{22}}{(1-|S_{11}|^2)(1-|S_{22}|^2)} \qquad (5.6\text{-}21)$$

$U = 0.0631$ for the example. The lower the value of U, the better the approximation to the real amplifier with $S_{12} \neq 0$. If U is less than 0.1, the errors in G_T will be under 1 dB (10).

The isolation of parameters in the unilateral gain equations is quite revealing. Using the same example as before, $G_{s\text{-max}} = 2.02$ dB while $G_{L\text{-max}} = 0.28$ dB. This immediately tells the designer that working to match the output port of the amplifier will produce only a slight increase in gain. However, a match at the input will pay larger dividends, up to 2-dB improvement.

The Smith chart may be used to graphically illustrate the effect of matching. This is done by plotting circles of constant gain on the chart. Consider the example with circles representing the gain obtained by matching the input.

The input reflection coefficient of the unilateral amplifier is S_{11}. Hence, the optimun match is S_{11}^*. This is the best that may be done with input match. Other source immittances will produce various degrees of mismatch with corresponding decreases in gain. A given degradation over $G_{Tu\text{-max}}$ is represented by points lying on a circle containing the optimum match, S_{11}^*, within it. This is shown in Fig. 5.23 for the input of the example amplifier. The circles shown represent mismatches producing gain reductions of 0.25 dB, 0.5 dB, 1 dB, 2 dB, and 6 dB.

Sec. 5.6 Scattering Parameters

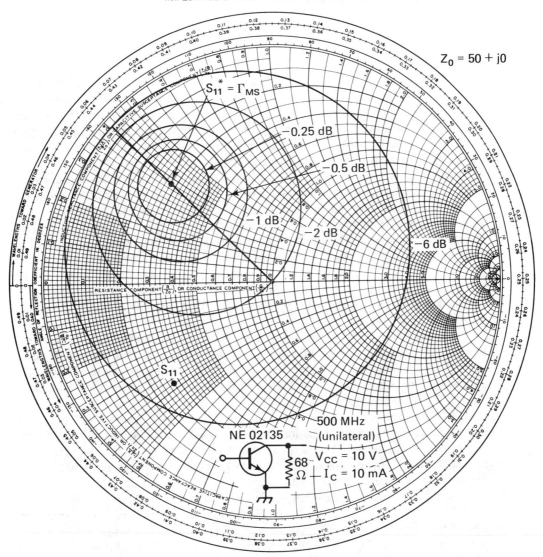

Figure 5.23 Constant gain circles showing the effect of mismatch at the input of a unilateral ce amplifier using the NE02135 at 500 MHz.

If the gain reduction in dB is G_0, a negative number with 0 dB corresponding to the optimum match, an algebraic gain, g_i, is given as $g_i = 10^{(G_0/10)}$. The position of the center of the constant gain circle is given by

$$d_i = \frac{g_i |S_{ii}|}{1 - |S_{ii}|^2 (1 - g_i)} \qquad (5.6\text{-}22)$$

while the radius of the circle is

$$R_i = \frac{(1-g_i)^{1/2}(1-|S_{ii}|^2)}{1-|S_{ii}|^2(1-g_i)} \tag{5.6-23}$$

S_{ii} is the s parameter associated with the port to be matched in both of the above equations. For the example of Fig. 5.23, $S_{ii} = S_{11}$. The centers of the circle lie on the line between the center of the chart and the optimum match point, S_{ii}^*. Similar circles could be drawn to represent the effect of output mismatch. Constant gain circles are especially useful in designing broadband amplifiers where purposeful mismatching is introduced to obtain flat gain with frequency.

The constant gain circles used in the example were applied to a unilateral amplifier. This is not necessary. Similar ones exist for the real ($S_{12} \ne 0$) case. The equations for the center positions and radii are found in the literature (11).

The Smith chart is especially convenient when scattering parameters are used to describe a network. However, it is not necessary to use scattering matrices in order to use the Smith chart. Carson has done considerable work with the chart on amplifiers described by admittance parameters (12). It doesn't really matter which matrix is used, for the devices and the phenomena are the same. Note that the results obtained in the example amplifier are identical with those obtained with the earlier y parameter analysis.

Drawing Smith chart contours to represent various conditions can be very enlightening when doing an amplifier design. Note in Fig. 5.23 that the -6-dB circle is very large, containing a sizeable fraction of the interior area of the Smith chart. So long as the input source immittance is within that circle, the gain will be at least 11.6 dB, assuming the output port is matched.

It is sometimes required to choose a source impedance which is different than that required for optimum gain, for example, to achieve an optimum noise figure. (Noise will be discussed in the next chapter.) Contours of constant noise figure may also be drawn. If superimposed with constant gain circles, the tradeoff between optimum gain and noise becomes readily available.

Stability may also be analyzed with stability contours on the Smith chart. Equations 5.6-8 and 5.6-9 give port reflection coefficients as a function of the S parameters and of the termination at the opposite port. If a reflection coefficient at the output port is set to unity (Eq. 5.6-9), the equation may be solved to provide values of Γ_s which produce this condition. The resulting circle may lie within the Smith chart if the amplifier is potentially unstable. The line through the chart will divide it into a stable and an unstable region. Evaluation of Γ_{out} at a point off the dividing line will reveal which side of the stability contour is stable and which is unstable. Source immittances in the unstable region should be avoided in an amplifier, or sought if an oscillator is being designed. Using these methods, it is possible to work with networks which are less than unconditionally stable while still ensuring system stability. Numerous examples of stability contours are found in the literature (13).

It is often useful to transform to other matrix types when working with S

parameters. The y parameters have the virtue that parallel networks have a composite matrix which is the sum of the individual matrices. A similar situation applies to series networks described by z parameters. The *ABCD* matrix form allows cascaded networks to be evaluated through matrix multiplication. The transformations from scattering parameters to other matrix forms are found in the literature (14).

It is sometimes desirable to renormalize a set of S parameters to a different characteristic impedance. This is done by transforming to, for example, a y matrix and then back to a scattering matrix of the new Z_o. Direct transformation is also possible.

REFERENCES

1. Carson, Ralph S., *High-Frequency Amplifiers,* John Wiley & Sons, New York, 1975.
2. Pease, Marshall C., III, *Methods of Matrix Algebra,* Academic Press, New York, 1965.
3. Shilov, Georgi E., *An Introduction to the Theory of Linear Spaces* (translated from the Russian by Richard A. Silverman), Prentice-Hall, Englewood Cliffs, N.J., 1961.
4. Kurokawa, K., *An Introduction to the Theory of Microwave Circuits,* Academic Press, New York, 1969.
5. See reference 1, p 7.
6. Hejhall, Roy, "RF Small-Signal Design Using Two-Port Parameters," Motorola Application Note AN-215A. Outstanding summary of y-parameter design.
7. Linvill, John G. and Gibbons, James F., *Transistors and Active Circuits,* McGraw-Hill, New York, 1961.
8. Stern, Arthur P., "Stability and Power Gain of Tuned Transistor Amplifiers," *Proc. IRE,* **45**, *3,* pp. 335–343, March 1957.
9. "S-Parameters . . . Circuit Analysis and Design" (a collection of published papers), Hewlett-Packard Application Note 95, September 1968.
10. *Transistor Designers Guide,* Microwave Associates, Burlington, Mass., 1978.
11. See reference 9.
12. See reference 1.
13. See reference 9.
14. Shea, Richard F. (editor), *Amplifier Handbook,* McGraw-Hill, New York, 1966.

SUGGESTED ADDITIONAL READINGS

1. Krauss, Herbert L., Bostian, Charles W., and Raab, Fredrick H., *Solid-State Radio Engineering,* John Wiley & Sons, New York, 1980, Chap. 4.
2. Van Valkenburg, M. E., *Network Analysis,* 3rd ed., Prentice-Hall, Englewood Cliffs, N.J., 1974, Chap. 11.
3. Shuch, H. Paul, "Solid-State Microwave Amplifier Design," *Ham Radio Magazine,* October 1976.

6
Practical Amplifiers and Mixers

Most of the circuits presented in previous chapters have used ideal elements. Some designs have used measured device data. Others have used models. However, the output of the networks have been replicas of the input. If the network was an amplifier, the output was larger than the input with a predictable difference in phase. There were no additional outputs.

This chapter will take us into a more realistic realm. Circuits are described with methods for characterizing the features that make them less than ideal. The factors that lead to noise will be considered. Distortion, covered briefly in Chap. 1, will be discussed again with emphasis on those types of vital interest to the rf designer.

Nonlinear circuit elements will also be discussed. Of special interest in this class are *mixers*, networks with three ports that produce outputs which contain sum and difference frequencies with respect to the inputs.

6.1 NOISE IN AMPLIFIERS

The ideal amplifier is one that has an output that is greater than the input. However, it will have no other outputs. The output power vanishes if the input signal is removed.

Using the assumption of being ideal constitutes a form of modeling, affording simplicity. The real network is not ideal. One output of great concern is noise. Any network, active or passive, will have an output unrelated to the input in the form of noise so long as the network is at some finite absolute temperature.

The simplest device we consider is a resistor. The noise output will be modeled as a generator with a characteristic impedance. The maximum noise power will occur when the load is conjugately matched to the source.

There are several noise components arising from assorted physical phenomena in the resistor. The most common one is thermal noise. Other noise sources are often termed excess noise; they constitute noise components in excess of the thermal term. The available thermal noise power from a resistor is given by

$$P_n = kTB \tag{6.1-1}$$

where P_n is in watts, k is Boltzmann's constant, 1.38×10^{-23} watts per kelvin, and B is the bandwidth in Hz in which the noise appears.

Equation 6.1-1 is basic and we will use it a great deal. A flat noise distribution with frequency is implicit. The total noise power in a given frequency segment will be proportional to the width of the segment in Hz. The noise power available is -174 dBm at "room" temperature, 290 K, in a 1-Hz bandwidth.

The power available from a resistor is a function only of T whether it's a 1-MΩ carbon-type or a piece of copper wire with a very small fraction of an ohm. It may be easier to extract the noise from one than the other.

We know from earlier work that a generator may be represented as either a voltage source with a series impedance or a current source with a parallel admittance. The power available is related to the open circuit voltage and the short circuit current by $P_a = V^2/4R = I^2R/4$. These relationships are used to form a noise model for the resistor, shown in Fig. 6.1. The strength of the sources is

$$V_n = (4kTBR)^{1/2} \tag{6.1-2}$$

or

$$I_n = (4kTB/R)^{1/2} \tag{6.1-3}$$

A typical network will contain series or parallel combinations of resistors. The model in Fig. 6.1 suggests that the voltage sources of a series connection would add. They do, but not algebraically. The device physics must be considered.

Thermal agitation in the conductor will lead to a distribution of energy associated with the free electrons. Because the electrons are free to move throughout the structure, at any one instant there may be a less than even spatial distribution. There will then exist potential differences between various locations, such as the two ends of a

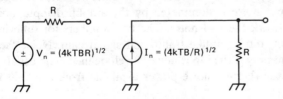

Figure 6.1 A resistor is modeled with either a series noise voltage or a parallel noise current.

resistor. The motion is completely random. The uneven distribution of charge will, itself, establish fields within the conductor that will cause the free charges to move back toward an even distribution.

The electronic motion in one resistor is generally unrelated to that in the next. Hence, there is no phase relationship between the noise voltages in two series connected resistors and the instantaneous voltages do not add algebraically. The available noise powers do add directly, though, leading to the rule that the voltages add as the root of the sum of the squares. Thermal noise voltage calculated in this manner is the same as obtained by applying the fundamental equations with a resistor equal to the total. Noise currents also will add as the root of the sum of the squares.

There will sometimes be a relationship between noise generators. An example would be two different resistors used to model part of a single device such as a transistor. When this occurs, the noises are said to be correlated.

A signal applied to the input of a network will be the sum of two components, a desired signal with information of interest and some noise. There exists a signal-to-noise ratio (snr) at the input, a ratio of the signal power to the noise power. When the composite signal is applied to the network, an output signal-to-noise ratio is defined. Because the network is noisy, the output snr will be degraded over that at the input. The ratio of the two is the noise factor (F), a vital parameter in the characterization of a noisy network. Formally

$$F = \frac{S_i/N_i}{S_o/N_o} = \frac{S_i N_o}{S_o N_i} \quad (6.1\text{-}4)$$

where S_i and S_o are the input and output signal powers and N_i and N_o are the corresponding noise powers.

Equation 6.1-4 may be written in a different form. The terms are power ratios, or gains. Hence

$$F = \frac{G_n}{G_s} \quad (6.1\text{-}5)$$

where G_s is the available power gain for signals while G_n is the so-called noise gain. The signal gain is well defined. Usually, it is the transducer gain when the network is operating with well-defined terminations. The noise gain is not so well defined, as least not from the equations.

If the input noise were small, the output noise would be the amplified input noise plus a component representing the thermal noise of the network. If input noise, however, is large, the output noise is dominated by the amplified input noise with little contribution from network thermal noise. The two numbers for differing input noise levels will be different, totally unlike traditional signal gain. Noise factor is a meaningful parameter if and only if the input noise is well defined. The usual assumption is that input noise is the thermal noise power available from a 290 K resistor.

The *noise factor, F,* is an algebraic ratio. The *noise figure* (NF) is the logarithmic equivalent, $NF = 10 \log F$.

Assume that a 20-dB gain amplifier is to be evaluated for its noise figure. A signal is driven into the amplifier from a proper source impedance, Z_s. The power delivered to a load is measured. G_s is confirmed at 20 dB. A resistor is then attached in place of the input generator. If required, a matching network is built so the impedance seen looking back from the amplifier input is Z_s. The output noise power is measured. If the amplifier was perfect with no internal noise and was impedance matched, the output would be -174 dBm $+ 10 \log B + G_s$, or -154 dBm in a 1-Hz bandwidth. Assume the power measured is actually -149 dBm. There is 5 dB more noise than there would be from a perfect amplifier; hence, the noise figure is 5 dB. The noise factor is 3.16.

Noise figure is a bandwidth invariant parameter so long as the bandwidth is reasonably narrow. A noise figure measured in a narrow bandwidth is termed a spot noise figure. If the bandwidth is doubled, the input noise will double as will that at the output, leaving the ratio in the equations unaltered.

If the bandwidth is very large a more appropriate noise figure is a broadband one, a frequency average of the spot noise figures over the band of measurement. The spot noise figure is the viable measure for a device for most receiver and instrumentation applications.

The noise figure concept has the drawback that it depends upon definition of a standard temperature, usually 290 K. A related concept that does not depend upon such a standardization is noise temperature, illustrated in Fig. 6.2. A "real" amplifier is replaced with an ideal amplifier having no internal noise. Attached to the input of the ideal amplifier is an ideal summing network. The first input to the summing network is usually a signal power and an input noise power. The other is an excess noise source at a temperature T_e. If there were no excess noise, the noise output of the amplifier would be GkT_0B where T_0 is the temperature of the input noise source and G is the power gain.

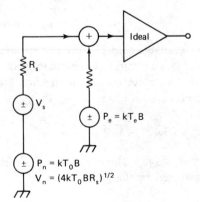

Figure 6.2 Model used to evaluate the noise temperature concept.

A real amplifier does have internally generated noise. Hence, the total output noise power will be $kBG(T_0 + T_e)$ where T_e represents the excess noise. Assume that the input is the usual room temperature resistance. The noise figure is then related to the noise temperature by application of Eq. 6.1-5. Noting that the input noise is kT_0B, the noise factor is given as $F = (1 + T_e/T_0)$.

The utility of noise temperature over noise figure is the independence from a standard. It is also a much more graphic measure of amplifier noise performance in some special applications.

Consider an amplifier with a 1-dB noise figure. This corresponds to $F = 1.26$ and $T_e = 75$ K. The output noise will correspond to a total noise input from a resistor of $290 + 75 = 365$ K if the amplifier is driven from a room temperature source. Efforts to improve noise figure (reduce the noise temperature) will have little effect upon the noise output of the amplifier. It is dominated by the high input noise temperature used during measurement.

Consider the same amplifier in a different environment. The amplifier is attached to an antenna that is pointed at a dark region of outer space. The antenna pattern is assumed to be good, allowing virtually no irradiation from the earth. This antenna will have a characteristic radiation resistance with a temperature equal to the source of energy impinging upon it. This might be 20 K for deep space. The noise output of the amplifier will now correspond to that coming from a 95 K source. Clearly, a few degrees improvement in the noise temperature of the amplifier will be significant in reducing the amplifier noise output. Noise temperature is best used for characterizing amplifiers (or receivers) used for reception of weak signals from quiet sources.

Most systems consist of several cascaded networks. Each is characterized by a gain and a noise factor or equivalent noise temperature. We now evaluate the effective noise factor of a cascade of two stages, shown in Fig. 6.3.

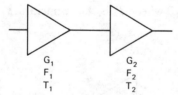

Figure 6.3 A cascade of two stages, each specified by a gain and noise figure or temperature.

The input stage has an available gain G_1 and a noise factor F_1. A related noise temperature is T_1. The second stage is similarly characterized. The input noise power to the combination is kT_0B. The noise output of the first stage is the input noise and the equivalent network noise, both increased by the gain

$$N_{o1} = k(T_0 + T_1)BG_1 \qquad (6.1\text{-}6)$$

The noise output of the second stage is the excess noise of that stage plus the input from the first stage, both increased by the gain

$$N_{o2} = k(T_0 + T_1)BG_1G_2 + kT_2BG_2 \qquad (6.1\text{-}7)$$

Dividing by the net gain, $G_1 G_2$, the equivalent input noise of the combination is $k(T_0 + T_1)B + kT_2B/G_1$. Hence, the excess noise temperature of the combination is that of the first stage plus that of the second diminished by the gain of the first. The equivalent noise factor is calculated using Eq. 6.1-7. Using Eq. 6.1-5

$$F_{net} = G_n/G_s = (1 + T_1/T_0) + T_2/T_0 G_1 \tag{6.1-8}$$

The first term is F_1 and T_2/T_0 is $F_2 - 1$. The final result is

$$F_{net} = F_1 + \frac{F_2 - 1}{G_1} \tag{6.1-9}$$

This result is used extensively in the analysis of low noise systems. The vital factor is the way the noise contribution from the second stage is diminished by the gain of the first. Consider an example where two amplifiers each have a noise figure of 3 dB. The corresponding noise factors are 2. Assume the gain of the first stage is 8 dB, or 6.31. The noise figure of the cascade is then 3.34 dB. The noise of the first stage is dominant in controlling the result. This is generally true if the gain of the first stage exceeds the noise figure of the second.

The noise figure of a multiple system is evaluated through repeated application of Eq. 6.1-9. Alternatively, equations with a large number of terms may be derived.

The previous equations presume that the noise from both amplifiers in a cascade are observed in the same bandwidth. This may not always be a valid assumption. Envision two amplifiers separated by a narrow bandwidth bandpass filter. The net result is then observed in a system with a bandwidth exceeding that of the narrow filter. This situation is typical of many receiver intermediate frequency amplifier systems. The noise contribution from the second stage is commensurate with the measurement bandwidth. The contribution from the first stage is, however, reduced, for the noise spectrum is restricted to a narrower bandwidth. The equations may be modified to account for the difference.

The noise bandwidth will differ from that of the filter. The details are presented later in this section.

The filter itself may alter the results over those predicted in the equations. The filter is a passive network with some insertion loss. As long as the temperature of the passive network is also close to the 290 K standard, it will create noise and must be considered as an element in the net cascade. Any passive network will have a noise factor equal to the reciprocal of the available power gain. For example, a 3-dB pad or attenuator will have a 3-dB noise figure and a gain of -3 dB, allowing evaluation with Eq. 6.1-9.

Noise figure measurement is straightforward, at least in principle. One method, shown in Fig. 6.4, uses the so called "Y factor" method.

The first element is a noise source. There are a number of types. Many modern ones use the noise from a carefully biased zener diode. A second type is a discharge tube filled with a noble gas. The electrodes in the tube are biased to cause a discharge

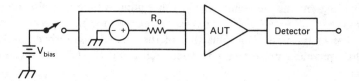

Figure 6.4 System for the measurement of the noise figure of an amplifier or a complete system.

to occur. The noise temperature of such a source is a well-defined function of the tube current.

A third is the so-called "hot-cold" noise source. Two resistors are used, each with the same resistance. One is placed at either room temperature or in boiling water. The other is placed in a Dewar flask filled with liquid nitrogen at a temperature of 77 K. Well-defined available noise powers are available for the input by switching the amplifier under test between cables connected to the two resistors. The hot-cold source is often used as a method for calibrating other noise sources.

Assume that the source used is a zener diode. With no bias, the noise temperature of the source is room temperature, about 290 K. The noise output increases by a known ratio when the diode is biased. A typical ratio is 15 dB or 31.6.

The next element in the system, Fig. 6.4, is the amplifier under test. This is followed by a detector. The detection system must be extremely sensitive, for the noise powers involved are small.

The detector is first turned on and adjusted to the frequency of interest. Power is then applied to the amplifier under test. The detected noise should increase. If it does not, the noise figure of the detector or the gain of the amplifier is insufficient. Additional amplification between the amplifier under test (AUT) and the detector may be needed.

A measurement may progress once the amplifier is biased on. The noise output of the amplifier is noted. The noise source is then turned on. The noise reaching the detector will increase. If the amplifier was perfect, the detected noise would change by an amount equal to the increased noise of the source. The detected noise change would be minimal if the amplifier had a high noise figure.

The noise output from the source is kT_0B with the source off. The source strength will increase to $kT'B$ when bias is applied where T' is the equivalent noise temperature. T' is the sum of T_0 and an excess noise, T_e. The ratio, T_e/T_0, is the excess noise ratio, a parameter usually specified by the manufacturer. This ratio is specified by E.

The output of an amplifier under test will change as the noise source is switched off and on. However, the output change will be less than E unless the amplifier is noisefree. The output power ratio is Y, or the Y factor. Manipulation of the fundamental equations (1) shows

$$F = E/(Y-1) \qquad (6.1\text{-}10)$$

Consider an example where E is 31.6 or 15 dB. Assume that the amplifier output noise increases by 12 dB when the source is biased on. Hence, $Y = 15.85$, or 12 dB. The noise factor is calculated from Eq. 6.1-10 as 2.13, or 3.28 dB.

The switching shown in Fig. 6.4 is either manual as presented, or automatic at a low rate. A typical switching frequency is 400 Hz. Automatic switching allows the Y factor to be displayed on a meter attached to the detector output. The meter is usually calibrated directly in noise figure.

All of the calculations presented have utilized B, the observation bandwidth of the noise. Usually, B cancels from the equations. The bandwidth must be considered if actual noise powers are being calculated through.

B is the noise bandwidth. This is slightly different from the bandwidth of a filter which is the half power (3-dB) width. A typical filter characteristic is presented in Fig. 6.5a. The response is down by 3 dB at a separation from the center of the bandpass equal to half of the bandwidth. As the filter is investigated at further separations, the response continues to drop. The noise will be attenuated at all frequencies except at the center if noise is applied to this filter. The input noise away from the center of the filter will still contribute to the total output.

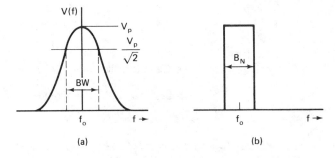

Figure 6.5 Comparison of the response of a typical filter (a) and an ideal one (b) for noise response. The noise bandwidth of (a) is found by integration to evaluate the total noise power. The bandwidth is then that of an ideal filter (b) with the same total noise power transmitted.

The response shown in Fig. 6.5b is an ideal "brick-wall" filter. It has no attenuation or ripple within the passband. The stopband attenuation is infinite. The width of this filter is B for all values of attenuation. The noise bandwidth of this filter is also the filter bandwidth.

The noise bandwidth of an arbitrary filter, such as that in Fig. 6.5a, is that of an ideal brick-wall filter that would allow an identical noise power to pass. Formally, this is given by the integral relationship

$$B_n = \frac{1}{V_p^2} \int_0^\infty V^2(f)\, df \qquad (6.1\text{-}11)$$

where $V(f)$ is the voltage transfer function as a function of frequency and V_p is the peak value of $V(f)$ (2). The integral is evaluated over all frequencies. Often a numerical integration is performed from a careful measurement of filter characteristics.

The usual filters are multiple resonator types with a Butterworth or Chebyshev response in communications equipment. They are close enough to the ideal brick-wall filter that the 6-dB width closely approximates the noise bandwidth. The differences may be calculated (3).

6.2 NOISE MODELS AND NOISE MATCHING

The active devices discussed have been modeled for signal behavior. The models have not contained any information about noise. We can now add these features.

The ideal amplifier has no excess noise. We model the perfect amplifier as one with a real input impedance, R_i, but with no other nonideal features. The input resistance is assumed to be at a finite temperature and thus capable of generating some noise. This amplifier is shown in Fig. 6.6 where it is driven with a real generator, one with both a signal and a noise component. The noise from the amplifier input resistance is modeled with a series noise voltage.

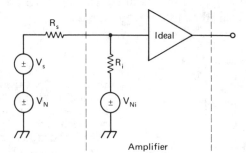

Figure 6.6 Evaluation of the noise figure of an ideal amplifier with a finite input resistance but no other internal noise.

The power available from each resistor, R_s and R_i, is kT_0B. The noise voltages are related to the resistances by Eq. 6.1-2. To minimize the noise figure of the ideal amplifier, the voltage, V_i, from the input resistor must be kept small. This is done by using a source resistor that is small with respect to the input resistance. Most of the available noise power in the input resistor is then shunted to ground by the source resistor. However, the input mismatch is severe. The transducer gain is well below that available from the amplifier.

The opposite extreme would be a match where the source impedance was high with regard to the input. Virtually all of the noise in the input resistor is then present at the input to the ideal amplifier following the input resistor, leading to a poor noise figure. Moreover, the impedance match is again poor, leading to a degradation in gain.

Consider the case where the source is power matched to the amplifier input. This would have been the goal in earlier analysis. The two noise voltages are equal. Each is matched to the other resistor, leading to a voltage at the amplifier input terminal from each source of $V = \frac{1}{2}\sqrt{4kTBR}$. The power from each is then that available from the respective sources, leading to a total noise power at the input of $2kTB$. The input noise power is kTB. The "noise gain" is 2. The amplifier gain vanishes from the calculations, leaving the conclusion that the noise figure of a perfect amplifier with a finite input resistance is 3 dB when the input is matched.

There are fallacies in this analysis which will be treated subsequently. It does, however, illustrate some vital factors in low noise amplifier design. First, noise figure is critically dependent upon source impedance. Moreover, the impedance required is rarely related to that required for a conjugate match. The optimum gain does not coincide with the best noise figure. This justifies the instrument described in the previous section where noise figure was displayed on a meter. Such an instrument allows the source "matching" to be adjusted for optimum noise figure, perhaps at the expense of gain.

The analysis used a noise voltage generator in series with the input resistor of the amplifier to model the noise. A parallel current generator could also have been used. The conclusions would have been identical for the case of a matched source. Which representation is the best? The answer is that both must be used.

Figure 6.7 shows a more general noise modeling of an amplifier. The amplifier is characterized as an ideal one with some input resistance. However, the noise is modeled with a noise current source in parallel with the input terminals and a noise voltage source in series with the input port. The optimum source impedance for producing the lowest possible noise figure is E_n/I_n where the parameters are the strengths of the noise generators. The magnitudes of both generators may be determined experimentally by measuring the lowest attainable noise figure and the source resistance required to achieve it.

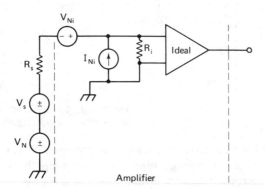

Figure 6.7 A "real" amplifier is modeled as an ideal one with noise sources added at the input.

The two generators are shown in Fig. 6.7 as being separate and isolated. However, they may well be correlated noise sources. The correlation must be taken into account if careful modeling is to be done.

The optimum source impedance for noise figure is determined using these methods. This impedance will be complex rather than the resistive one shown in the figure in the more general case. The impedance may be located on a Smith chart. Then, circles drawn about this point, although not with the point as a center, will be contours of constant noise figure. Such displays are useful in evaluating the precise effect of source match (or mismatch) on noise figure (4). Circles of constant gain were described in the previous chapter. When the two are drawn on the same chart, the tradeoff between stage gain and noise figure is graphically illustrated.

The conclusion was reached that an amplifier that is impedance matched at the input, no matter how noisefree it may be, cannot deliver a noise figure under 3 dB. This is not strictly true. It is a reasonably accurate statement when the simplest of matching methods are used. However, if feedback is used, the input impedance may be changed while not altering the impedance seen between the "input" terminals of the amplifier, the factor determining noise performance.

Figure 6.8 shows a familiar circuit, the common emitter amplifier with feedback in the form of combined emitter degeneration and parallel feedback. Shown with the amplifier are two additional noise sources. These represent the noise generated in the feedback elements. The presence of this noise can degrade the performance of the amplifier. However, it need not be severe. The transistor noise characteristics are not generally modified by feedback.

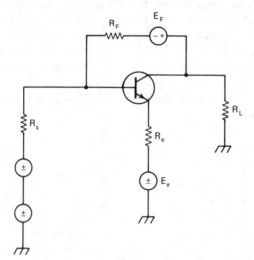

Figure 6.8 Evaluation of the effects of feedback on noise figure.

The voltage associated with a given resistance was presented in Eq. 6.1-2. This is an open circuit value. The actual voltage appearing at an amplifier input from a resistor will depend upon the resistance seen at that input. This is essentially a matter of impedance matching. For the resistive feedback amplifier of Fig. 6.8, a typical input resistance seen looking into the base might be 50 Ω. Most of the available noise voltage is decreased at the input by voltage divider action if the feedback resistor from the collector is large with regard to 50 Ω. The noise figure degradation is then not usually severe.

Some special connections will further aid the maintenance of noise figure. For example, the feedback resistor, R_F, in Fig. 6.8 is directly from the collector. The feedback action depends upon developing a current in the resistor that is proportional to the output voltage. The same current that comes from the collector connection could just as well come from a transformer connected to the output. The resistor could be twice as large while still delivering the same current to the base if the transformer had a 2:1 turns ratio. However, the impedance match between the base and the feedback resistor would then be even worse, yielding a reduced noise current being introduced to the base.

A transistor biased to a few milliamperes will have an "input" resistance of just a few ohms when looking into the emitter. A comparable value external emitter resistor will be well matched to that input, and can severely degrade the noise figure. Again, transformer action at the output can be used to reduce the noise problem while still maintaining the same essential feedback properties. The emitter resistor serves the role of sampling the output current and developing a proportional voltage to be summed with the input. A small value resistor in the emitter, much smaller than would normally be used, could be substituted. Then this is augmented with a large resistor leading from the emitter to a transformer that has a voltage proportional to that at the collector. The feedback voltage produced at the emitter is the same. However, the noise voltage introduced may be smaller.

Reactive feedback is often used to great advantage in building low noise amplifiers with a good input impedance match. One form is through the application of an emitter inductor. This may be nothing more than the lead length of the transistor, or a small section of microstrip transmission line, at microwave and uhf. An example is shown in Fig. 6.9. The inductor in the emitter will have some resistance. The noise power available from that resistor is still kTB. However, the resistance is so small that the noise voltage is also vanishingly small.

Assume that the transistor is biased to a current of 5 mA and that it is described by a simple hybrid-pi model. Assume further that the operating frequency is 500 MHz, the transistor has a 5-GHz F_t, and a low frequency beta of 50. A unilateral analysis is used. Under these conditions, the input impedance will be $(\beta + 1)Z_e$. Z_e is the series combination of r_e, the internal emitter resistor $(26/5 = 5.2\ \Omega)$, and the inductive reactance, X_L. β will be complex, $\beta = \beta_0/(1 + j\beta_0 F/F_t)$. For the frequencies assumed, the quantity $(\beta + 1) = 2.923 - j9.62$. For convenience, this is defined as $A - jB$. Then, the input impedance will be $(A - jB)(r_e + jX_L) = (Ar_e + BX_L) + j(AX_L - Br_e)$. X_L may be chosen to produce an input impedance

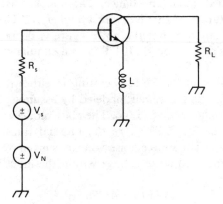

Figure 6.9 Evaluation of the effect of emitter inductance in changing the input resistance of an amplifier while contributing no additional noise. This is but one mechanism that is used to build low noise amplifiers (NF less than 3 dB) while maintaining a good input match.

with a 50-Ω real part. The value is $X_L = 3.62$ Ω, a value resulting from an inductance of only 1.2 nH. With this emitter inductance, the input impedance is still capacitive with a reactance value of 39.4 Ω. A series inductance will produce a perfect input match.

The quality of the input match may be very important in low noise amplifiers. There is usually a preceding bandpass filter in low noise receivers. The filter must be properly terminated to provide the desired frequency response.

The example presented uses the simplest possible model with the needed salient features, the hybrid-pi. The model would be replaced in practice with measured S parameters or those provided by the manufacturer. It is, however, both surprising and rewarding to find that the inductor predicted is not all that different in most applications. A complete design would, of course, include an analysis of the amplifier's stability, not only at the operating frequency but at all frequencies where the device is capable of gain. The operating gain of the amplifier is also evaluated.

We considered noise from feedback elements. The same care should be given to the resistors that might be used to bias the transistor. Some amplifiers like that shown in Fig. 6.8, where a shunt feedback resistor is used, also employ a resistor from the base to ground. This resistor, in combination with that from the collector, serves to properly bias the stage while also providing the required broadband feedback. Unfortunately, this resistor may frequently have a value that is close to the input impedance seen looking into the base. The impedance match is so "good" that the noise figure is measurably degraded by the bias element. An inductor in series with the resistor will destroy this undesired "noise power match" while leaving the biasing unaltered.

Many types of feedback are used with low noise amplifiers with a goal of realizing low noise, controlled and stable gain and well defined terminal immittances. The concepts will now be illustrated with a FET amplifier.

The common gate field effect transistor has an input resistance equaling the reciprocal of the common sources transconductance. For a typical small signal FET, the g_m value might be around 5000 μS, leading to $R_{in} = 200$ Ω. The FET will deliver reasonably good noise figure in the common gate mode with a well-defined, although usually mismatched, input impedance. The match may be improved with a transformer or by substitution of an FET with a higher g_m. The match to a 50-Ω line is excellent when $g_{ms} = 0.025$, but the noise figure is degraded over that seen with a mismatch. As an aside, a large g_m JFET is well simulated experimentally by paralleling a number of small FETs.

The noise characteristics of the FET, operating in either the common gate or the common source configuration, are well modeled by assuming a resistor in series with the source lead with a value equal to $1/g_m$. The noise model applied to a common source amplifier is shown in Fig. 6.10. R_t is the internal transistor noise resistor leading to a noise voltage V_{tn}. The source noise voltage is given by the earlier expression. Using rms addition, the total noise voltage within the input loop is

$$V_{ng} = (4kTBR_t + 4kTBR_s)^{1/2} \qquad (6.2\text{-}1)$$

Sec. 6.2 Noise Models and Noise Matching

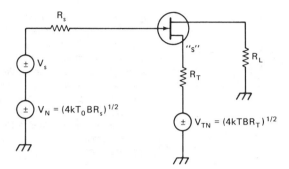

Figure 6.10 A noise model for an FET.

This is the net noise voltage appearing at the gate with respect to the "internal source." However, so far as the external world is concerned, this voltage appears to originate from a source impedance of R_s. The noise power available is then

$$N_t = kTB(R_t + R_s)/R_s \qquad (6.2\text{-}2)$$

Dividing by the input available noise power, kTB, the noise factor is

$$F = 1 + R_t/R_s \qquad (6.2\text{-}3)$$

This expression is, of course, a simplification resulting from application of the simplest possible model having the desired salient characteristics. Still, it does a good job of describing the noise figure of either a JFET or MOSFET in the hf region. The noise figure becomes closer to that of an ideal amplifier as the source impedance driving the gate increases. The equation suggests that there is no limit. This is a practical conclusion to the extent that practical details usually define the limit. The usual way of performing the input noise-match transformation is through an LC network, often an L- or pi-type. The network Q_L increases as the impedance transformation increases. Noting that the insertion loss of a single resonator is given as $(1 - Q_L/Q_u)^2$, the insertion loss will climb to a part of 1 dB with typical unloaded Q values of 250 when the impedance is a few thousand ohms. This insertion loss will subtract directly from the noise figure obtainable and is often the dominant limitation. Noise figures close to 1 dB are not unusual in the hf spectrum with a number of dual-gate MOSFETs such as the 3N140 or RCA type 40673.

The common source FET, when driven with an L-network at the input, will present a very poor match. Generally, matched amplifiers are preferred. Figure 6.11 shows a scheme that will allow the input to be matched while still preserving a low noise figure. Note that this is neither a common source nor common gate configuration. Neither terminal is grounded.

The circuit used for analysis is shown in Fig. 6.12. The input is driven from a current source. The FET is replaced with the simplest available model, a current generator controlled by the gate-to-source voltage, V_{gs}. The gate is assumed to be an open circuit. The transformer is assumed perfect, being described by the relationship

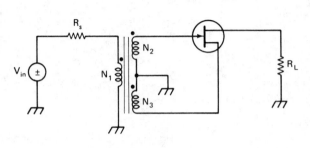

Figure 6.11 An FET amplifier which is neither common source nor common gate. The coupling between the input winding (N_1 turns) to that driving the source (N_3) provides the desired input resistance. The winding driving the gate increases the voltage between gate and source. This provides a high driving source resistance, allowing for low noise while maintaining a good impedance match.

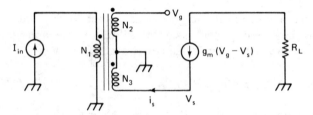

Figure 6.12 The small-signal model used to evaluate input resistance. The input is driven from a current source.

$$\sum_{1}^{3} N_i i_i = 0 \tag{6.2-4}$$

The input current is related to that at the FET source by

$$i_{\text{in}} = i_s \frac{N_3}{N_1} \tag{6.2-5}$$

Using the voltage relationships of an ideal transformer

$$V_g = \frac{-V_s N_2}{N_3} \tag{6.2-6}$$

Combining the equations

$$i_s = -g_m V_s (1 + N_2/N_3) \tag{6.2-7}$$

The equations are combined to calculate the input resistance, the ratio of the input voltage to the input current

$$R_{\text{in}} = \frac{N_1^2}{N_3^2} \frac{1}{g_m(1 + N_2/N_3)} \tag{6.2-8}$$

If the amplifier is driven from a source with an impedance R_0, the transducer gain may be evaluated

$$G_t = 4g_m^2(1 + N_2/N_3)^2 \frac{N_3^2}{N_1^2}\left(\frac{R_{in}}{R_{in} + R_0}\right)^2 R_L R_0 \quad (6.2\text{-}9)$$

To estimate the noise performance of the amplifier, we must determine the impedance "seen" by the gate-source port of the FET. The external load from the driving generator is transformed by the transformer to an equivalent R_0' between the gate and the source, $R_0' = R_0(N_2 + N_3)^2/N_1^2$. The gate-to-source port is now driven by a current source as shown in Fig. 6.13. Calculation of the voltage at the generator will allow evaluation of the resistance seen at the port.

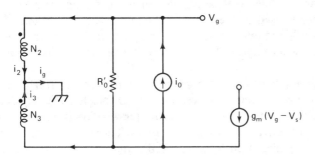

Figure 6.13 The small-signal model used to evaluate the driving source resistance presented to the FET gate-source port needed to estimate noise figure. A finite source impedance is simulated by R_0'.

Nodal equations are written with currents having directions shown by the arrows in Fig. 6.13. Note that the current in the ground leg of the transformer, i_g, is not zero. The result of this calculation is Z_{gs}, the resistance seen by the gate-to-source port of the FET

$$Z_{gs} = \frac{1 + N_2/N_3}{g_m + \dfrac{1}{R_0'}(1 + N_2/N_3)} \quad (6.2\text{-}10)$$

The equations collapse to the traditional common gate amplifier if $N_2 = 0$ and $N_1 = N_3$. The results are much different if N_2 is made large.

If a FET with $g_m = 0.004$ S is used with a 50-Ω source resistance and a 450-Ω load, a matched amplifier is realized with the proper transformer. The input and source windings are set at 4 turns each. Using Eq. 6.2-8, the input resistance is found to be 50 Ω if $N_2 = 16$ turns. The transducer gain is then 9.5 dB while the resistance seen by the gate-to-source port is 625 Ω. Using Eq. 6.2-3, a noise figure of 1.5 dB is predicted. This amplifier will have an excellent input impedance match with a good noise figure. The output match will be poor.

It is useful to ignore the equations and look at the circuit intuitively. The FET is a voltage driven device with a very high input impedance. Hence, any common source amplifier will have an infinite input vswr. A common gate amplifier can be matched through transformer action. The current flowing will still be determined by the difference between the gate and the source potentials. With transformer input coupling, the source current is transformed so that a match is seen. However, the large number of turns of the gate winding will present an increased gate voltage and an increased impedance seen from the gate-to-source port.

Amplifiers like that described for the FET are also realized with a bipolar transistor. The FET model is so simple that we have used that device for analysis. Qualitatively, the same things happen with a bipolar transistor. Often, with either FETs or bipolar transistors, the input transformer is tuned. The tap in the secondary, that defining N_2 and N_3, can be simulated with series capacitors which can be adjusted.

Other amplifiers can be modified with transformer feedback, providing both low noise and good impedance matches at both ports. One common one is a modification of the familiar bipolar amplifier with a parallel feedback resistor and emitter degeneration. A transformer is used in the output. The input lead of the amplifier, that leading to base, is routed through the transformer in a 1- or 2-turn winding of the proper phase. The result is an improved match as well as lower than usual noise figure.

Figure 6.14 shows an amplifier that uses a bipolar transistor and transformer feedback exclusively. This amplifier was first described by D. Norton (5). Equation 6.2-4 describes the essentially ideal transformer. The emitter winding is a single turn while the other two windings have m and n turns, respectively. The transistor, operating as a common base amplifier, is biased on to several milliamperes, producing an impedance seen looking into the emitter which is vanishingly small.

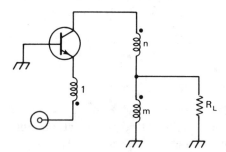

Figure 6.14 A transformer feedback amplifier providing low noise, excellent impedance matching, and low distortions.

The amplifier is analyzed easily from an intuitive viewpoint. Assume that drive is from a current source. Virtually all input emitter current appears at the collector. The m and n turn windings form an autotransformer so that the collector load is larger than the output termination, R_L. Some of the voltage appearing across the $m + n$ collector winding occurs across the single turn emitter winding. By phasing the emitter winding properly, the voltage developed is just the right amount that, when evaluated with the input current, a desired input impedance is produced. No extra noise is introduced because all of the feedback is from a transformer.

The analysis is easily formalized. The results are that the input impedance is identical to the output termination if

$$n = m^2 - m - 1 \tag{6.2-11}$$

and the associated insertion transducer gain is

$$G_T = m^2 \qquad (6.2\text{-}12)$$

The output is also matched. If the impedance driving the amplifier is R_0, the impedance seen looking into the output is also R_0.

Using the equations with $m = 3$, $n = 5$ for a single-turn emitter winding, this amplifier will have a 9.5-dB gain. In models built by the author, noise figures as low as 1.5 dB have been obtained simultaneously with input and output return losses in excess of 25 dB over the frequency range of 5 to 100 MHz.

This amplifier is an excellent one, but it is still not ideal. The problem is a lack of input-to-output isolation. Any change in one termination at either end will severely change the match at the other port. The "ideal" amplifier would have all of the characteristics of this one plus $S_{12} = 0$.

It is possible to further extend the concept of lossless feedback. Some workers (6, 7) have built amplifiers using directional couplers like those described at the end of Chap. 4. Some of the output power is sampled and applied to the input as feedback. It is done so that only the forward power, that going from the amplifier to the output termination, is sampled and used as feedback. Excellent isolation is produced with these amplifiers, with little compromise in noise figure or input match.

Low noise design is an interesting and important aspect of rf work. Considerable work has been published on the subject, much of it of an empirical nature. Although not aimed at rf design, an excellent treatment of the subject is given in the text by Motchenbacher and Fitchen (8).

6.3 DISTORTION IN AMPLIFIERS AND THE INTERCEPT CONCEPT

The previous sections presented methods for treating noise in amplifiers. Noise is the factor that limits the smallest signals that may be processed with a network. This section looks at the other extreme, the factors which limit the maximum signals that may be processed.

Distortion was treated using simple scalar transistor models in Chap. 1. The predominant interest was to find the limitations for small signal modeling. The work with the simplified Ebers-Moll model showed some of the distortion characteristics of the bipolar transistor. Shifts in bias with high drive and the modified time domain shape of current waveforms all resulted from that analysis. The work presented in this section is more generalized; it is applicable to virtually any network that displays a nonlinear behavior. Also, the characterizations are in the frequency domain. That is, the analysis will show not only the magnitude of distortions, but the frequencies which appear from that distortion.

Any network is characterized by a transfer function, the output voltage (current) as a function of an input voltage (current). The assumption was made in Chap. 5 that the networks were linear. An output was linearly related to the inputs. This is equivalent to assuming that only small signals are present. The assumption of linearity is not needed in a more general sense. If voltages are chosen as the variables, a network may be characterized as a power series

$$V_{\text{out}} = K_0 + K_1 V_{\text{in}} + K_2 V_{\text{in}}^2 + K_3 V_{\text{in}}^3 + \cdots$$
$$= \sum_{0}^{\infty} K_n V_{\text{in}}^n \qquad (6.3\text{-}1)$$

This is a Taylor series expansion centered about the zero input signal condition. If a large-signal model exists, such as the simplified Eber-Moll model presented in Chap. 1, the series may be evaluated analytically. Even if the model is not available for mathematical evaluation, Eq. 6.3-1 may still be used as a tool for analysis.

Two types of input signal will be considered in this analysis. The first is a single frequency input or "tone," $V_{\text{in}} = E \sin \omega t$. The other is a pair of unrelated inputs added to form a "two-tone" input, $V_{\text{in}} = E_1 \sin \omega_1 t + E_2 \sin \omega_2 t$. A practical amplifier may, of course, be subjected to much more complicated inputs. Most of the salient features of the distortions or departures from linearity are suitably characterized with the single-tone or two-tone inputs.

The first term in the series of Eq. 6.3-1, K_0, is a constant, unrelated to the input signal. This term always arises from a formal Taylor expansion and has the significance of specifying bias conditions.

The next term is the linear one, $K_1 V_{\text{in}}$. This term will be the dominant one in most of the networks of interest. It is the basis of all of the two-port network analysis performed earlier. Considering only the linear term, the output is merely a replica of the input. For a two-tone input, $V_{\text{out}} = K_1 E_1 \sin \omega_1 t + K_1 E_2 \sin \omega_2 t$.

Next in the series is the quadratic or second order term. This is the dominant term in the simple FET model presented in Chap. 1. When the second order term is considered alone with a single tone input, the output voltage will be

$$V_{\text{out}} = K_2 (E \sin \omega t)^2 \qquad (6.3\text{-}2)$$

With the application of trigonometric identities, this becomes

$$V_{\text{out}} = \frac{K_2 E^2}{2} (1 - \cos 2\omega t) \qquad (6.3\text{-}3)$$

Two terms arise. The first one is proportional to the amplitude of the input but contains no time dependence. This is an offset in bias. The same result was predicted for a bipolar transistor in the time domain analysis presented in Chap. 1. Here, however, we see that the bias shift is a general characteristic of a second order curvature to the transfer function of the amplifier and is not restricted to a specific model. This distortion is used in practice for the detection of signals. If a network with a suitably large value for K_2 is driven with an input signal, a low pass filter at the output will allow the dc shift to be observed.

The second term in Eq. 6.3-3 is at twice the input frequency. This is termed *second harmonic distortion* and is a direct result of second order curvature in the transfer function. This distortion is used for frequency doublers.

If a two-tone input is applied to the network, considering still only the second order term, the output is

$$V_{out} = K_2(E_1 \sin \omega_1 t + E_2 \sin \omega_2 t)^2$$
$$= K_2(E_1^2 \sin^2 \omega_1 t + E_2^2 \sin^2 \omega_2 t + 2E_1 E_2 \sin \omega_1 t \sin \omega_2 t) \quad (6.3\text{-}4)$$

The first two terms are replicas of those found with single-tone excitation. Each will lead to a bias shift and to the presence of the second harmonic of each input tone in the output waveform.

The third term in Eq. 6.3-4 is the product of two sine waves at the two input frequencies. This is the basis of circuits termed multipliers. Focusing only on this term and applying a trigonometric identity

$$V_{out} = K_2 E_1 E_2 [\cos(\omega_1 - \omega_2)t - \cos(\omega_1 + \omega_2)t] \quad (6.3\text{-}5)$$

The outputs are at frequencies that are the sum and difference of the inputs. A filter may be used at the output to select, for example, the difference. This frequency may then be further amplified and, if needed, detected or processed. This is the basis of the superheterodyne concept. The circuit which produces the output at the difference (or sum) frequency is called a multiplier, mixer, or in some specific applications, a product detector.

The output of a mixer is usually a desired response. Hence, the mixer operation is not termed as a distortion. Still, it is the result of nonlinear characteristics.

The presence of second order distortion in an amplifier is often a problem of little significance. These distortion products occur at frequencies well removed from the desired output. Hence, they may be removed from the output by filtering. Second order products are called intermodulation distortion (imd) when they arise from the interaction of a two-tone input. Second order imd is of greatest significance in broad band amplifiers covering more than one octave of frequency span.

The third order coefficient in Eq. 6.3-1, K_3, is the one that is often the greatest culprit in causing unwanted distortion products. If the third order term is analyzed with a single tone input, the resulting output is

$$V_{out} = K_3 E^3 \sin^3 \omega t \quad (6.3\text{-}6)$$

Application of trigonometric identities leads to

$$V_{out} = \frac{K_3 E^3}{4} (3 \sin \omega t - \sin 3\omega t) \quad (6.3\text{-}7)$$

The output contains a term at the fundamental drive frequency. It is not linear though, for it is proportional to the third power of the driving amplitude, E. The other term is a third harmonic, a sine wave at three times the frequency of the input. This is the basis of a frequency tripler, a circuit used in transmitters or in local oscillator chains in receivers or measurement instruments.

The application of a two-tone input while considering only the third order term produces the output

$$V_{out} = E_3(E_1^3 \sin^3 \omega_1 t + E_2^3 \sin^3 \omega_2 t \\ + 3E_1^2 E_2 \sin^2 \omega_1 t \sin \omega_2 t + 3 E_1 E_2^2 \sin \omega_1 t \sin^2 \omega_2 t) \qquad (6.3\text{-}8)$$

Four terms are found. The first two are the results of the two input tones acting independently and reduce to the frequency components described by Eq. 6.3-7. The last two terms are further altered with trigonometric identities. Consider only the third term as V'_{out}

$$V'_{out} = \frac{3E_1^2 E_2 K_3}{2} \{\sin \omega_2 t - \tfrac{1}{2}[\sin(2\omega_1 + \omega_2)t - \sin(2\omega_1 - \omega_2)t]\} \qquad (6.3\text{-}9)$$

There are two output forms, both of interest. The first part of V'_{out} is a sine wave at ω_2. However, it is a distortion product and often an annoying one, for it has an amplitude that is proportional to both of the driving input tones. If the first input tone has an amplitude which is a slowly varying function of time as would be found with amplitude modulation, that modulation will appear on the output at ω_2. This is termed *cross-modulation*.

The second term in V'_{out} is intermodulation distortion, in this case, third order imd. The frequencies appearing are related to the inputs as $f_{out} = 2f_1 - f_2$ or $2f_1 + f_2$. If the input tones are close to each other in frequency, the sum term is close to the third harmonic and is no more of a problem than would be a harmonic distortion, for it may be filtered from the system. However, the difference term is quite the opposite. The output frequency is very close to that of the input tones. It is not removed easily by filtering. V'_{out} was the result of manipulation upon the third term in Eq. 6.3-8. Closely spaced products were found at $2f_1 - f_2$. If the fourth term of Eq. 6.3-8 is considered, output terms will arise at $2f_2 + f_1$ and $2f_2 - f_1$. The second of these is, again, a closely spaced one.

Consider as an example an amplifier with input signals at 100 and 101 MHz. There will be outputs at the harmonics, 200, 300, 202, and 303 MHz. Addition-type imd products will occur at 201 and 1 MHz from second order curvature and at 301 and 302 MHz from third order imd. None of these is severe in a narrow band application. A filter following the amplifier will remove all of these products. However, the "close-in" third order imd products cannot be filtered from the system. They will occur at frequencies of 99 and 102 MHz. A sharply tuned filter following the amplifier would remove these products. However, assume that a desired signal was at 102 MHz. Undesired signals at 100 and 101 would also produce a signal at this frequency. No amount of filtering *after* the amplifier would remove these products—any filtering must be ahead of the offending amplifier if it is to reduce the third order imd problems.

Assume that the two input tones during a two-tone measurement are of equal amplitude, $E_1 = E_2$. We see from Eq. 6.3-9 that the output amplitude of the distortion products is then proportional to the third power of the input signals. Hence, if the two input signals are reduced by 1 dB, the distortion products will drop by 3 dB. This strongly nonlinear behavior is of great importance. By proper distribution of gain in a system such that the desired signals do not become too strong prior to filtering, and by careful amplifier design to keep K_3 small with respect to K_1, third order imd products may be maintained at a suitably low and predictable level.

The analysis may be continued. The higher order terms will create harmonics of like order with single tone input signals. Other components will also arise. For example, a single tone input with the fourth order term of Eq. 6.3-1 will produce not only a fourth harmonic but a second order output term and a zero order term, additional dc bias shift. Two-tone inputs will create distortion products beyond the usual harmonic distortions. For example, the fifth order term will generate undesired close-in imd signals at $3f_1 - 2f_2$ and $3f_2 - 2f_1$.

It is sometimes useful to consider more complicated input signals, containing three or more input tones. The analysis is like that presented, although it becomes predictably messy (9).

It is seen from all of the equations that harmonic distortion is related to imd. If the amplitudes of the harmonics are measured, the K_n values may be inferred. They may then be used to predict imd levels. This is rarely done though. The analysis is complicated by the fact that a given order of distortion will contribute not only to that order of harmonic, but to that two orders lower. That is, third order curvature leads to both third harmonic output and to a fundamental component.

Because most of the various distortion products occur at different and predictable frequencies, they may be evaluated experimentally. The usual test setup is shown in Fig. 6.15. Two signal generators are used for two-tone testing. The outputs are added in a hybrid combiner, discussed in Chap. 4 as a return loss bridge. The combiner is

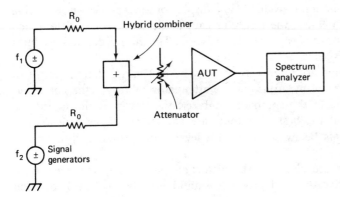

Figure 6.15 The system used to measure the intermodulation distortion of an amplifier or other two-port network.

extremely important. It provides isolation between the signal generators. If this isolation is not present, one signal generator may be modulated by the signal from the other, creating products at the same frequencies as predicted from amplifier distortion.

The hybrid combiner output is applied to a step attenuator. This serves two purposes. First, it ensures that the combiner output is terminated in something close to the proper impedance. An improper termination will degrade isolation. Secondly, the attenuator provides a convenient means of varying the strength of both tones simultaneously. Equal input tones are used in most experiments.

The attenuator is followed by the amplifier under test (A.U.T.). This could be virtually any network that might display nonlinear characteristics. The same general methods are used to evaluate filters, mixers, and electronically controlled attenuators.

The final element in the test set is a spectrum analyzer. The spectrum analyzer is a frequency selective instrument that allows the evaluation of outputs at a number of frequencies. The instrument is calibrated, allowing the output powers at each frequency of interest to be measured. Also, the spectrum analyzer is swept, allowing an entire part of the amplifier output spectrum to be displayed in one presentation. The usual output form of an analyzer is as a display on the face of a cathode ray tube. Most spectrum analyzers have a logarithmic output. Hence, the powers available are presented in dBm or other convenient logarithmic units.

Figure 6.16a shows the output spectrum of an amplifier during a two-tone test to evaluate third order imd. The input frequencies are 99 and 101 MHz. The output consists of two predominant tones at these frequencies plus the third order imd products at 97 and 103 MHz. The baseline, the output at frequencies other than the inputs and the imd products, is the result of noise from the amplifier and the spectrum analyzer. This noise level may be decreased by reducing the bandwidth of the spectrum analyzer.

The display of Fig. 6.16a was obtained with a Tektronix 492P Programmable Spectrum Analyzer. It was controlled by a Tektronix 4052 Computer and the output display shown was plotted on a Tektronix 4662 X-Y Plotter. The signal generators were both Hewlett-Packard HP608's and the hybrid combiner was home-built. The amplifier being evaluated was a cascade of two Motorola HWA-120 broadband hybrid integrated circuits.

A great deal of information is available in a display of the type shown. Knowing the amplitude of the input tones, the insertion gain of the amplifier is determined immediately. The output power of each tone is +10 dBm. The third order imd products are at −25 dBm, or 35 dB below the output tones.

Additional experiments are often performed. The first varies the amplitude of the input signals. This is done by adjusting the attenuator following the combiner. Usually, the reference level of the spectrum analyzer, the power at the top at the display, is adjusted so that the desired outputs are at the top of the screen. It is advisable to keep all signals below the reference level. This prevents distortion in the analyzer.

As the input signals are changed, the distortion products should vary by a predictable amount. Third order imd products should increase by 3 dB for each

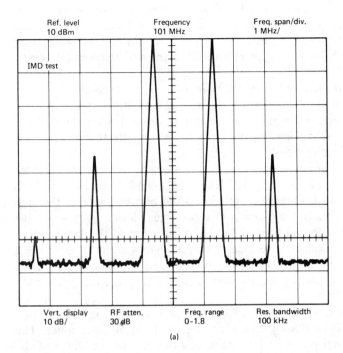

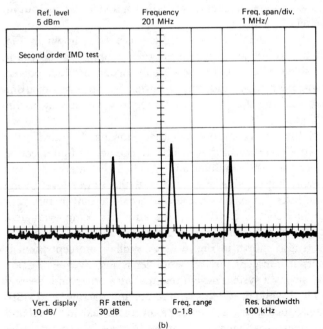

Figure 6.16 Intermodulation distortion measurements on a two-stage wideband amplifier. The third order imd is shown in (a). A fifth order imd product is also visable in the display. Second order imd is shown in (b). The center peak is the intermodulation distortion product. The other responses are the second harmonics of the input tones. This data was obtained with a Tektronix 492P Programmable Spectrum Analyzer. Control was provided by a Tektronix 4052 computer.

increase of 1 dB in the two reference tones. The difference between the desired outputs and the distortion products will then change at a 2 dB per 1-dB input level rate change. If the distortion products change at the predicted rate, the network being tested is said to be "well-behaved." Typical amplifiers are well-behaved. However, many mixers are not.

Another characteristic that may be evaluated is gain compression. This may be performed with a single input tone. The gain is measured at a number of input levels. As the input is increased, a point is eventually reached where the output does not increase in proportion. Usually, the output power where the gain is 1 dB less than that at low levels is specified.

Another measurement of interest is evaluation of second order intermodulation. The same input tones are used, 99 and 101 MHz. The second order imd products will appear at 2 and 200 MHz. Figure 6.16b shows the plot obtained at the sum frequency. Three output signals are shown. The two equal levels away from the center of the display are the second harmonics of the input tones. The larger signal near the center is the imd product. The desired output signals are at a level of +5 dBm during this measurement. The second order imd responses will drop in proportion to the square of the input signals. Hence, the imd ratio will drop in direct proportion to the input levels.

Figure 6.17 shows a plot of output amplitudes as a function of drive signal strength. This plot is obtained during a two-tone test with equal input tones. The main curve shows the output power of each tone as a function of the power of the individual input tones. Gain compression occurs with this amplifier at an output power per tone of approximately +11 dBm.

The output powers of the third order imd products are also shown in Fig. 6.17. The products are plotted only for levels where no gain compression occurs in the amplifier. Note that the slope of the curve showing distortion product powers is 3. This is a logarithmic plot. Both input and output powers are in terms of dBm. Hence, a slope of 3 indicates that the third order imd products are varying as the third power of the input voltages.

The two curves, that of the desired output and that showing the third order distortion products, may be extrapolated until they intersect. The point of intersection is a very valuable figure-of-merit for the amplifier and is called the intercept point. Note that the amplifier is usually not capable of operating at this power level owing to gain compression. Still, the intercept is a valuable number. The output intercept is +28 dBm while the input intercept is 0 dBm for the amplifier being evaluated. The input and output intercepts differ by the gain of the network.

Second order imd data is also shown in Fig. 6.17. A similar intercept point is defined. The second order output intercept for this amplifier is just over +40 dBm.

Most distortion analysis during a system design regards third order imd. Second order distortion is neglected. This is valid in systems with reasonably narrow bandwidth filters following the distortion producing stages. It can be a severe mistake in broadband systems though. It is common in broadband applications to have a second order distortion that exceeds the third order component. This is evident in the intercept

Sec. 6.3 Distortion in Amplifiers and the Intercept Concept

curves for the example amplifier (Fig. 6.17) at input power levels below −20 dBm.

The utility of the intercept concept is best seen with an example. Assume that an amplifier is available with a third order output intercept of +20 dBm. Analysis with a spectrum analyzer shows 1-dB gain compression at an output of +5 dBm; the circuit is otherwise well-behaved. Let the amplifier be operated with output tones

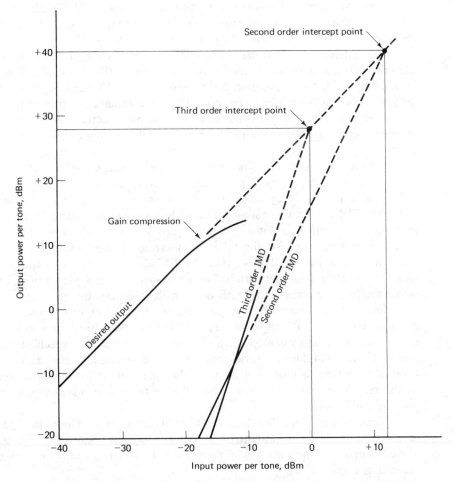

Figure 6.17 Plot of output power per tone versus input power per tone during a two-tone test. The desired output is shown with a slope of unity. Distortion products are also presented with a slope commensurate with the order of that distortion. Extrapolation of the curves defines the intercept points of the amplifier. The second order intercept is much higher than the third order value, suggesting that third order imd is the dominant distortion effect. However, this is not true at low signal levels. Note that the second order imd exceeds the third order products for inputs below −20 dBm. This is especially important in broadband systems.

of −10 dBm. The difference (ratio) of the intercept and the output is 30 dB. Hence, the third order imd products will be twice this amount below the output, 60 dB down at −70 dBm. If the outputs are decreased by 10 dB to −20 dBm, the products will be 80 dB below the desired outputs at −100 dBm.

Measurement of intercept is implicit in the concept. Consider the data of Fig. 6.16a. That spectrum shows that each tone has an output power of +10 dBm and that the third order imd products are below the output by 35 dB. Hence, the output intercept is the desired output plus half of the difference between the desired output and the distortion products, or +10 dBm + ½(35 dB) = +27.5 dBm.

The intercept concept may be applied to virtually any network. In some cases the output frequencies will be different than those of the input. An example would be a mixer. Still, the application of two input tones will result in two output tones with an identical frequency separation. The third order curvature of the devices in the mixer will lead to third order imd. The output spectrum of the mixer may be examined with a spectrum analyzer to determine the level of imd and a corresponding intercept.

Considerable care must be used with the intercept concept. It is viable only when it is useful for predicting network performance. Several "high level" mixers are available which display excellent imd. For example, one examined produced third order imd 70 dB below the desired output with two inputs of 0 dBm. This is outstanding performance. From this data, one would predict an input intercept of +35 dBm. This may not be valid for the mixer in question. Careful examination reveals the third order imd just a bit more than 70 dB below the desired outputs with input tones of −10 dBm. If the network had been well-behaved, the +35-dBm input intercept would imply third order imd 90 dB down with −10-dBm inputs.

Care must also be used when specifying an intercept value. A proper specification is "input intercept" or "output intercept." Many workers are sloppy in their description, giving only an "intercept point." Generally, mixers are specified by an input intercept while amplifiers are described by the output parameter. Either is just as valid as the other so long as it is clear what the worker (or manufacturer) means. In this text, intercepts will often be modified with an i or o subscript to clarify the point of specification.

Noise figures were discussed in the previous section. The noise figure of two cascaded amplifiers was evaluated and found to be dependent upon the noise figure of each amplifier. The intercept of cascaded networks is also of great interest when evaluating a system.

Figure 6.18 shows a cascade of two amplifiers. The gain and third order output intercept of each is presented. Assume initially that the second amplifier is distortion-free, but the first one has some imd. If the output intercept of the first amplifier is I_{o1} dBm, the output intercept of the cascade is $I'_{o1} = I_{o1} + G_2$ where G_2 is the gain of the second stage in dB and the prime indicates that the output intercept of the first stage has been normalized to a different plane. Because the second stage is assumed to be distortion-free, the output tones and the distortion output from the first amplifier are both amplified without further distortion.

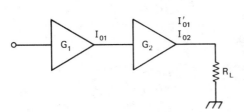

Figure 6.18 A cascade of two amplifiers, each with a known output intercept. I'_{o1} is the output intercept of the first stage renormalized to the output plane, achieved by increasing I_{o1} by G_2, the second stage gain. If the distortion products are assumed coherent, and all intercepts are normalized to one plane, the equivalent intercept is calculated just as the net resistance of parallel resistors is evaluated.

Consider now the more general case where both amplifiers have finite output intercepts. The analysis will be confined to third order imd although the approach is easily extended to distortion of any order. Assume that the intercepts of both stages have been normalized to the same plane in the cascade. The intercepts will be designated by I_n where the subscript n denotes the stage. D_n will refer to a distortion power while P_n will describe the desired output power of the nth stage normalized to the plane of interest.

If the fundamental defining concepts of the intercept are invoked in algebraic terms instead of logarithmic units, the distortion power of the nth stage is $D_n = P_n^3/I_n^2$. This power appears in a load resistance, R. Hence, the corresponding distortion voltage is $V_n = (RD_n)^{1/2} = (P^3R)^{1/2}/I_n$. The total distortion will come from the addition of the distortion voltages.

As was the case with noise voltages, distortion voltages must be added with care. If the voltages are phase related, they should be added algebraically. However, if they are completely uncorrelated, they will add just as thermal noise voltages do, as the root of the sum of the squares. There is usually a well-defined phase relationship between signals with amplifiers. The worst case is when distortions from two stages add exactly in phase. This will lead to the largest distortion. Some cases may exist where distortion voltages are coherent (phase related) and cancel to lead to a distortionless amplifier. Like most physical phenomena, this is unusual and not the sort of thing that a designer can depend upon. We will take the conservative approach of choosing the worst possible case, that of algebraic addition of the distortion voltage, assuming them to be in phase.

Using the worst case assumption, the total distortion voltage is $V_T = V_1 + V_2$

$$V_T = \left(\frac{1}{I_1} + \frac{1}{I_2}\right)(P^3R)^{1/2} \qquad (6.3\text{-}10)$$

The corresponding power is then

$$D_T = \frac{V_T^2}{R} = P^3\left(\frac{1}{I_1} + \frac{1}{I_2}\right)^2 \qquad (6.3\text{-}11)$$

From the earlier definition, the net or total intercept at the plane of definition is

$$I_T = (P^3/D_T)^{1/2} \qquad (6.3\text{-}12)$$

Further manipulation yields the final result

$$I_T = \left(\frac{1}{I_1} + \frac{1}{I_2}\right)^{-1} \qquad (6.3\text{-}13)$$

Equation 6.3-13 has a familiar form with an easy to remember analogy. If intercepts are normalized to a single plane and are expressed as powers in milliwatts or watts rather than logarithmic units, the total intercept at the plane of definition is a sum similar to that for resistors in parallel. This applies only for the case of coherent addition of distortion voltages for third order imd. Not only is this analysis conservative to the extent that it is "worst case," but it works well in practice, predicting measured results with reasonable accuracy.

Consider an example, two identical amplifiers with a gain of 10 dB and an output intercept of +15 dBm. If the two intercepts are normalized to the corresponding ones at the output, they are +15 and +25 dBm. Converting to milliwatts, the two intercepts are 31.62 and 316.2. Application of the resistors-in-parallel rule yields an equivalent output intercept of 28.75 mW, or 14.59 dBm. Essentially, the imd is completely dominated by the output stage.

A more realistic design would be one with a "stronger" second stage. Assume that the output intercept of the second stage is increased to +25 dBm. That of the first stage is still +15 dBm, while both gains remain at 10 dB. The result is an ouput intercept of +22 dBm. The output intercept of the first stage equals the input intercept of the second to yield equal distortion contribution from each and a 3-dB degradation over the intercept of an individual stage.

Generally, the last stage in a chain will determine the third order imd performance. This will be maintained so long as the output intercept of the previous stage is greater than the input intercept of the last.

Some generalizations may be made about the intercepts of some amplifiers. Consider first the question of gain compression in a common emitter bipolar amplifier. From an intuitive viewpoint, we would expect the gain to begin to decrease significantly when the collector signal current reaches a peak value equaling the dc bias current. The signal current will then be varying from the bias level to twice that value and to zero on negative-going peaks. This assumes that the supply voltage is high enough that no voltage limiting occurs. The load also effects the possibility of voltage limiting.

It is found experimentally that the 1-dB gain compression point is well approximated by the current limiting described. Gain will still be present at higher levels and the continued gain compression is gradual until a "saturated" output is reached. Distortion is severe at high levels above the point of 1-dB compression. A bipolar transistor with a 50-Ω collector termination will have a 1-dB compression point of

+10-dBm output when biased to a collector current of 20 mA. The compression point is proportional to I_c^2.

A general rule of thumb for bipolar transistors terminated in collector loads near 50 Ω is that the output intercept will be 13 to 16 dB above the 1-dB gain compression point. This assumes that the transistor is a good one, having relatively low capacitances. Also, the better transistors (low imd) will be those with a current gain relatively independent of current level.

Intermodulation distortion may often be reduced by matching at the output of an amplifier. If an impedance larger than 50 Ω is presented to the collector, smaller current excursions are required for a given output power.

Feedback can change the imd performance of an amplifier. However, except at rather low frequencies, the changes are slight so far as output intercept is concerned. The main role of negative feedback is to reduce the gain of the amplifier and to control terminal immittances. Hence, negative feedback which reduces stage gain can have a dramatic effect on input intercept while changing the output intercept little.

If resistance feedback is utilized, some of the available output power is dissipated in the feedback elements, decreasing that available to the load. For this reason, feedback schemes utilizing lossless feedback offer both improved noise performance and better imd characteristics. The amplifier shown in Fig. 6.14 offers excellent results for both noise and imd.

A useful figure-of-merit for amplifiers, especially when battery operation is contemplated, is intercept efficiency (10). Formally, it is the ratio of output intercept to bias power. From the earlier rule of thumb, we found that the output intercept of an amplifier tends to increase as the square of the bias current. However, most amplifiers are operated in a system with a single power supply (or perhaps two) available for biasing. Increasing current will cause only a proportional increase in bias power. To this extent, intercept efficiency of a bipolar transistor will increase linearly with increasing bias power.

Field effect transistors are harder to characterize in a general way. Some offer very high third order intercept efficiencies owing to a dominance of a square law curvature in the transfer function. However, small-signal FETs often have low enough I_{DSS} values that they may not be biased to high currents, limiting their output intercept. FETs still find great acceptance in systems requiring low power consumption with moderate output intercepts.

Some large signal VMOS-FETs offer high output intercepts. However, they often require very high bias power to achieve this, perhaps $V_{dd} = 25$ V and $I_d = 400$ mA. Examination of the low frequency characteristics are initially puzzling, for a curve of I_d versus V_{gs} is extremely linear, lacking in both second and third order curvature. However, the parasitic capacitances are high. Moreover, they are strongly dependent upon the drain voltage, perhaps accounting for the poor intercept efficiencies.

Intermodulation distortion is an important characteristic of rf networks. It is often, although not always, the limiting factor which determines the largest signal that a system may handle. It is useful to the extent that those design efforts aimed at reducing imd will often reduce other nonlinear effects such as harmonic distortion and cross-modulation. It is easy to measure imd as well, making it a convenient tool.

Having discussed distortion and, in the preceding section, noise, we are now in a position to examine systems. This will be postponed to Chap. 8 where receiver measurements are presented. In simple systems, distortions are easily measured without the aid of a spectrum analyzer. The measurement methods as well as the parameters to be measured are presented later.

6.4 MIXERS

Mixer operation was mentioned briefly in the previous section. When a two-tone input was applied to a device with square law curvature, the output contained sum and difference frequencies of the two input tones. Mixers are extremely important elements in most rf systems and will be presented in more detail here.

A symbolic representation of a mixer is shown in Fig. 6.19 with two inputs. The larger input is termed the *local oscillator* (lo). Usually this signal will be at a power level of from 1 to 100 mW. The other input is an rf signal. Its level may be quite low. The upper limit on the rf signal strength is determined by gain compression properties. If the rf input is a collection of signals as one might find at the input of a receiver or rf instrument, the upper limit is then determined by intermodulation requirements for the system.

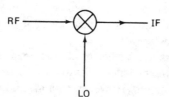

Figure 6.19 Schematic representation of a mixer.

The mixer output is an intermediate frequency (i.f.). This is usually a single frequency with a relatively narrow bandwidth in receiver applications. In other applications, the mixer i.f. port may be very wide in frequency response, perhaps covering several octaves.

If the lo and rf frequencies are assumed to be constant, single-tone inputs, there will be two possible i.f. outputs. One is the sum of the input frequencies and is termed the upper sideband while the other is the difference frequency, called the lower sideband. This terminology is historic, having arisen from modulator definitions.

The only outputs of interest are the sum and difference frequencies in normal mixer applications. The other outputs that may occur are then termed spurious outputs. These will include the lo and the rf frequencies as well as their harmonics. The harmonics may be present in the input signals or they may be created by high order curvature in the transfer function of the mixer. Other spurious outputs will be at frequencies that are the sum or difference of harmonics of the input frequencies. If the inputs are f_{lo} and f_{rf}, possible outputs are at $f_{out} = nf_{lo} \pm mf_{rf}$ where m and n are both integers from 0 upward. The magnitude of the spurious responses will be different for various mixer types. The responses always decrease as m and n become large. However, the rate of decrease may be small for some mixers.

Implications of the spurious response problems can be great. If a system is to operate with small distortions, the frequencies must be chosen with great care. Alternatively, one can sometimes make use of the spurious responses. For example, diode-ring–type mixers typically have a 6-dB loss so far as 1,1 products are concerned, frequencies where m and n both equal 1. However, 3,1 products are also strong with a typical loss in the mixer of 18 dB. This could be used to advantage in working with different input frequencies. An example would be a system with an i.f. of 500 MHz. A filter at the output of the mixer will allow only 500-MHz signals to continue. A lo tuning over the range of 2 to 4 GHz will cause input rf frequencies over the range of 1.5 to 3.5 GHz to yield a difference output of 500 MHz, resulting from a 1,1-type product. A 3,1 response will allow input rf components in the 5.5- to 11.5-GHz range to produce an output at 500 MHz, again from a difference response. While the response will be attenuated over that obtained with a 6- to 12-GHz lo, it is still usable. This is termed harmonic mixing and is the basis of many microwave spectrum analyzers.

The spurious responses mentioned assumed a single frequency input at both the lo and rf ports of the mixer. Other spurious responses will arise from the mixer's ability to respond to other inputs. Consider an example of a receiver with an intermediate frequency of 20 MHz. The receiver is to be used to receive signals at 50 MHz and the lo frequency is 70 MHz. The mixer will produce a 20-MHz i.f. response for rf inputs at either the desired 50 MHz or at 90 MHz. These responses will be equal in strength with typical mixers. A filter must be used in the path to the rf port of the mixer to select the proper input frequency. Because it precedes the input, this filter is called a *preselector*. The other input frequency, 90 MHz in the example, which can also yield an i.f. output is called the *image frequency*.

An amplifier is used in the chain preceding the mixer in some receivers. The rf amplifier may have a preselector filter ahead of it. The filter between the amplifier and the mixer is then termed an *image stripping* filter. It serves the same role as the preselector in suppressing the image frequency. One might reason that the filter would be redundant if a suitable filter preceded the rf amplifier. This is not true. Noise will be generated within the rf amplifier, not only at the desired frequency (50 MHz), but at the image (90 MHz). The noise figure of the complete receiver will be degraded if the undesired noise is not removed from the mixer input.

The term "image" is usually used in reference to input frequencies that will also yield a mixer output. However, the term is also applied to what we have described as the opposite sideband, a mixer output. Care must be taken when reading the literature or describing a system phenomenon.

Additional subtleties of receiver and instrument system design will be presented in Chap. 8. In this section, mixers will be viewed from a fundamental viewpoint, showing how mixing action occurs and presenting the configuration of some practical mixers.

Mixers may be classified in a number of ways. One would be to group mixers according to the type of device they utilize. The classification we shall use is based upon the mode of operation. The two are *square-law* and *switching-mode* mixers.

The square-law-type is typified by the mixer using a JFET as shown in Fig. 6.20. Square-law mixers are those which emphasize the second order curvature of the device transfer function.

From earlier work, we learned that the drain current in a junction FET is $I_d = I_{DSS}(1 - V_s/V_p)^2$ where I_{DSS} is the drain current with the gate and source at the same potential, V_s is the source potential with the gate at ground, and V_p is the pinch-off voltage. A normalization is applied for simplicity, setting $I_{DSS} = 1$ and $V_p = 1$. Then, $I_d = (1 - V_s)^2$.

Examination of Fig. 6.20 shows that the rf signal is injected onto the gate through a tuned circuit. The resonator serves the role of transforming the impedance of the source to something higher, not only to improve gain but to reduce noise figure. The local oscillator is injected onto the source while the drain is tuned with a resonator at the intermediate frequency. The role of the drain resonator is twofold. It provides an impedance transformation, allowing higher gain. It also provides a relatively low impedance path to ground for signals at frequencies other than the i.f.

Noting that the control of the FET drain current is actually from the difference between the gate potential and that at the source, the signal voltages will effectively add, even though one, the lo, is applied to the source while the rf signal is at the gate. Hence, we may assume a value for V_s of

$$V_s = v_L + v_s + v_b \qquad (6.4\text{-}1)$$

where v_L is the local oscillator voltage, v_s is the rf voltage appearing at the gate, and v_b is the bias voltage resulting from a dc current flow through the source resistor.

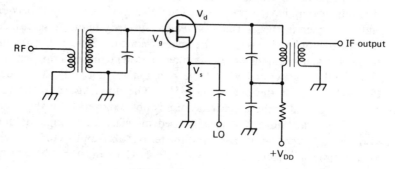

Figure 6.20 A mixer using a JFET.

The multiple term expression for V_s may be substituted into the normalized equation for drain current, yielding

$$I_d = 1 - 2(v_s + v_L + v_b) + v_s^2 + v_L^2 + v_b^2 + 2(v_s v_b + v_L v_b + v_s v_L) \qquad (6.4\text{-}2)$$

Of all of these terms only one, the last, is significant. This is the only one that may have frequency components at the intermediate frequency. All the rest are either at the signal or lo frequencies or are dc terms. The viable drain current is then

$$I_d = 2 v_s v_L \qquad (6.4\text{-}3)$$

We will assume that each of the signals is sinusoidal, resulting in

$$I_d = 2 E_s E_L \cos\omega_s t \cos\omega_L t \qquad (6.4\text{-}4)$$

where E_s and E_L are the peak amplitudes of the rf signal and lo inputs. Using the trigonometric identity $\cos(A \pm B) = \cos A \cos B \mp \sin A \sin B$, and a bit of manipulation

$$I_d = E_s E_L [\cos(\omega_L - \omega_s)t + \cos(\omega_L + \omega_s)t] \qquad (6.4\text{-}5)$$

We have outputs at the sum and difference frequencies.

The tuned drain circuit will allow only one of the frequencies to appear as a voltage at the drain. Assuming the drain is tuned to the difference frequency

$$I_d = E_s E_L \cos(\omega_L - \omega_s)t \qquad (6.4\text{-}6)$$

The mixer may be characterized by a conversion transconductance, the ratio of the drain current flowing at the intermediate frequency to the rf signal voltage appearing at the gate

$$G_m = E_L \qquad (6.4\text{-}7)$$

Recall that the square-law model used for the JFET applies only in a specific region. The source-to-gate voltage must always lie between 0 and 1 if the normalized form is used. However, to maximize the conversion transconductance the local oscillator voltage should be as large as possible while operating the device within the allowed region. Hence, the bias should be ½ and E_L should be ½. A normalized transconductance of ½ results with this bias and lo level.

If the normalized equation for drain current, $I_d = (1 - V_s)^2$, is differentiated with respect to V_s, the amplifier transconductance is determined as $g_m = 2(1 - V_s)$. The maximum amplifier gain will occur at the lowest source bias, producing a peak amplifier g_m of 2. The ratio of the amplifier transconductance to the conversion value is then 4. The gain from a FET mixer is then 12 dB less than that expected from the same device operated as an amplifier with maximum gain biasing. This assumes identical impedances presented to the gate and drain in the comparisons. A 12-dB difference is typical of carefully designed practical FET mixers.

Optimum design depends upon many factors. Using the earlier models for noise in FETs and noting the effects of external resistors, the lo should come from a generator with low characteristic impedance, not only at the lo frequency, but at that of the signal and the i.f.

The characteristic assumed for the JFET was that of a square-law device. There was no higher order curvature to the transfer function. Hence, the ideal FET is completely free of third order intermodulation distortion. This is, of course, an idealization. The parabolic nature of the FET is not perfect and there will be imd, although it is not severe. It can become very bad though if the biasing conditions are violated. If the FET ever goes into the pinch-off region or if the gate-source diode becomes forward biased, distortion can become severe.

Some workers have built JFET mixers in the hf spectrum with a noise figure of 4 dB and an input intercept of around +10 dBm with an associated gain of a few dB (11).

Another problem to avoid with FET mixers is gain compression from excessive drain voltage excursion. This will occur if the drain impedance is made too high, a result of poor transformer design, or terminating the output tuned circuit with a high Q filter. Ideally, the output would be terminated in a crystal filter at the i.f. The filter will present a reasonable load to the mixer output within the passband. However, when moving into the skirts of the filter, the transition between the passband and the stopband, the impedance may become high. This is reflected through the output transformer as a high drain load, allowing excessive drain voltage excursions. This will occur for a signal that is away from the frequency to which the receiver using this mixer is tuned. The best, and most elegant, solution to this problem is the replacement of the simple resonator with a network with impedance inverting properties. A properly designed pi-network serves this function. Such networks were described in Chap. 4.

A significant problem with JFETs is poor specification. The I_{DSS} and V_p values can both vary by a factor of 2 or more for a given FET type. This is not a severe problem in amplifier service. It can be disastrous to a mixer though. For this reason, JFET mixers are not routinely used in many receiver or instrumentation applications even though they are potentially capable of excellent performance.

MOSFETs are similar in performance to the JFET. They are generally enhancement-mode devices, requiring a forward bias on the gate to establish a current flow. The drain current as a function of gate voltage is a square law similar to that of the junction device. With the exception of the biasing, the single gate MOSFET mixer is identical to the one using a JFET. Performance is similar, although the noise figures are often not as low.

The dual gate MOSFET is often used as a mixer. A typical schematic is shown in Fig. 6.21 where the local oscillator signal is injected onto gate 2 of the device. The gate 2 potential has the effect of varying the transconductance in an approximately

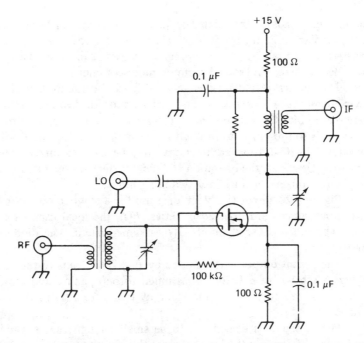

Figure 6.21 A mixer using a dual gate MOSFET.

linear way. Hence, it is exactly analogous to varying the instantaneous source potential of the JFET, leading to square-law mixing. The rf is injected onto gate 1 through an impedance transforming network while a similar network at the output feeds the i.f. amplifier system. The drain load resistor shown is sometimes used to ensure a reasonably constant output impedance, a requirement for driving a filter. The low reverse capacitance of the dual gate device (low S_{12}) leads to the device having a large output resistance, much larger than the JFET. The dc bias on gate 2 is set equal to the source voltage and the local oscillator level is adjusted for about 5 V peak-to-peak on gate 2.

Typical MOSFET performance is a noise figure of about 8 to 10 dB and an output intercept of +20 dBm. The conversion gain is also 12 dB less than can be expected from the same device operated as an amplifier. The dual gate MOSFET has the virtue that there are no diodes to be forward biased by lo or rf signals. Hence, circuits like that of Fig. 6.21 are much more tolerant of variations in device parameters, making such mixers more reproducible. The performance is generally not as good as that achieved with carefully designed JFET mixers though.

The bipolar transistor has an exponential large signal characteristic. It also has very large transconductance when compared with the usual small-signal field effect transistors. As such, it will function as a mixer. It is biased into conduction in the normal manner with the rf signal usually being injected onto the base. The lo is injected at the emitter, causing the transconductance to vary. The variable transconductance leads to the mixing action. The presence of large high order curvature in the bipolar transistor produces considerable distortion at large signal levels, making the device unsuitable for most applications except in inexpensive consumer applications. Even there, FETs are becoming more common.

The mixers just described have utilized the square-law characteristics of the device to produce a desired second order distortion, leading to mixing. The devices were biased into the active region. The relative lack of higher order curvatures with FETs leads to immunity to imd. Other devices, including those with very high order curvatures in their transfer functions, may be used as mixers, and ones that work very well. However, the operation is different; they are operated in a switching mode. Most diode mixers operate in a switching mode.

Figure 6.22 shows the block diagram for a single diode switching-mode mixer. Two input signals are injected in series. $F(t)$, the local oscillator, is assumed to be large. The other is the rf input, $G(t)$, assumed small. The diode is typically a hot-carrier type.

The circuit output is dominated by $F(t)$. This leads to the circuit model shown in Fig. 6.22b where a switch is assumed to be opening and closing in accordance with the lo. We then assign the value of 1 or 0 to $F(t)$ to represent the switching action.

The rf signal is considered to be small enough that it cannot cause forward conduction of the diode alone. Hence, the circuit output will be affected by $G(t)$ only when $F(t)$ is unity. The lo and rf waveforms are shown in Figs. 6.22c and d, although the magnitude of $G(t)$ is increased for illustration. The resulting output is shown in Fig. 6.22e.

The square wave drive signal can be described analytically as a Fourier series

$$F(t) = \frac{1}{2} + \frac{2}{\pi} \cos \omega_L t - \frac{2}{3\pi} \cos 3\omega_L t + \cdots \qquad (6.4\text{-}8)$$

When this lo waveform is multiplied by the rf signal, $G(t) = \cos \omega_s t$, the resultant output, neglecting $F(t)$, is

$$V_0(t) = \frac{1}{2} \cos \omega_s t + \frac{2}{\pi} \cos \omega_s t \cos \omega_L t - \frac{2}{3\pi} \cos \omega_s t \cos 3\omega_L t \qquad (6.4\text{-}9)$$

Both amplitudes are assumed to be unity for simplicity in analysis.

The first term in Eq. 6.4-9 is the feedthrough of the rf signal appearing in the load resistor. The second term is a familiar one, the product of two sine waves. With the trigonometric identity quoted earlier, this term produces outputs at the sum and difference frequencies

$$V_0(t) = \frac{1}{\pi} \cos(\omega_L + \omega_s)t + \frac{1}{\pi} \cos(\omega_L - \omega_s)t \qquad (6.4\text{-}10)$$

Sec. 6.4 Mixers

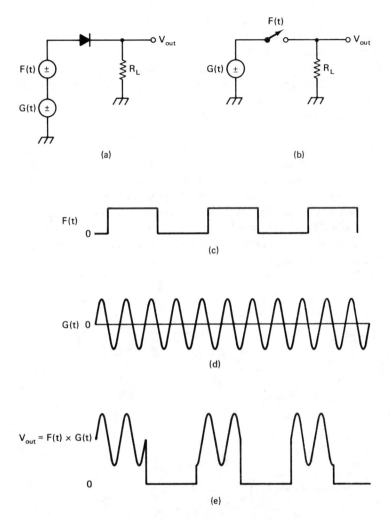

Figure 6.22 A single diode mixer, the simplest of the switching-mode mixers. The idealized circuit (a) is modeled as a switch opening and closing at the lo rate, F(t), in series with the rf signal, G(t). The two input waveforms are shown at (c) and (d) with the resulting output at (e). The rf component is normally very small with respect to the lo amplitude. It is shown larger for illustration.

The next term in Eq. 6.4-9 will create sum and difference frequencies between the rf input and the third harmonic of the lo. This and higher terms are the basis of the harmonic mixing.

The single diode mixer described is not common at lower frequencies. However, it is used in some microwave applications, especially in mixers for rf signals above 20 GHz.

The greatest deficiency of the mixer of Fig. 6.22 is the presence of all input signals with virtually no attenuation at the output. The large lo signal will appear at the i.f. ports as will the rf signal. This imposes a severe problem on the circuitry following the mixer. The problem is virtually eliminated through the use of balance circuitry. Figure 6.23a shows a singly balanced diode mixer while a doubly balanced one is shown at Fig. 6.23b.

The virtues of the singly balanced mixer are understood if the diodes are assumed to be identical, perhaps each containing some series resistance. When the lo is positive, current will flow into the dot of the primary transformer. Hence, it will be flowing out of the dots of the two secondaries. This forward biases the diodes. There will be no lo voltage at the i.f. port if the diodes are identical in every respect. The switching action still occurs with the diodes. The rf signal is injected at the center tap of the transformer. Because the two secondary windings are identical, usually being bifilarly wound, the lo voltage at the junction of the two secondaries is zero. There is, however, nothing to prevent rf energy from appearing at the i.f. port of the singly balanced mixer.

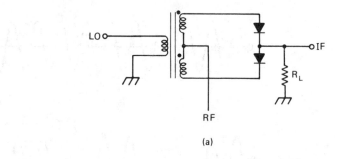

(a)

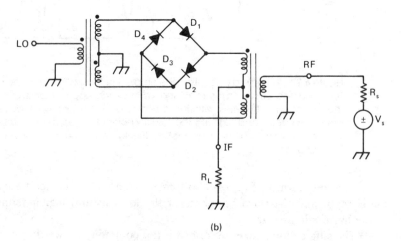

(b)

Figure 6.23 Balanced mixers. A singly balanced design is presented at (a) with a doubly balanced diode ring at (b).

Operation of the doubly balanced mixer is understood with the aid of Fig. 6.24. The lo waveform is assumed to be positive. This forward biases diodes D_1 and D_2 of Fig. 6.23b, but reverse biases D_3 and D_4. The rf voltage from the primary appears directly across half of the rf transformer. Hence, the rf voltage with its source impedance, R_s, appears in series with the i.f. load, R_L, for half of the lo driving cycle. The diodes are shown in Fig. 6.24 as series resistors.

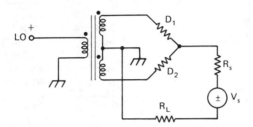

Figure 6.24 Model of the diode-ring mixer of Fig. 6.23b at an instant when D_1 and D_2 are conducting. Diodes shown as resistors.

When the lo changes polarity, the opposite action occurs. Diodes D_1 and D_2 are reverse biased while D_3 and D_4 conduct. There is a reversal of polarity brought about by the switching action. With the original condition shown in Fig. 6.24, the center tap of the "on" diodes is connected to the positive polarity of the rf waveform owing to the winding sense of the rf transformer. When the lo polarity changes, forcing D_3 and D_4 to be the conducting diodes, the negative reference polarity of the rf waveform is connected to the junction of the diodes. The connection to complete the loop is always through the "on" set of diodes. The doubly balanced mixer is often called a commutating mixer owing to the polarity reversing action.

We see that the rf port is balanced from Fig. 6.24. Energy from the rf port is applied in the loop with the center tap of the lo transformer. Hence, it will appear with two opposite phases in the primary of the lo transformer, causing cancellation of rf energy at the lo port. Similarly, lo energy is removed from the rf port. Both lo and rf energy are inhibited from appearance at the i.f. port by action similar to that of the singly balanced mixer, Fig. 6.23a.

Assuming that the mixing action is from fundamentals of the lo waveform, a 1,1-type response, the question might well be asked how the operation differs from that of the square-law mixers considered earlier. Examination shows that only one term in the expansion of the square-wave lo drive leads to mixer action while the higher order terms contribute only to higher order mixing. The difference is that the lo currents are considered to be so high that the details of the curvature vanish from consideration. The diode currents for an on set of diodes are so high that any rf that is applied will constitute a "small signal."

The extent to which this is valid may be examined both theoretically and experimentally with the circuit shown in Fig. 6.25. The inductors are large, serving as isolating rf chokes while the capacitors are large enough that they are short circuits for the applied rf input. The diodes are biased on with a dc current source. Using the diode equation, the terms in the transfer function of the composite circuit may be evaluated. The imd is evaluated with the distortion analysis presented in Sec.

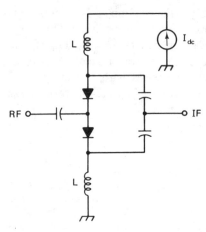

Figure 6.25 A circuit used for evaluation of mixer imd resulting from diode nonlinearity. Measurements and calculations both show third order imd to be unrelated to diode nonlinearity. Third order imd in diode mixers is dominated by modulation of the local oscillator switching action by the incoming rf signal. See the text and the work of Walker cited at the end of this chapter.

6.3. The intercepts may be calculated. The same circuit can also be used for experimental evaluation of the intercepts. Walker (12) has performed this analysis and measurement, showing that the intercepts are quite large so long as the current exceeds a few milliamperes. The loss through the circuit of Fig. 6.25 is small, making the input and output intercepts virtually identical. Essentially, there is no imd contributed by the "on" diodes, so long as they are really on.

There is still imd present in diode-ring mixers. The mechanism is, however, not diode nonlinearity. Instead, imd is created by modulation of the on–off nature of the lo waveform by the rf signals. Walker has studied this phenomenon, both experimentally and theoretically with the results reported in the previously referenced paper.

Local oscillator modulation by the rf signal occurs predominantly when the lo is near a zero voltage condition, the switching point. The usual lo is from a sinusoidal source. An increase in available lo power causes a corresponding increase in the rate of change of the switching voltage at this critical point. The mixer is subjected to the effects of strong rf signals for a reduced time period. Improved imd performance with increased lo power is often observed experimentally.

Walker has also shown that imd performance is enhanced by application of a square-wave lo. The power required to achieve a desired output intercept may then be decreased by as much as a factor of 15. The harmonic distortion introduced is of little consequence, for the diode ring itself will create considerable harmonic content from the diode limiting action. A square-wave lo is generated by a high speed differential amplifier operated in an overdriven mode, as shown in Fig. 6.26.

Another source of imd in diode-ring mixers is the voltage controlled nature of the capacitance of the reverse biased "off" diodes. If at some instant a finite charge exists on an ideal capacitor and the voltage is then altered, the movement of charge will be linear with applied voltage. However, if the capacitance is itself a function of voltage, the movement of charge through the capacitor will not be linear. Any semiconductor junction that is reverse biased will have a capacitance which

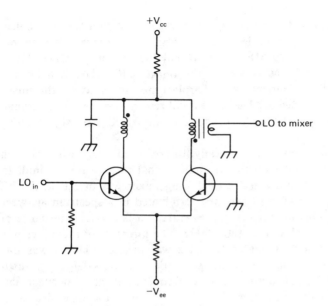

Figure 6.26 An overdriven differential amplifier used to provide a square wave lo drive for a diode-ring mixer.

varies with applied voltage. The reverse capacitance of the diodes should be small and should be matched just as the forward current characteristics are matched.

Some mixers are modified by the insertion of resistors in series with the diodes. This has the effect of increasing the voltage drop across the "on" diodes which, in turn, causes an increase in reverse voltage across the "off" diode pair. The resistors dissipate some of the rf energy passing through them, increasing insertion loss. Balance and imd performance are usually improved though. Another method of improving imd performance in diode-ring mixers is the replacement of the individual diodes with series pairs. Mixers that have a modified topology, either with resistors or additional diodes, are termed "high level mixers." The usual diode ring utilizing hot-carrier diodes is specified for an lo drive power of +7 dBm. High level mixers usually require +17 dBm. The values specified are the lo drive powers available from the source into a 50-Ω load. Power delivered to the mixer may often be much different owing to the limiting nature of the diodes.

A modern communications receiver tuning the hf spectrum, essentially from quite low frequencies up to 30 MHz, might have a first i.f. at 70 MHz. The lo would then tune the range from 70 to 100 MHz. The desired output would be the difference between the lo and the rf incoming frequency. The sum frequency, the lo plus the incoming rf, will produce outputs in the spectrum above the lo frequency. This opposite sideband is often called an image response in the literature.

There is little to preclude movement of signals in different directions through switching mode mixers. The lo is the waveform that causes the switching to occur. Once this happens, signals at the rf port may be converted to appear at the i.f. port as desired. Signals at the i.f. port, however, can also be mixed with the lo to appear back at the rf port.

The bidirectional nature of a switching mode mixer is often a severe problem when considering the image response described. Consider the receiver example described with an input signal at 20 MHz. The lo will then be at 90 MHz. Mixer outputs occur at the desired 70 MHz and at the image, 110 MHz. If a filter is placed at the i.f. port of the mixer which is sharply tuned to 70 MHz, the image response at 110 MHz will be reflected. The energy will return into the mixer through the i.f. port where it will mix with the lo, producing another 20-MHz signal, along with other responses.

The effects of the image are varied. Usually, the reflected image will return in such a phase that it will interact with the incoming signal to increase the imd. In one experiment performed by the writer, a diode-ring mixer was investigated with an output i.f. of 10 MHz. The i.f. port was first terminated in a spectrum analyzer with an input attenuator. This instrument had essentially a flat input impedance of 50 Ω at all frequencies from below 1 to 1800 MHz. The intercept of the mixer was measured. Then, a series tuned circuit at 10 MHz was inserted. The Q was low and the conversion loss of the mixer was changed little. The impedance presented at the desired i.f. was still 50 Ω. However, the image frequency at the i.f. port encountered a severe mismatch; a reflective termination. The intercept of the mixer decreased by 10 dB! What would normally have been acceptable performance was degraded to a completely unacceptable level.

Generally, a mixer should be terminated at the i.f. port by a broadband match. The desired i.f. output should be properly matched. Of equal, if not greater significance, the image frequency should also be terminated. Generally, the results of higher order conversions should also be terminated. An improper termination will cause reflected signals that will usually degrade mixer imd.

A proper match is achieved in many ways. One is to place a 50-Ω attenuator between the switching mode mixer and the filter that will eventually follow. Usually, a 6-dB pad is adequate. However, the loss introduced by the pad along with the loss of the mixer may degrade system noise figure to an unacceptable level. An alternative is to terminate the mixer in a broadband amplifier. The amplifier should have an input impedance that provides proper termination at all frequencies. This is the usual method in receivers using diode mixers. A third method is to terminate the mixer in a diplexer network. These are combinations of filters which are configured to present a constant impedance in at least one of the ports. Diplexers were described in Chap. 4. Care must be taken with a diplexer to assure that the impedance variations of a following filter will not couple back through in such a manner that improper mixer terminations occur at frequencies close to, but not exactly at, the desired i.f.

In some mixers, usually designed by diabolically clever people, the image response can be used to advantage. If the image is reflected at the i.f. port such that it returns to the mixer with a "proper" phase, it will be reconverted to the original rf frequency by the switching action of the lo. Part of the extra rf energy in the mixer will now be reconverted to the desired i.f. with such a phase that it adds to the original signal, resulting in a mixer with a reduced conversion loss. Additionally, the imd response may even be improved. Image-reflection–type mixers, if supplied with lots of lo power and carefully terminated, can even be made to oscillate! This is rare. Generally, the more conservative approach to design is to passively terminate mixers with broadband resistive loads.

The traditional diode-ring mixer such as that shown in Fig. 6.23b will have a conversion loss of 6 to 7 dB. Little extra noise is added at lower frequencies by the diodes; hence, the noise figure is the same numerical value as the loss. The noise figure may be higher than the loss at uhf. The usual bandwidth is very wide. A diode-ring mixer using four hot-carrier diodes and two transformers wound on high permeability cores will usually have a bandwidth that extends from 0.5 to 500 MHz for the rf and lo ports. The response at the i.f. port will extend from dc up to 500 MHz. There is approximately 40 dB of balance over much of the frequency range. For example, the ratio of the lo power available from the generator to that delivered from the rf port is 40 dB. The limiting factor on balance is the match between diodes and the nature of the transformers. The balance may be improved by placing a psuedobalun transformer in the line with the input windings of the two transformers.

It's hard to generalize the imd characteristics of a diode-ring mixer. A reasonable rule of thumb is that the output intercept is comparable with the lo drive in the mixer. This is adequate as a starting point in analysis, but eventually measurements must be performed. High level mixers will have improved imd with an output intercept of +20 to +25 dBm being possible. Care must be used when applying high level mixers though, for they are often not well-behaved. Some are ill-behaved enough that it is questionable whether the intercept concept may even be applied. Again, careful measurements must be performed.

Switching-mode mixers usually use diodes. However, other types may be constructed. MOSFETs have been used in switching-mode mixers. They are configured such that the channels replace the diodes in the traditional ring. Local oscillator power is applied to the FET gates such that they are turned on in pairs just as the diodes would be. Some mixers of this type have been reported to have an input intercept as high as +35 dBm, a very strong mixer indeed.

Some workers have also built switching mode mixers using bipolar transistors. Many integrated circuit, doubly balanced mixers use four bipolar transistors as switches. While these devices offer excellent balance, they suffer with a high noise figure and poor imd characteristics. They should not be ignored for mixer applications though, for they have probably not been optimized. Much of the noise and imd may be the result of the usual input differential pair rather than the switching devices.

One doubly balanced mixer of considerable interest was reported in the amateur literature (13). Four bipolar transistors were used in a doubly balanced configuration. The transistors each had negative feedback applied. There was no forward bias other than that supplied by the local oscillator. This mixer was reported to have an intercept (presumably an input value) of +40 dBm and a noise figure of 10 dB. The lo power required to achieve this performance was +13 dBm.

Singly and doubly balanced mixers are not limited to switching types. JFETs will function well in such configurations. However, the FETs must be carefully matched. Some manufacturers supply duals and quads of FETs in a single package.

The devices are matched for both V_p and I_{DSS} such that all device parameters are within 10% of the others. One mixer of this type using a U350, a quad of matched Siliconix U310 JFETs, was reported by Oxner to have an input intercept of +30 dBm, a noise figure of 8 dB, and a gain of 4 dB. The oscillator power required was +15 dBm and the bandwidth was from 50 to 250 MHz (14).

Oxner has also reported on a similar mixer using only two matched U310 transistors in a singly balanced configuration (15). The results were nearly as good as those obtained with the quad configuration. As was the case with the single JFET mixer, the adjustment of lo power level and bias are interrelated and critical. Operation beyond pinch-off and with forward bias of the gate-source diode must be avoided.

DeMaw and Collins (16) have described a singly balanced mixer using a pair of VMOS transistors. An output intercept of +39 dBm was realized with an associated gain of 15 dB. The lo power required was only +15 dBm.

The reader is urged to review the literature on mixers. There is considerable room for further design and analysis work, leading to further improvements.

6.5 INTERMEDIATE FREQUENCY AMPLIFIERS

A superheterodyne places much of the system gain at an i.f. This has a number of virtues. First, distortion is reduced by early use of selective filters. The distortion within the filter bandwidth is usually of less significance, for the output contains the dominant desired output.

The frequencies used for i.f. amplifiers will vary depending upon system requirements. A modern communications receiver will often have a multiplicity of i.f.'s, the first being in the low vhf spectrum, even if the coverage of the receiver is restricted to the hf spectrum. Most modern spectrum analyzers, especially those designed to operate at uhf and microwave frequencies, will have a first i.f. of several GHz.

After passing through a first i.f. that may be high in frequency, a second (or third) i.f. is eventually used, typically from a couple of hundred kilohertz up to perhaps 20 MHz. This is usually the highest gain amplifier used in either a communications receiver or a spectrum analyzer and will be the subject of the design methods presented in this section.

In many respects the design of an i.f. amplifier is no different than any other operating at radio frequencies. Low noise is sometimes required. A predictable level of distortion and signal handling ability without gain compression is needed. However, there is one characteristic which differentiates the i.f. amplifier from others, the ability to vary the gain. The gain in a receiver must accommodate the tremendous range of signals arriving at the input while still keeping the system output approximately constant. In a measurement instrument, the gain must be varied to measure a wide range of input signals. The amplifiers for the two applications differ, though. A receiver must have a gain variation that is very large and is accomplished with a single control. This may be from a potentiometer on the front panel of the receiver, or from an automatic gain control (agc) circuit. The typical range required may exceed 120 dB. The instrument i.f. usually requires only that the gain be variable in steps, usually small compared with the total range of the instrument.

The difficulty of design is similar with the two applications. Receiver i.f. distortion is rarely critical and gain stability is not of paramount importance, for it is usually controlled by an automatic control system. System gain must be stable with variations in temperature and signal level in a measurement application.

The traditional i.f. amplifier is one using a bipolar transistor with variable bias, shown in Fig. 6.27. The output is obtained with a transformer, either tuned or broadband.

The amplifier operation is analyzed with the aid of the simplest of models, a beta generator with an emitter resistor. Amplifier gain is well approximated by R_L/r_e where R_L is the load resistance presented to the collector. As the dc amplifier bias is reduced from some nominal level, the gain will decrease. The rate at which this change occurs with changing control voltage will depend upon the magnitude of the external emitter resistor, R_e.

The amplifier of Fig. 6.27 has a number of problems. First, gain is reduced by decreasing the standing current. However, gain reduction is usually needed when signals become large. An amplifier operating with a low standing current and large signal output represents a worst case for distortion and gain compression. This amplifier also has an input impedance that varies with dc current. This can adversely affect the performance of a filter that might precede such an amplifier. Circuits like this are used only in casual applications such as consumer equipment.

Stability is a possible problem with this amplifier. The y parameters (or whatever

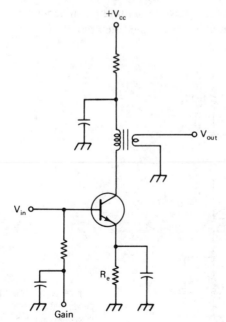

Figure 6.27 A common emitter amplifier with variable base bias, allowing the gain to be electronically controlled.

is used for analysis) will vary with dc current. As they vary, the stability parameters will also change. Care must be taken to ensure that stability is maintained at all gain levels.

The circuit of Fig. 6.27 presumes that a simple and traditional model is adequate for device description. This would predict a decrease in gain with decreasing emitter current. Some transistors designed specifically for i.f. application have significantly different characteristics. They display a high frequency gain which decreases with increasing emitter current. A change of I_e from 2 to 10 or 20 mA will be accompanied by as much as a 30-dB decrease in gain. This is termed *forward gain control* while operation of the more traditional amplifier is described as reverse gain control. The forward gain control amplifier has the virtue of larger standing current when required for large signals. Impedance variations still persist.

Field effect transistors are frequently used in i.f. amplifiers. A JFET will function in a circuit much like that of Fig. 6.27. For maximum gain, the gate voltage is close to ground and the external source resistor is small. Gain is decreased by reverse biasing the gate, making the control voltage negative with respect to ground. This decreases the standing current and decreases the common source transconductance. Amplifiers of this sort have the virtue of presenting a relatively constant input impedance, a characteristic of the open circuit input nature of the FET. However, a decrease in gain is also accompanied by a decrease in standing current.

Figure 6.28 shows a popular i.f. amplifier with controllable gain, one using a dual gate MOSFET. The input to gate 1 is tuned while the output is extracted from the drain in an untuned transformer. If the bias voltage, V_{dd}, is about 12 V, a typical value, maximum gain occurs with a control voltage of about 4.

The dual gate MOSFET will have a maximum gain dependent upon the transconductance and the impedances presented to the gate 1 and drain ports. For low frequency

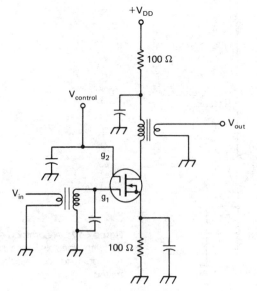

Figure 6.28 A dual gate MOSFET amplifier with variable gain. Distortion increases as gain is reduced.

(hf) analysis, a simple model consisting of nothing more than a voltage controlled current generator provides accurate results. A typical gain maximum might be 25 dB. The gain is usually decreased by 20 dB by decreasing the gate 2 bias from 4 V to ground potential. The gain will decrease further with application of a negative voltage to gate 2. At maximum gain, a typical output intercept would be +20 dBm. When the gate 2 bias is reduced to zero, the output intercept will decrease, often by an amount almost as much as the decrease in gain. Terminal impedances remain fairly constant with application of gain control. The noise figure is also low.

Monolithic integrated circuits are frequently used for amplifiers requiring electronic control of gain. The methods used will vary, some being nothing more elaborate than those already described. Some use a current "robbing" scheme in a cascode configuration. This method is shown in Fig. 6.29 with a discrete transistor circuit.

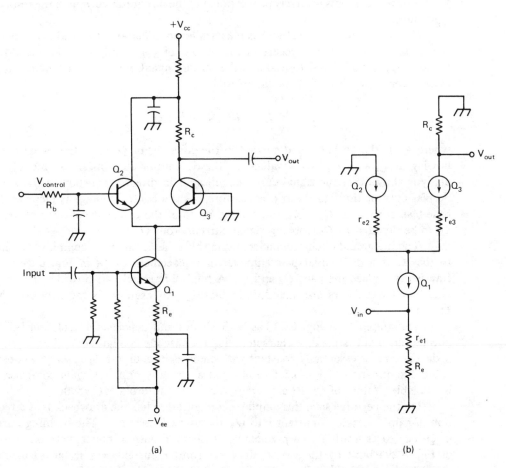

Figure 6.29 A cascode amplifier with "current-robbing" by Q_2 to allow for reduction of gain. A practical circuit is shown at (a) with a small-signal model for analysis at (b).

The basic amplifier is shown in Fig. 6-29a. Three transistors are used with a dual power supply in this example. Single power supplies are more common with the integrated circuit versions.

Input is applied to the base of Q_1. Some emitter degeneration in this stage will establish gain, setting a known g_m for the input device. The biasing establishes a constant dc emitter current in Q_1.

The control line is open circuited for maximum gain. Hence, all of the current flowing in Q_1, both dc bias and signal, appears in Q_3. An output signal is extracted from R_c, a collector load resistor.

Gain is reduced by applying a positive bias to the control line, causing current flow to occur in Q_2.

The gain variation is analyzed with the small-signal model shown in Fig. 6.29b. Each emitter resistor is inversely proportional to the dc emitter current in the respective transistor.

The collector load seen by Q_1 is the parallel combination of r_{e1} and r_{e2}. However, the signal current which appears at the output of Q_3 is only that portion flowing through r_{e3}. This is used to evaluate the output signal current as a function of the bias current in each half of the differential pair.

$$I_{c3} = I_s I_{e3}/(I_{e2} + I_{e3}) \qquad (6.5\text{-}1)$$

where I_{c3} is the output signal current in the collector of Q_3, I_s is the signal current flowing in Q_1 and the two I_e values are the dc emitter currents in Q_2 and Q_3. The analysis shows that the signal currents will split in direct proportion to the split in dc bias current through the two transistors. As the control voltage is increased, the base bias current in Q_2 increases, forcing more of the constant collector current of Q_1 to be diverted to Q_2, robbing signal current from Q_3.

The integrated circuit amplifiers using this scheme are more complex. The input to the IC is a differential pair amplifier, a replacement of Q_1 in Fig. 6.29a. Four transistors replace the pair, Q_2 and Q_3. A final differential pair provides the output. Additional transistors are included for biasing. An example is the Motorola MC-1590G.

The amplifier of Fig. 6.29 has both virtues and deficiencies. Isolation between output and input is good, a characteristic of a cascode configuration. The input impedance remains essentially constant with changes in gain, for I_{e1} is set at a constant value. Examination of Eq. 6.5-1 shows that a virtually unlimited gain control range is available. Values of 65 dB are reported for the Motorola MC-1590G.

On the negative side, this amplifier configuration has the drawback that a reduction of gain diverts bias current into Q_2, the unused transistor. The standing current in Q_3 decreases, leading to a predictable decrease in output intercept. In one version of Fig. 6.29a built by the writer, the total emitter current was set at 10 mA and the output load at the collector of Q_3 was 200 Ω, provided by a ferrite transformer to match to a 50-Ω termination. At maximum gain the output intercept was +25 dBm. When the gain was reduced by 8 dB, the output intercept had decreased to +17 dBm. The amplifier was well-behaved.

Another problem with the circuit is the sharp nature of the control. The exponential nature of the bipolar transistor becomes apparent as the base voltage is increased. The reduction of gain is fast as Q_2 begins to conduct. This rapid change may be reduced by driving the Q_2 base from a current source, realized by making R_b as large as reasonable.

Another method of improving the "smoothness" of control is with negative feedback. This refinement is shown in the partial schematic of Fig. 6.30. A PNP transistor, Q_4, is added to the amplifier as "wraparound" biasing for Q_2. The dc current in Q_2 is sampled with a collector decoupling resistor, R_s, to produce a control voltage. This is compared with the voltage at the base of Q_4 and forced to be equal with negative feedback. The result is a well-predicted variation in gain. This relatively simple refinement changes an amplifier generally useful only in receiver applications to one suitable for instrument applications.

The other problem with the amplifier, the degraded output intercept with application of gain reduction, is reduced with a similar refinement. The degradation in imd performance results from decreased Q_3 collector current as gain is reduced. The collector current in Q_3 may be sampled with another PNP transistor. The collector of the added device is routed to the base of Q_1 so the current in the output device, Q_3, is maintained constant. This was done in the experimental version. The value of I_{e3} was maintained constant at 10 mA. The output intercept changed only slightly with a gain reduction of 8 dB.

This further refinement has a limited range of application owing to the large increase in the bias current in Q_2 that must be provided to realize a given level of gain reduction. Also, the input impedance of Q_1 will now be dependent upon gain.

The refinements described illustrate the nature of the problems encountered in designing an i.f. amplifier. Solutions do exist, although usually at the cost of increased

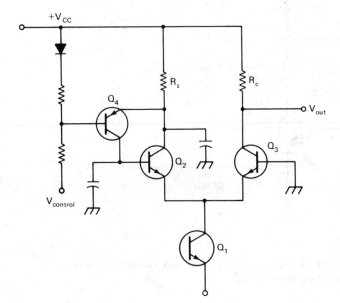

Figure 6.30 Addition of a dc feedback loop via Q_4 to provide the agc amplifier of Fig. 6.29 with a more predictable reduction in gain with control voltage.

complexity and power consumption. The configuration to be used will depend upon the application requirements. The salient features of the amplifiers may usually be analyzed with relatively simple models.

A different approach to gain variation is with a PIN-diode network. The PIN diode is one with an intrinsic (nondoped) layer of silicon between the P and N regions, producing a very slow diode response resulting from charge storage. A parameter of interest in the PIN diode is the charge lifetime, the period required to remove charge from the junction through recombination and migration. So far as rf energy is concerned, the PIN diode is a resistor whose value depends upon the dc current flowing.

There are two different types of PIN diodes resulting from differences in the dopings. One is for switching. The resistance to rf is high with no dc current flow. However, with only a few milliamperes of forward current, the resistance drops to a low value, sometimes less than 1 Ω. These devices are used for switching filters or similar applications. The other type is that intended for use as a current controlled resistor. These are generally described by $R = KI^{-b}$ where K is a constant of proportionality and b is a parameter, usually close to unity.

There are many ways to apply PIN diodes in i.f. amplifiers. One, shown in Fig. 6.31, is as a variable resistor in a common emitter amplifier. The gain is low with $V_{control} = 0$, being predicted by the unbypassed emitter resistor. As the control voltage increases, the resistance of the diode decreases. The lessened feedback causes gain to increase.

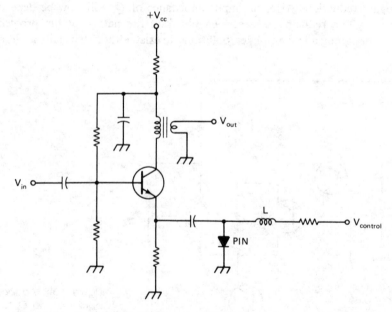

Figure 6.31 Use of a PIN diode as a voltage controlled emitter degeneration resistance, providing low distortion agc.

Sec. 6.5 Intermediate Frequency Amplifiers

A PIN diode is sometimes capacitively coupled to the collector of an amplifier like that of Fig. 6.31. The action is similar except that an increase in control voltage decreases gain. Both methods are applied simultaneously in some amplifiers. The advantage of using elements isolated from the transistor is that the bias conditions can be established to preserve reasonable distortion and power output performance.

It is not always necessary to vary gain continuously. An example would be an i.f. amplifier found in a spectrum analyzer where gain is added in steps of 10 dB. This may be done with switching-type PIN diodes or with transistors. The latter method is shown in Fig. 6.32. The gain of Q_1 is changed by altering the emitter degeneration when applying a voltage to saturate Q_2. The potentiometer allows the size of the gain step to be adjusted precisely. The 100-kΩ resistor attached from Q_2 to the positive supply ensures that the switching transistor has a reverse biased collector-base junction during operation at the lower gain status. The value is not critical. The transistor switch is often preferred to a PIN diode since it is usually less expensive.

The methods presented, either with current controlled PIN diodes or with switches, work well through the hf spectrum. They are not as effective at vhf and higher. The higher frequency amplifiers must usually be designed with more care, for there is not as much excess gain to be deleted. Stability is less easily achieved

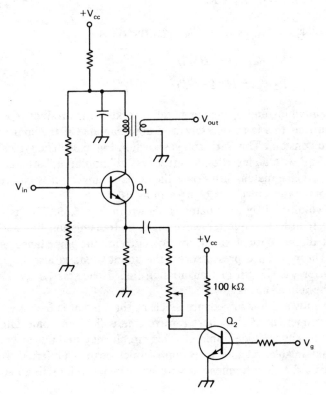

Figure 6.32 Application of a transistor switch, Q_2, to alter the gain of an amplifier in a discrete step.

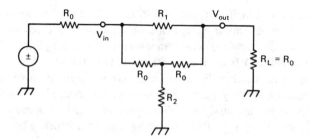

Figure 6.33 A bridged-T attenuator. This topology is useful when R_1 and R_2 are replaced with PIN diodes.

owing to the higher F_t transistors used. A preferred design method is the construction of stable, matched amplifiers with PIN-diode attenuators in cascade. These attenuators are also useful in applications where no gain is desired.

There are a number of attenuator circuits in common use. One is the bridged-T type shown in Fig. 6.33. The adjustable resistors are R_1 and R_2; they will eventually be replaced by diodes. From inspection, we see that the attenuation will be small if R_1 approaches a short circuit while R_2 becomes large. Conversely, the attenuation is large with R_2 very small and R_1 large, and, moreover, the input impedance is R_0, the characteristic value of the pad. The attenuator is matched if

$$R_1 R_2 = R_0^2 \tag{6.5-2}$$

If the gain of the attenuator is defined as $H = V_{\text{out}}/V_{\text{in}}$, the resistors are

$$\begin{aligned} R_1/R_0 &= (1 - H)/H \\ R_2/R_0 &= H/(1 - H) \end{aligned} \tag{6.5-3}$$

The condition for maintaining an impedance match is to keep the product $R_1 R_2$ constant. It may be shown that this is moderately well approximated if the sum of the currents are maintained constant. This variation is shown in Fig. 6.34. The typical variation in gain is 10 to 12 dB with an insertion loss (the loss at "minimum" attenuation) of 2 to 3 dB. The impedance match varies over the control range, but is within reason at all settings, a worst case return loss being around 10 dB.

A variation of the two-diode PIN attenuator is shown in Fig. 6.35. This is a narrow bandwidth design. It utilizes a quarter wavelength of transmission line and identical currents in each diode. With a small control current, the impedance of the series diode is high. The resistance presented by the shunt diode is also high, but it is isolated by a quarter wavelength of transmission line. Hence, as far as the attenuator is concerned, the impedance is low.

Both diodes will move toward low impedance levels as the current is increased. The impedance inversion properties of the quarter wave line will again come into play, making the effect of the shunt diode minimal. This circuit may be built at vhf with a short length of coaxial cable. At uhf, a section of microstrip is preferred. A quarter wave line is synthesized at low frequencies with an impedance inverting network.

Sec. 6.5 Intermediate Frequency Amplifiers

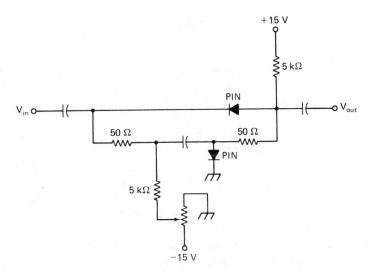

Figure 6.34 A broadband PIN-diode attenuator based on the bridged-T circuit. A typical gain range is over 10 dB while maintaining reasonable input and output impedance matches.

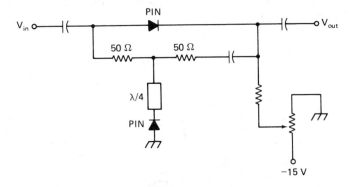

Figure 6.35 A narrow band version of the PIN-diode bridged-T attenuator. The transmission line provides an impedance inverting function and could be simulated with an *LC* network at lower frequency.

Many other circuits use PIN diodes as current controlled resistors. Many contain up to six diodes, offering very large variations in attenuation. These are found in the literature (17).

PIN diodes are not distortionless components. As sufficiently low frequencies are reached, the charge lifetime is short with regard to a single cycle of the rf. The devices then display diode action and distortion occurs according to the diode equation. Most inexpensive PIN diodes available are for vhf applications. They often perform poorly at the lower end of the hf spectrum. FETs are sometimes used as voltage controlled resistors in PIN-type circuits. Also suitable are some optical devices. These are resistors whose value depends upon the amount of light falling upon them. They are packaged with a light emitting diode whose current controls the amount of resistance.

Amplifiers are often required as gain blocks in an i.f. system which must be carefully controlled. Gain must be predictable, stable, and, more often than not, variable. An amplifier for such a system must use extensive feedback to control characteristics.

Figure 6.36 shows one amplifier that works well in these applications, a differential pair with transformer coupling. The output is obtained from a link on the collector transformer. Negative feedback back to the base of Q_2 establishes a well-defined voltage gain and keeps the impedance at the link output low. A resistor (46 Ω) returns the net output impedance to 50 Ω. The input is matched to 50 Ω with a transformer and a 200-Ω terminating resistor.

Gain is controlled by the two feedback resistors, R_f and R_i. If the gain must be varied, R_i may be made variable with a PIN diode or, alternatively, with a resistor that is switched with a transistor or PIN switch.

A traditional nodal analysis of this circuit using simple models works well at low gains. A suitable model is a beta generator with an emitter resistor.

The amplifier of Fig. 6.36 loses some of its desired characteristics as the feedback network is changed to increase the gain. Output impedance is not as well defined and gain stability degrades.

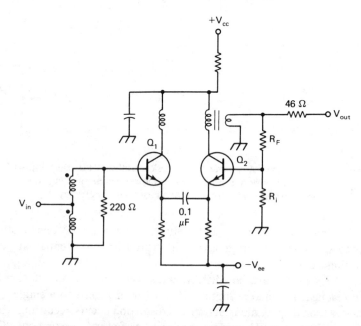

Figure 6.36 A broadband, transformer coupled feedback amplifier suitable in instrumentation i.f. systems. Individual bias resistors keep the emitter currents equal. Gain is controlled by R_f and R_i.

The circuit will show a noise figure of 6 to 8 dB with an output intercept of +36 to +40 dBm if the gain is set at about 10 dB and the transistors are biased to a current of around 40 mA in each emitter. The intercept would be higher except that much of the available power is dissipated in the output resistor and feedback elements. Reverse isolation is good with this amplifier as long as the gain is kept low. The results quoted were measured at 10 MHz.

A modification of the differential pair amplifier is shown in Fig. 6.37. Q_1 and Q_2 function as a differential pair while Q_3 is an emitter follower with high standing dc current. The emitter follower provides a low output impedance, even without feedback. Hence, the output impedance of the amplifier, when modified with a series resistor, is close to 50 Ω at all gain settings. Negative feedback is applied to the base of Q_2 from the output to control the voltage gain. This amplifier will show similar performance to the one of Fig. 6.37 with a bias current of about 40 mA in the output stage and only 10 mA in Q_1 and Q_2.

A vital feature of the three transistor amplifier, Fig. 6.37, is the compensation network. The simple models work well for analysis at low frequencies. However, as the F_β of the transistors are exceeded, the capacitive nature of the transistors contribute extra phase shift. When a loop is closed by negative feedback, the extra phase shift may lead to oscillations. This tendency is controlled with the 50-pF capacitor and 100-Ω resistor. The added network is transparent at low frequencies. At vhf the Q_2 collector load resistance becomes lower, reducing the gain without feedback to a level where stability is assured. An excellent method of testing is to sweep the amplifier over a large frequency range. The gain should be flat, eventually decreasing smoothly. A significant peak in the gain is indicative of possible instability. The measurement

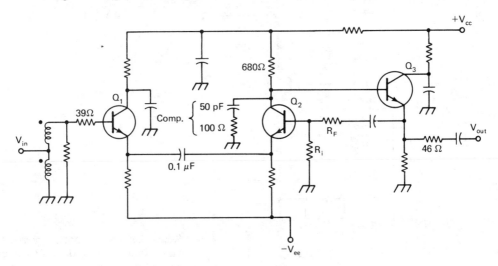

Figure 6.37 An extension of the circuit of Fig. 6.36. This circuit has higher gain before application of negative feedback. Port immittances are then better controlled. A "compensation" network provides a dominant pole in the frequency response to ensure stability.

should be performed to much higher frequencies, at least 200 MHz if the amplifier is to operate at 10 MHz. The compensation should be adjusted so that the high frequency gain is decreasing when well above the operating frequency. The evaluation should be performed at all possible gain settings if the gain is made variable through replacement of R_i by a PIN diode.

The amplifier of Fig. 6.37 has one additional component that may not have an obvious function, the resistor in series with the base of Q_1. This is used to stabilize that stage. The instability noted experimentally was related to the high F_t of the transistors used for the differential pair and was unrelated to the application of negative feedback.

There are clearly a number of variations to the previous amplifiers. Stability is always a consideration. The output intercepts and the reverse isolation may also be critical performance parameters. Figure 6.38 shows an amplifier that has features similar to the previous ones, but also has unique characteristics.

The input stage, Q_1, operates as a common emitter amplifier with a slight amount of emitter degeneration. A compensation network at the collector provides the "dominant pole" required for vhf stabilization. The output stage, Q_2, is similar to the output of the previous amplifier, Fig. 6.37. Feedback is obtained from the emitter and applied to the base of the input. This path also provides dc bias for the composite amplifier.

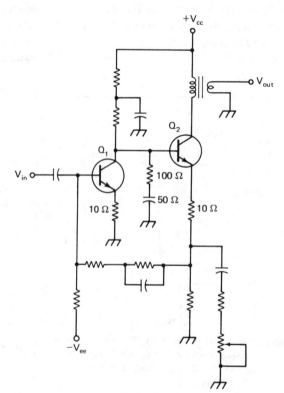

Figure 6.38 A feedback amplifier for instrumentation providing reasonably low noise and distortion and good intercept efficiency. The output impedance is not as well defined with this circuit as with those of Fig. 6.36 and 6.37.

The different feature of this circuit is the extraction of the output. It is obtained from the collector of Q_2 via a transformer. This improves reverse isolation while negative feedback ensures gain stability. The input impedance is controlled by the negative feedback. Output impedance is not well defined for the amplifier of Fig. 6.38 though. However, the intercept may be higher, for no output power is dissipated in an output resistor or feedback network.

REFERENCES

1. "Instruction Manual for AILTECH 75 Precision Automatic Noise Figure Indicator," AILTECH, Farmingdale, N.Y.
2. Blinchikoff, Herman J., and Zverev, Anatol I., *Filtering in the Time and Frequency Domains*, John Wiley & Sons, New York, 1976.
3. Engelson, Morris and Telwski, Fred, *Spectrum Analyzer Theory and Applications*, ARTECH House, Dedham, Mass., 1974.
4. "S Parameters . . . Circuit Analysis and Design," Hewlett-Packard Application Note 95, September 1968.
5. Norton, David E., "High Dynamic Range Transistor Amplifiers Using Lossless Feedback," *Microwave Journal*, pp. 53–57, May 1976. *Also see* U.S. Patent 3,891,934.
6. Meade, Hansel B. and Callaway, Gary R., "Broad Band Amplifier," U.S. Patent 4,042,887.
7. See reference 5.
8. Motchenbacher, C. D. and Fitchen, F. C., *Low-Noise Electronic Design*, John Wiley & Sons, New York, 1973.
9. Simons, Kenneth A., "The Decibel Relationships Between Amplifier Distortion Products," *Proc. IEEE*, **58**, 7, pp. 1071–1086, July 1970.
10. "The Amplifier Factor," Applications Note from Anzac Electronics, Waltham, Mass.
11. Sabin, William, "The Solid-State Receiver," *QST*, **54**, 7, pp. 35–43, July 1970.
12. Walker, H. P., "Sources of Intermodulation in Diode-Ring Mixers," *The Radio and Electronic Engineer*, **46**, 5, pp. 247–255, May 1976.
13. Rohde, Ulrich L., "High-Dynamic Range Active Double-Balanced Mixer," *Ham Radio Magazine*, pp. 90–91, November 1977.
14. Oxner, Ed, "Active Double-Balanced Mixers Made Easy with Junction FET's," *EDN*, pp. 47–53, July 5, 1974.
15. Oxner, Ed, "FETs Work Well in Active Balanced Mixers," *EDN*, pp. 66–72, January 5, 1973.
16. DeMaw, Doug and Collins, George, "Modern Receiver Mixers for High Dynamic Range," *QST*, **65**, 1, pp. 19–23, January 1981.

SUGGESTED ADDITIONAL READINGS

1. Cheadle, Dan, "Selecting Mixers for the Best Intermod Performance," *Microwaves*, November and December 1973.

2. Will, Peter, "Reactive Loads — The Big Mixer Menace," *Microwaves,* April 1971.
3. Friis, H. T., "Noise Figures of Radio Receivers," *Proc. IRE,* **32,** *7,* pp. 419–422, July 1944. *Also see* paper by same writer in *Proc. IRE,* **33,** *2,* pp. 125–127, February 1945. (Classic papers dealing with noise figure of cascade of stages with differing bandwidths.)
4. Munford, W. W. and Scheibe, E. H., *Noise Performance Factors in Communications System,* Horizon House-Microwave, Inc., Dedham, Mass., 1968.
5. Wilson, Stuart E., "Evaluate The Distortion Of Modular Cascades," *Microwaves,* **20,** *3,* pp. 67–70, March 1981.
6. Hayward, Wes and DeMaw, Doug, *Solid-State Design for the Radio Amateur,* Chap. 6, ARRL, Newington, Conn., 1977.
7. Strid, Eric W., "Measurement of Losses in Noise-Matching Networks," *IEEE Trans. on Microwave Theory and Techniques,* **MTT-29,** *3,* pp. 247–252, March 1981.

7

Oscillators and Frequency Synthesizers

The literature, both in the form of textbooks and journal articles, is filled with information on amplifier design. Oscillators are not as well covered. More often than not, oscillators in communications and radio frequency test equipment have been "designed" from a cookbook, or were built and tested empirically with no real effort toward design.

There is reason for this lack. Small-signal theory may be applied, but only partially. An oscillator is a nonlinear circuit. The nonlinearity dominates the behavior. The linear two-port network is easily analyzed. The same network, when operating without the linear assumption, becomes much more difficult.

Oscillators are of extreme importance. Virtually any piece of communications or rf test equipment will have at least one oscillator. The oscillators are not perfect and contain output components beyond that desired, a simple sinusoid. The "extra" outputs consist of harmonics and noise. The harmonics are rarely of concern. The noise is of great importance, though, being the limiting factor for most equipment performance.

Frequency synthesizers are extensions of oscillators. They add stability and accuracy as well as electronic control to an otherwise simple circuit. However, the process of synthesis is a complicated one. If it is not done with great care, the resulting output may be spectrally much worse (noisier) than expected of a simple oscillator.

This chapter will treat oscillators primarily from a small-signal viewpoint. However, the nonlinear details will be covered and used to obtain methods which allow small-signal analysis to estimate oscillator operation.

The oscillator analysis is extended to obtain a glimpse at the problems of frequency synthesis. Generally, the more refined methods used for synthesis will not be covered. Instead, the emphasis is on the fundamental factors which limit the performance and application of frequency synthesizers.

7.1 OSCILLATOR CONCEPTS

An amplifier provides an output that is a replica of the input. An oscillator shows an output at a specific frequency with no input signal required. Output frequency and power are both controlled through the choice of components.

Figure 7.1 shows a block diagram of an oscillator. There are three fundamental parts, an amplifier or active device, a resonator, and an output load. The amplifier is a device capable of gain at the frequency of interest. The resonator is a frequency selective element. It may contain transformers or other impedance transforming components. Some output loading is required, for the circuit is of no use if no output is extracted. The loading shown in Fig. 7.1a is a resistor at the amplifier output. The loading may actually be obtained from any point in the oscillator circuit. The effect of the external load may be quite small depending upon the design.

There is no output other than some noise when power is initially applied to

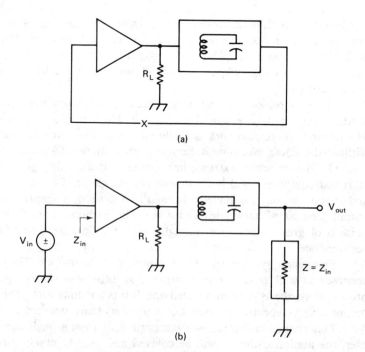

Figure 7.1 Block diagram of an *LC* oscillator. The feedback loop is broken at an arbitrary position for analysis, shown as "x" in (a). The broken loop is shown in (b).

Sec. 7.1 Oscillator Concepts

the circuit of Fig. 7.1a. Even if the amplifier was perfect and noiseless, there would be resonator noise from the loss components. This is amplified. Part of the resulting output noise appears in the load. The rest is applied to the resonator which acts as a filter. Noise within the tuned circuit bandwidth passes with finite attenuation. Noise outside of the passband is attenuated more.

The output from the resonator is again applied to the amplifier input. If the resulting signal is greater than the original noise power and is in phase with the original signal, it will be further amplified. If the phase is wrong, the noise will not grow.

When confronted with an oscillator circuit, the first question to ask is if oscillation will actually occur. The method for evaluation is shown in Fig. 7.1b. The oscillator feedback loop is broken somewhere, although it does not really matter what point is chosen. In this example, the break is at the point marked with an x in Fig. 7.1a.

The amplifier is characterized by a small-signal input impedance. The loop input is driven by a voltage source. The output is terminated in an impedance equal to that at the amplifier input.

The output voltage appearing across the terminating impedance is evaluated for a given input voltage. The voltage is mathematically resolved into two components. One is in phase with the input while the other is in phase quadrature (± 90 degrees out of phase) with the input. The in-phase part is compared with the input. If the net loop voltage gain is positive and greater than unity, oscillation is possible at the frequency of the input signal. If the gain is less than $+1$, oscillation cannot occur, at least not at the frequency of the input drive.

Assume that the evaluation has been performed at the tuned circuit resonant frequency and oscillation is predicted. We now examine the original closed loop of Fig. 7.1a. Noise present at the instant when power is applied is amplified and filtered and reapplied to the input. It will grow in amplitude. The analysis predicts that power will increase forever, without bound if the amplifying device and resonator are described with the usual small-signal models. This is, of course, impossible. Energy must be conserved, implying that the amplitude will eventually stop growing. The physical mechanism that provides amplitude stabilization is usually the nonlinear amplifier behavior. Amplifier gain decreases as signals grow, at least at the operating frequency. The amplitude will stabilize when the gain around the loop is exactly unity.

The evaluation may also be performed with an input current source. The output current is then evaluated and an appropriate gain is calculated. The results will be the same.

A nonlinear amplifier is not required for a stable oscillator. An external detector could be used in conjunction with some means of electronically changing amplifier gain. When the amplitude reaches a predetermined level, the gain is reduced until unity loop gain is achieved. Stabilization then occurs as a result of automatic gain control (agc).

The magnitude of amplifier nonlinearity may be evaluated by examining the harmonic content. There will be no harmonic distortion if the amplifier is perfectly linear. Even when agc is used for stabilization, there may be some harmonic distortion, indicating some amplifier nonlinearity. Oliver (1) has shown in an elegant treatment of the Weinbridge oscillator that some nonlinearity is desirable. If the circuit is exceptionally "clean" with very low distortion, the amplitude will fluctuate considerably, especially when the circuit is perturbed, such as occurs when the "resonator" frequency is changed.

Many rf oscillators use some agc. However, agc is usually applied to an amplifier with an inherent nonlinear behavior. The agc action determines a proper operating point so amplitude stabilization is obtained from intrinsic amplifier nonlinearity. It is clear from the foregoing that both small- and large-signal device behaviors are critical in describing oscillator behavior. Small-signal models were discussed extensively in Chap. 5. Large-signal models for the bipolar junction transistor and junction field effect transistor were presented in Chap. 1.

The bipolar transistor is described by an exponential equation. If the emitter-base port is driven with a voltage, emitter current is given by $I_e = I_{es} e^{(qV/kT)}$. The collector current is approximately equal to that in the emitter. This is a highly simplified large-signal model, but is often adequate to describe the salient features of oscillator behavior.

The collector current may be evaluated as a function of signal amplitude and bias if the junction voltage is considered to be a sum of a bias, V_0, and a sinusoidal signal, $V_s \sin \omega t$. Small-signal conditions are maintained for a normal bias and signal amplitudes of less than about 10 mV. Current waveforms become highly nonlinear for larger signal levels, containing not only the fundamental signal but harmonics. A shift in dc bias also results.

Most amplifiers and oscillators are biased for a constant emitter current. Harmonic content in the collector current increases as input signals become large. This occurs only at the expense of reduced amplitude of the fundamental current. Hence, the transconductance *at the fundamental frequency*, the ratio of the collector current at ω to the drive voltage, is a decreasing function of drive voltage. This is the usual current limiting mechanism in bipolar transistor oscillators. Voltage limiting may also occur as a result of transistor saturation. The details of current limited operation are presented in an excellent treatment by Clarke and Hess (2). Also, several pertinent curves are presented in Chap. 1.

Large-signal JFET operation is similar. The idealized JFET is described by a quadratic with drain current given by $I_d = I_{DSS}(1 - V_s/V_p)^2$. If the device is biased with a constant current source small with regard to I_{DSS}, the quiescent source voltage, V_s, is close to V_p. Large-signal behavior is evaluated if the source potential is replaced with the sum of a dc bias and a sinusoidal signal. As signal amplitude increases, the positive-going source voltage will cause the current to decrease. Negative-going source voltage peaks will increase the drain current. Because of the quadratic device nature, the current increases will be larger in magnitude than the decreases. The dc source voltage will shift as the increasing signal amplitude "tries" to increase the current, owing to constant current biasing. The shift will be toward the pinch-off region, decreasing the average transconductance.

7.2 THE COLPITTS OSCILLATOR

A small-signal equivalent circuit for the Colpitts oscillator is presented in Fig. 7.2a. The amplifier operates in the common base configuration. This will be shown later to have no bearing upon circuit operation.

The basic Colpitts oscillator is biased with a constant emitter current. The standing current determines the impedance seen looking into the emitter, $R_{in} = 1/g_{me} = 26/I_e(\text{mA})$. The common emitter transconductance is g_{me}. The model used is

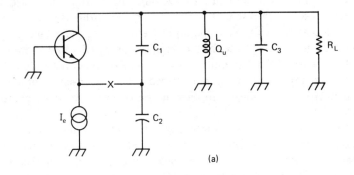

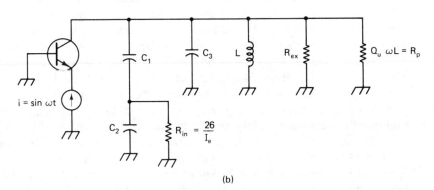

Figure 7.2 The Colpitts oscillator. The basic configuration is shown in (a) while (b) shows the broken loop for analysis.

the simplest possible for the common base amplifier, a device with an emitter resistor and a noninverting current gain of unity.

The inductor is shown going from the collector to ground. In practice, this inductor often serves a dual role, providing a path for collector bias voltage. The grounded end is actually attached to ground through a large bypass capacitor. The resonant frequency is determined by the inductance and the net capacitance, C_3, in parallel with the series combination of C_1 and C_2. R_L accounts for the loss associated with the parallel loss resistance of the inductor, $R_p = Q_u \omega L$, and for external loading that may be present.

Starting is analyzed with the circuit of Fig. 7.2b. The loop has been broken at the point marked with an x in the closed loop oscillator. The input signal comes from a current source with a strength of sin ωt. The position in the loop formerly occupied by the transistor input is terminated in a resistor equaling the small-signal common base input resistance. An external resistor is included in the line from the tapped capacitors to the emitter in some versions of the Colpitts oscillator. The value for R_{in} shown in the figure will also include this resistance. In the opened loop, both components of resonator loading are shown, that from an external load (R_{ex}) and R_p representing the inductor losses. The input frequency used for analysis is usually the tuned circuit resonant frequency, although this is by no means mandatory. In later examples, we will allow the input frequency to vary, sweeping the response.

Before going through the formal details, it is worthwhile to examine the circuit intuitively. The amplifier has unity current gain. Because the input is driven from a current source, input current is not altered by the transistor input resistance. The current injected at the emitter appears at the collector with no change in phase or amplitude.

Assume that external loading is negligible. The circuit at the collector is essentially resonant. Hence, a resistive collector impedance is seen, the parallel combination of the resonator load, R_p, and a feedback resistance, R_f. The feedback component is the result of R_{in} transformed by the $C_1 - C_2$ network. This impedance is usually high, comparable with R_p. A signal voltage appears at the collector.

The collector voltage is divided by the series capacitors, C_1 and C_2. If the capacitors are large enough, the reactance in parallel with R_{in} will be small with respect to that resistance. Hence, there is a minimum of phase shift associated with the voltage division. The current flowing in R_{in} will have approximately the same phase as the input drive. Moreover, the magnitude will be higher. Although the transistor has unity current gain, the collector current appears across a larger load impedance. The voltage division action of C_1 and C_2 is accompanied by a current multiplication action. Oscillation is predicted.

The analysis may now be formalized. The so-called Colpitts network is first evaluated. This is the network associated with C_1, C_2, and the resistive load, R_{in}. There are two things that must be known about the network. What is the immittance seen looking into the network? This is shown in Fig. 7.3a. The other characteristic that must be calculated is the voltage transfer function, H, shown in Fig. 7.3b.

If we redefine some of the parameters, the analysis is simplified slightly. C_1 is

Sec. 7.2 The Colpitts Oscillator

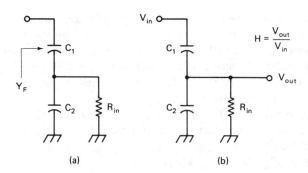

Figure 7.3 The vital networks of the Colpitts oscillator. The input admittance of the capacitive divider network is evaluated at (a) where the load, R_{in}, is the input resistance of the common base amplifier under small signal conditions. The transfer function of the capacitive divider is presented at (b).

replaced by C while C_2 is normalized to C with a constant, K; specifically, $C_2 = KC$. R_{in} is replaced by the equivalent conductance, $G = 1/R_{in}$. Then, the admittance seen at the collector from the Colpitts network is

$$Y_f = \frac{\omega^2 C^2 G}{G^2 + \omega^2 C^2 (K+1)^2} + j\omega C \frac{[G^2 + \omega^2 C^2 K(K+1)]}{G^2 + \omega^2 C^2 (K+1)^2} \qquad (7.2\text{-}1)$$

where the f subscript tells us that this is the result of a feedback component.

Using the same notation, the voltage transfer function from the collector to the load, R_{in}, is given by

$$H = \frac{\omega^2 R_{in}^2 C^2 (K+1) + j\omega R_{in} C}{1 + \omega^2 R_{in}^2 C^2 (K+1)^2} \qquad (7.2\text{-}2)$$

The gain of the oscillator may now be evaluated. The voltage at the collector is the net impedance appearing at that node for an emitter input of 1 A. The net admittance is

$$Y_{col} = \frac{\omega^2 C^2 G}{G^2 + \omega^2 C^2 (K+1)^2} + \frac{1}{R_{ex}} + \frac{1}{Q\omega L} \\ + j\left\{ \frac{\omega C[G^2 + \omega^2 C^2 K(K+1)]}{G^2 + \omega^2 C^2 (K+1)^2} + \omega C_3 - \frac{1}{\omega L}\right\} \qquad (7.2\text{-}3)$$

The impedance, Z_c, is the complex reciprocal of this admittance. The voltage appearing across R_{in} is HZ_c. The current flowing in the resistance, and hence, the current gain, is then GHZ_c. It is convenient to evaluate the phase angle of the gain and to multiply the cosine of this angle by GHZ_c, the result being the gain component in exact phase with the input.

The equations presented can be used to write a computer or calculator program for calculation of gain and phase angle. While the equations look formidable, they are complete and contain a number of similar terms. The programs are not particularly complicated.

An alternative approach is perhaps somewhat more direct for a numerical analysis. This is to use the ladder method. Both, of course, give exactly the same results.

The preceding equations provide little intuition. Hence, a numerical example is considered, a 10-MHz oscillator. The inductor is 0.5 μH with $Q_u = 250$. The Colpitts capacitors are chosen to be 100 and 1000 pF. The extra capacitance required for resonance is then 416 pF. The circuit is shown in Fig. 7.4.

The emitter current is varied from 0.01 to 10 mA for the first numerical experiment. The in-phase portion of current gain in dB and the phase angle are both plotted as functions of current. A logarithmic current scale is used in the graph. The results are also shown in Fig. 7.4 for a 10-MHz input.

The gain is high at high current, approaching 20 dB for $I_e = 10$ mA. The phase angle is over 10 degrees when the gain is high. It is intuitively reasonable that the angle of the gain should not be zero at high emitter current. This phase difference results from the low values of R_{in} that correspond to higher I_e values. The low resistance partially shunts C_2 (1000 pF), causing the net Colpitts network capacitance to be larger than it would be with no loading. This decreases the resonant frequency. Circuit behavior is further complicated by a phase shift associated with H, the voltage transfer function of the Colpitts network.

As the current decreases, the gain follows. The phase angle also becomes closer to the ideal zero degrees. Unity gain occurs for $I_e \simeq 0.04$ mA. This provides information about the eventual operating level of the oscillator. Since the emitter current is 0.04 mA for unity gain, *operating* transconductance will be $g_{me} = 1/R_{in} = 0.00154$ S. The associated value for R_{in} is 650 Ω.

The gain at starting is nearly 18 dB with a phase angle of 9 degrees if the oscillator is biased to 1 mA. The transconductance at 10 MHz must decrease by a factor of $1/0.04 = 25$ to achieve stable operation. The transistor must be operating then in a very nonlinear mode. The collector current will contain large amounts of harmonic current, most of it shunted to ground by the resonator. A lower starting gain may be chosen if low harmonic distortion is desired. This is realized by rebiasing the oscillator or by inserting resistance between the capacitor tap and the emitter. Both will have the desired effect on starting gain. The series resistor, however, allows the oscillator to operate at a higher power level.

For example, the oscillator could be biased at 1 mA with a series resistor of 500 Ω. The net resistance looking into the emitter is 526 Ω, the same as the transistor with no resistance and $I_e \simeq 0.05$ mA. Starting gain is then 1.7 dB with an associated phase angle of 1.4 degrees. Only a small excursion into the nonlinear region is required to achieve current limiting at unity gain.

There can be a significant difference in the phase of the gain between starting and operation. Assume that the oscillator is operating at the stable unity gain point and there is some noise in the transistor. The emitter current is the bias plus a small noise component. The instantaneous change in emitter current will modulate the gain. The operating level must change to achieve stable operation. This is accompanied by a slight shift in phase. The net result is a phase modulation of the output signal. This is a general characteristic for all oscillators.

Sec. 7.2   The Colpitts Oscillator

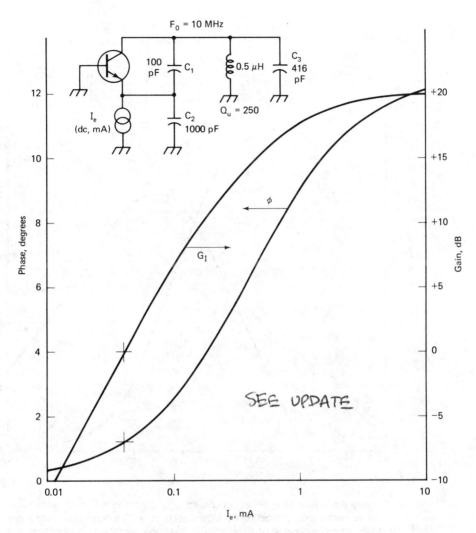

Figure 7.4 Evaluation of the open loop gain and phase angle of a 10-MHz oscillator as a function of the bias current.

An emitter resistor can reduce the magnitude of the phase modulation. If the preceding example is considered, the difference between the phase at starting and that during operation is only 0.26 degrees. The phase difference between starting and operation without emitter degeneration would be 7.88 degrees. The oscillator with the emitter degeneration will have a more restricted frequency tuning range though.

The C_1 and C_2 values chosen for the preceding example were arbitrary. Consider another example, a 10-MHz Colpitts oscillator with $I_e = 2$ mA and $L = 0.5$ μH and $Q_u = 250$. External loading is neglected. Figure 7.5 shows the effect of changing

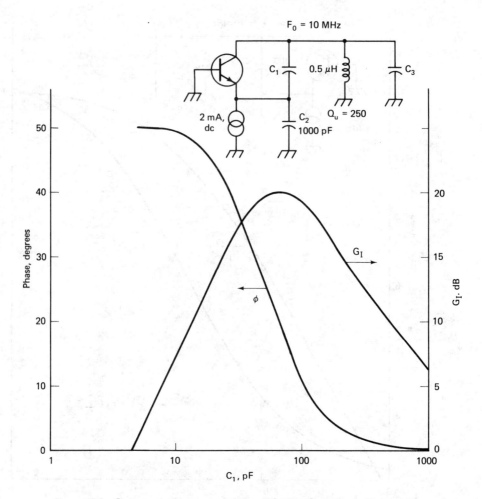

Figure 7.5 Gain and phase angle for a 10-MHz Colpitts oscillator where the upper Colpitts capacitor, C_1, is allowed to vary. C_3 is adjusted accordingly to maintain resonance at 10 MHz. There is a considerable amount of flexibility in the choice of components.

the upper Colpitts capacitor, C_1. C_3 is changed to preserve resonance at 10 MHz. C_1 varies from a few pF up to 1000 pF, the value where C_3 nearly vanishes. The curves of Fig. 7.5 show both the variation in gain and associated phase.

The results are rather dramatic with current gain greater than unity for all values of C_1 considered. It is not surprising that working oscillators may be built using nothing more than empirical design methods.

The gain peaks at C_1 close to 75 pF. This is a point where impedances are matched. The inductor has a parallel loss resistance of $Q_u \omega L = 7854\ \Omega$. The gain is maximum when the resistance presented to the resonator from the transistor via the Colpitts network, R_f, equals this same value. R_f is the reciprocal of the real part of the admittance given in Eq. 7.2-1.

Although starting gain peaks at a low value for C_1, there is considerable phase associated with the gain. The phase difference drops considerably as C_1 becomes large. For this reason, the better designs use large values for the capacitors. The upper limit occurs when $C_3 = 0$. This suggests that the better oscillators will be those with lower inductance values. Again, there is a practical limitation, although this may be overcome through modifications to the basic Colpitts design that are presented later.

Consider an open loop oscillator with "input" frequencies other than just the desired resonant one. Figure 7.6 shows phase and gain curves for the circuit, also shown in the figure. Two different emitter currents are used, 2 mA to represent starting and 0.041 mA to represent operation near current limiting. The input is swept 20 kHz below and above the 10-MHz resonant center frequency.

The response is broad at starting with $I_e = 2$ mA. Indeed, the peak is below the frequency range shown. The initial frequency will be lower than the final one.

The response is nearly that expected from an ideal resonator at the lower emitter current. The gain peak occurs at the resonant frequency and the phase differs from the ideal by only one degree. The phase response of a single resonator should be 45 degrees when the response is down by 3 dB from the peak. This is not shown in Fig. 7.6. The difference results from the specific gain plotted, the part in phase with the driving input.

The loaded resonator Q is much higher for the lower emitter current. The load presented to the resonator is 3.5 kΩ for $I_e = 2$ mA. The resistance presented to the collector increases to 77 kΩ as limiting is approached. The Q is essentially the unloaded value. This does not exactly represent an operating oscillator, for the average current is still that at starting. Generally, though, the Q will increase as limiting occurs. The more conservative design approach maintains a high loaded Q, even during the starting condition. Again, emitter degeneration may be used to advantage.

We have considered limiting only from an approximate viewpoint. The emitter current corresponding to unity gain has been evaluated and compared with the starting current. This provided some information about the degree of nonlinear operation required to achieve limiting. Nothing has been said about the actual operating level, the voltage at the collector during stable oscillation.

Consider the circuit shown in Fig. 7.7. Assuming an emitter-base offset voltage of 0.65, the emitter current will be 2 mA. Slight changes in emitter potential will not alter this value significantly. It is clear from the curves of Case A of Fig. 7.6 that this circuit will oscillate. The starting gain is about 19 dB with a phase of 10 degrees. Some emitter degeneration would reduce both. The task now is to evaluate the operating level.

Amplifier distortion was discussed in Chap. 6. A rule of thumb was presented to estimate the level where gain compression begins. It was when the signal current was large enough that the instantaneous value decreased to zero at one point in the

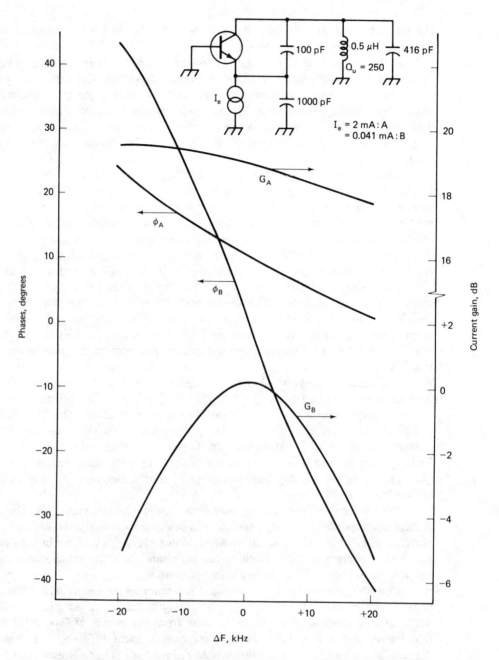

Figure 7.6 Gain and phase angle versus frequency of the broken loop of a Colpitts oscillator. Case A represents a bias current of 2 mA, a starting condition. Case B is a lower current, 0.041 mA, leading to a sharper response. Case B is representative of the oscillator during operation.

Sec. 7.2 The Colpitts Oscillator

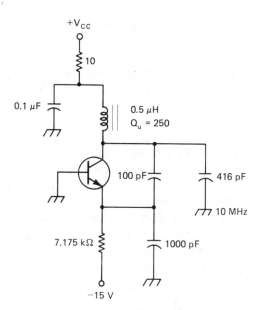

Figure 7.7 A practical 10-MHz Colpitts oscillator biased to $I_e = 2$ mA. This circuit is used in the text discussion of operating level.

operating cycle. Signal current would grow to twice the dc value at another point in the cycle. This does not represent the maximum output power of the amplifier, but the point of 1-dB gain compression.

This critical level depends upon two factors, the standing dc current and the impedance at the collector. Consider the rule of thumb as applied to the oscillator of Fig. 7.7.

The curves of Fig. 7.6 and related calculations showed that, in limiting, the impedance presented to the collector circuit from the Colpitts network was very high. The net load resistance was determined essentially by the unloaded resonator Q. The 0.5-μH inductor has a parallel resistance at 10 MHz of $Q_u \omega L = 7854$ Ω. Using this value with a peak signal current of 2 mA, the peak voltage predicted at the collector is 15.7 V.

The predicted voltage is that associated with the beginning of gain compression. An even higher value might be expected if the circuit was driven into saturated output. Measurements by the author have shown that a reasonable rule of thumb for an hf oscillator is that peak collector voltage is given by $V_c = 2I_e R_c$ where R_c is the resonator impedance resulting from the Q. Using this rule, a collector voltage of over 30 V peak, or 60 V peak-to-peak is predicted for the circuit of Fig. 7.7.

These values seem high. However, if a transistor with a high breakdown voltage was used and the power supply, V_{cc}, was in excess of 30 V, this estimate would be reasonable. The presence of emitter degeneration will not severely affect the result unless starting gain is very low, perhaps under 2 dB. The accuracy of this rule of thumb is about 2 dB.

The rule would not apply if the power supply was under 30 V. Current limiting would not occur. Instead, the transistor would go into saturation over part of the operating cycle. The transistor then no longer behaves as a current source. Instead, it becomes a relatively low value resistor. Energy from the resonator is absorbed in the saturation resistance, effectively lowering the collector impedance, R_c. The oscillator is then voltage limited. This is acceptable if a well-defined voltage is desired. However, there is a compromise. As the collector impedance drops, so does the loaded resonator Q. This will have adverse effects upon oscillator noise performance.

The oscillator of Fig. 7.7 is not a total loss. If the emitter bias was decreased to 0.5 mA and the collector supply was set at 10 V, the predicted current limited level is $V_c = 7.8$ V peak. Resonator Q remains high during operation.

Another change which could improve the circuit is to decrease the inductor value. The resonator impedance decreases accordingly, allowing current limited operation with higher emitter bias. This, however, presents many practical problems.

Consider a circuit modification, the Clapp oscillator, or series tuned Colpitts shown in Fig. 7.8a. The design is reasonably straightforward, although the real virtue of the circuit is not generally appreciated.

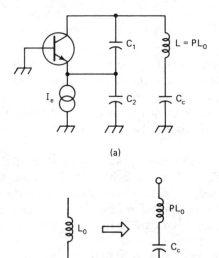

Figure 7.8 The Clapp variation of the Colpitts oscillator. The basic configuration is shown at (a) while the "Clapp transformation," the replacement of the Colpitts inductor, is shown at (b).

Initially, a traditional Colpitts oscillator is designed with an inductor L_0 and capacitors C_1 and C_2 being the only ones required for resonance. The inductor is then replaced with one that is larger by a factor P. This is shown in Fig. 7.8b. The extra inductive reactance introduced is eliminated by a series capacitor with such a value that the total reactance is identical to the original with only L_0. The conditions imposed on P and the required Clapp capacitor are

$$j\omega L_0 = j\left(\omega L_0 P - \frac{1}{\omega C_c}\right) \qquad (7.2\text{-}4)$$

Sec. 7.2 The Colpitts Oscillator

Solving for the capacitance

$$C_c = \frac{1}{\omega^2 L_0 (P-1)} \qquad (7.2\text{-}5)$$

Consider the impedance seen looking into the new inductor–capacitor combination. The Q_u of the inductor will be represented by the series resistor form, $R_s = \omega L/Q_u$. The impedance seen is

$$Z = \frac{\omega P L_0}{Q_u} + j\left(\omega P L_0 - \frac{1}{\omega C_c}\right) \qquad (7.2\text{-}6)$$

Using the condition imposed by resonance, this reduces to

$$Z = \frac{\omega P L_0 + j\omega L_0 Q_u}{Q_u} \qquad (7.2\text{-}7)$$

This is inverted to obtain the admittance

$$Y_{\text{in}} = \frac{Q_u(\omega P L_0 - j\omega L_0 Q_u)}{\omega^2 P^2 L_0^2 + \omega^2 L_0^2 Q_u^2} \qquad (7.2\text{-}8)$$

Our concern is primarily with the resistive part of the parallel admittance. Taking the reciprocal of the real part of Y_{in}

$$R_p = \frac{\omega^2 L_0^2 (P^2 + Q_u^2)}{\omega L_0 P Q_u} = \omega L_0 \left(\frac{P}{Q_u} + \frac{Q_u}{P}\right) \qquad (7.2\text{-}9)$$

This is to be compared with the usual parallel resonator impedance found in the simpler Colpitts oscillator, $R_p = Q_u \omega L_0$. Multiplication of the first term in Eq. 7.2-9 by Q/Q yields

$$R_p = Q_u \omega L_0 \left(\frac{P}{Q_u^2} + \frac{1}{P}\right) \qquad (7.2\text{-}10)$$

The second term is completely dominant with typical values ($Q_u = 250$ and P between 1 and 20), leaving the final conclusion that the resistance seen looking into the modified inductor is the resistance of the original resonator with L_0 diminished by the "Clapp factor," P.

An earlier Colpitts oscillator studied used a 0.5-μH inductor at 10 MHz, yielding $R_p = 7854$ Ω. Resonance is maintained if the inductor is replaced with the series combination of a 5-μH inductor and a suitable capacitor, 56.3 pF. The resistance decreases by a factor of $P = 10$ to 785 Ω. It is now possible to increase the emitter

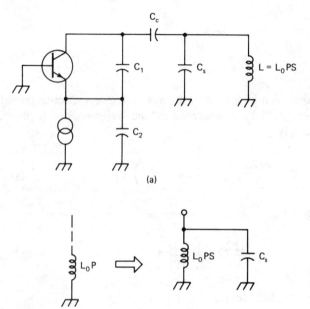

Figure 7.9 A further variation of the Colpitts oscillator, the Seiler. The inductor of a Clapp circuit is replaced with one of smaller value and a paralleled capacitor, as shown in (b).

current by a factor of 10 while still maintaining current limited operation. The oscillator gain equations should be reevaluated using the lower load resistance. The net resonator Q is left virtually unchanged at Q_u so long as the circuit is not significantly loaded by the transistor.

A second modification is in the addition of yet another capacitor. This variation is called the Seiler oscillator, shown in Fig. 7.9a, and may be regarded as a refinement of the Clapp circuit.

The Clapp circuit is limited from a practical viewpoint. The inductor may not be increased arbitrarily to very large P values. The resulting inductor would have a larger number of turns than desired, leading to stray capacitance effects. Also, the Clapp capacitor becomes small as P grows, making it more subject to the effects of other stray capacitance.

The large inductor of the Clapp type is replaced with a smaller one which is then paralleled with a capacitor, C_s, to establish the original resonance. The Clapp inductor is decreased by the factor S. While P was greater than unity, S is less than 1. This is shown in Fig. 7.9b.

The Clapp inductor on the left of Fig. 7.9b has an admittance

$$Y = \frac{1}{Q_u \omega L_0 P} - \frac{j}{\omega L_0 P} \qquad (7.2\text{-}11)$$

The Seiler variation has a similar admittance, with the factor S now factored into the equation

Sec. 7.2 The Colpitts Oscillator

$$Y = \frac{1}{Q_u \omega L_0 PS} + j\left(\omega C_s - \frac{1}{\omega L_0 PS}\right) \qquad (7.2\text{-}12)$$

Resonance is preserved if the reactive parts of the two admittances are equal, leading to the relationship for the Seiler capacitor

$$C_s = \frac{1-S}{\omega^2 L_0 PS} \qquad (7.2\text{-}13)$$

Inserting the resonance condition into the admittance equation (Eq. 7.2-12) yields

$$Y_{in} = \frac{1}{Q_u \omega L_0 PS} - \frac{j}{\omega L_0 P} \qquad (7.2\text{-}14)$$

Rearranging terms and inverting leaves the impedance looking in as

$$Z_{in} = \frac{Q_u \omega L_0 PS}{1 - jQ_u S} \qquad (7.2\text{-}15)$$

Having the immittance, the impedance of the Clapp capacitor, C_c, may be added. Collecting terms yields the expression looking into the position originally occupied by L_0, the Colpitts inductor

$$Z = \frac{Q_u \omega L_0 PS}{1 + Q_u^2 S^2} + j\left(\frac{Q_u^2 \omega L_0 PS^2}{1 + Q_u^2 S^2} - \frac{1}{\omega C_c}\right) \qquad (7.2\text{-}16)$$

This equation is suitable for numerical evaluation. The impedance is calculated, converted to an admittance with the reciprocal of the real part being the equivalent parallel resistance. Note that both the Clapp and Seiler factors, P and S, are included. If they are independently specified, L, the new inductance is known in terms of L_0, the original Colpitts inductor. If L is used as an input, only one parameter, P or S, must be specified. The other is predetermined by $L = L_0 PS$.

Consider an example. Again, $L_0 = 0.5$ μH with Q_u constant at 250 at 10 MHz. An inductor $L = 2$ μH is arbitrarily chosen and P is set at 15. The result of applying this data with the equations presented above is $C_c = 36.2$ pF, $C_s = 92.9$ pF, and $R_p = 146$ Ω. R_p is very low when compared with the original value of nearly 8000 Ω! Current limited operation is assured with very large emitter current values.

The parallel resistance presented to the collector may be evaluated as a function of P if all parameters except P are held constant. A minimum will be found. A change in Q_u will alter the position of the minimum.

The analysis presented is not exact. Resonance is calculated by comparing only imaginary parts of immittances. With high Q circuits and large impedance transformations, the reactive components are affected by the loading. This alters the resonant frequency of the composite network. More refined calculations or empirical work will be required to sort out these details. The qualitative behavior of the circuits is, however, clear. The Clapp and the Seiler oscillators are both variations of the fundamental Colpitts type. They utilize transformations to reduce the impedance presented to the collector. This allows preservation of resonator Q and operation at a higher power level.

The Clapp and Seiler circuits present a biasing problem. There is no longer a dc path through the tuned circuits. A practical version of the Seiler oscillator is shown in Fig. 7.10. The Seiler capacitor, C_S, is made variable, allowing the frequency to be tuned. Collector bias is applied to the transistor through a radio frequency choke with a series resistor, R_c. This resistor is often of critical importance.

Examination of the circuit reveals two resonant frequencies. The desired one results from L, the usual inductor. This inductance is essentially a short circuit at low frequencies. However, the inductance of the choke resonates with C_1 and C_2 at a frequency well below the operating range of the oscillator. This assumes that the inductance of the choke is large with regard to L. The circuit may oscillate at the lower resonant frequency as a traditional Colpitts oscillator if the Q of the choke is high enough.

Oscillation at the lower frequency could have two possible effects. The oscillation could dominate the circuit behavior, forcing the gain to be compressed so that high frequency oscillation could not occur. The other possibility, a common one, is that oscillation occurs at both frequencies. When this happens, the time domain output of the oscillator will be a modulated waveform. In the frequency domain, the higher

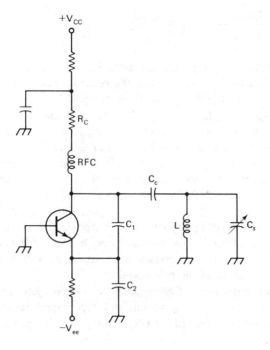

Figure 7.10 A practical version of the Seiler oscillator. Frequency tuning results from C_s. A series resistor, R_c, is added to the radio frequency choke. This degrades the Q at the lower frequency where the choke becomes resonant with circuit capacitances to prevent oscillation at an undesired frequency.

frequency will occur with sidebands separated from the desired carrier output by the frequency of the lower resonator. The low frequency itself will also be present in the output.

The solution is straightforward. The resonant frequency associated with the choke is calculated. Then, the gain for oscillation starting is evaluated at this frequency. If the gain is greater than unity, R_c is increased, effectively decreasing the unloaded Q of the choke at the lower frequency. Generally, the value of R_c and the inductance of the choke may be chosen to have a minimum affect upon the higher frequency resonance.

7.3 FURTHER LC OSCILLATOR TOPICS

Most of the information presented for the Colpitts oscillator and variations have been theoretical. The examples were at 10 MHz. The methods are, however, quite general.

The methods are applicable from very low frequencies up to the vhf and low uhf spectrum. Experiments by the author at frequencies up to 100 MHz show that the gain equations presented are accurate to within a fraction of 1 dB. That is, the conditions where oscillation just begins can be predicted to the stated accuracy. The absolute accuracy of the open loop gain is perhaps not as well defined, especially considering the nonlinear nature of an LC oscillator.

The best predictions of operating level are done with refined computer modeling methods. The rule of thumb seems to hold well through the hf spectrum and at vhf when good transistors are used, ones with an F_t well above the operating frequency.

The model used in the analysis was extremely simple. In spite of this, the predicted results agree well with observation. The methods outlined could certainly be extended to use a more refined model. This could be a hybrid-pi type or actual measured small-signal parameters. The opened loop consists of a common base amplifier with a following network that may usually be represented as a ladder. As such, composite small-signal parameters may be calculated using appropriate cascading operations.

The simple two-port analysis methods are lacking at frequencies above about 500 MHz. More often, a "one-port" method of analysis is used. This is covered in a later section.

All of the examples considered have used the common base connection. Many oscillators, however, are not really connected, or don't appear to be connected as a common base amplifier. In the most general sense, it does not matter what point in an oscillator circuit is grounded.

The lack of importance of the grounded point is emphasized in Fig. 7.11. Figure 7.11a shows the topology considered so far. The circuit is the "common base" amplifier, the ground point being the transistor base. However, the ground symbol could be removed from the circuit of Fig. 7.11a and it would still be a valid small-signal representation. The ground may now be moved to any point that might be convenient.

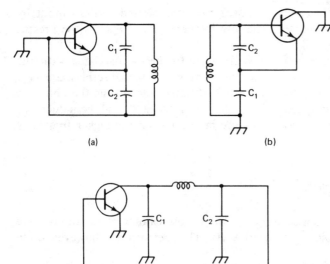

Figure 7.11 Three variations of the Colpitts oscillator with various points grounded. The topology used for analysis, the cb, is shown at (a). A cc variation is shown at (b) with the common emitter variation at (c). All configurations are identical, although some are more practical than others.

Figure 7.11b shows a common collector topology while Fig. 7.11c shows the grounded emitter configuration. All three are identical.

It may become subtle to demonstrate that all topologies of Fig. 7.11 are indeed identical. The loop was broken at the emitter in the analysis of the previous section. A signal was injected between the emitter and ground while the output load was placed between the Colpitts capacitor tap and ground. Consider the grounded collector version, Fig. 7.11b. The loop is again broken at the emitter and a generator is inserted between the emitter and the base. The positive polarity drives the emitter. The termination is then placed across C_2. The results would then be identical to those achieved with the previous analysis.

The loop might be broken at a different place. For example, the base could be opened in Fig. 7.11b. A driving voltage (or current) is applied between base and ground with a termination placed from the resonator to ground. The resistor value is the small-signal input resistance of the transistor. The gain numbers obtained with this analysis will be different than those obtained with the previous analysis. However, if the emitter current is varied as was done before, oscillation will cease at the same I_e value.

Other differences occur in the modified topologies. Some are of practical importance while others are only details of analysis. For example, C_2 is the capacitor between the base and emitter. This is usually large with respect to C_1. Sometimes just the opposite is seen in practical oscillators using a grounded collector. This can have adverse affects upon gain and phase characteristics. A detailed analysis is required.

The grounded collector version has a virtue that is both practical and experimentally interesting. The collector appears as a current source, at least when the simple models are used. Hence, it may be lifted from ground and a series resistor inserted. This provides a convenient place to obtain an output that is moderately well isolated from the rest of the circuit. Changes in the loading have little impact upon oscillator operation. Moreover, collector current may be monitored with a high frequency oscilloscope. The highly nonlinear nature of the current is then easily observed.

Operation is viewed differently for various connections. Both the grounded base and collector oscillators utilize amplifiers which are noninverting. To the contrary, the common emitter circuit, Fig. 7.11c, uses an inverting amplifier. If the reactive part of the circuit is examined as a pi-network, it is found that the network shows a phase shift of 180 degrees if the Q is high. Oscillation is again predicted intuitively, although the reasoning process is quite different.

Field effect transistors are used frequently in oscillator applications. Two versions of the Colpitts oscillator using a JFET are presented in Fig. 7.12. Both are virtually identical. The design is the same as used with the bipolar transistor with R_{in} replaced by $1/g_{ms}$, where g_{ms} is the common source transconductance evaluated at the bias point. R_s is chosen so the FET is biased close to pinch-off. As oscillation begins to build up, the negative-going excursions of the source will increase the channel current in proportion to the square of the source voltage. Positive-going source voltage changes will decrease current. The negative-going excursions dominate, leading toward increased current. However, this increases the dc voltage drop across the bias resistor, moving the bias point further toward pinch-off and reduced average transconductance.

The FET must use nothing more than a square-law characteristic to achieve limiting. This is contrasted with the bipolar transistor with its stronger, exponential characteristic. If proper current limited operation is to be maintained, the biasing must be carefully adjusted. If the starting gain is too high, resulting in an excess g_{ms}, the gate-source diode may become forward biased during part of the cycle. This results in voltage limited operation, degraded resonator Q, and, often, excessive noise. Circuits like those shown are possible if the biasing is chosen so that the starting gain is sufficiently small, usually under 1 dB when analyzed in the "common gate"

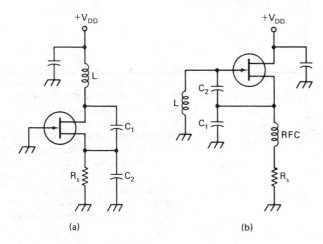

Figure 7.12 Colpitts oscillators using a JFET. Bias must be chosen carefully to achieve current limited operation without causing conduction in the gate-source diode.

mode. It is wise to measure the gate and source potentials with a dual trace oscilloscope to ensure that the diode is never forward biased. This is often difficult, or even impossible; oscilloscope probes usually load the resonator so heavily that resonator Q is degraded enough to prevent oscillation.

Figure 7.13 shows a modification of the previous circuits. The oscillator is converted to the Clapp configuration. A series drain resistor provides a convenient place for output to be obtained. The biasing method is changed significantly as well. The presence of the Clapp capacitor, C_c, removes the gate from dc ground. As such, the gate may take on other dc potentials. The gate is initially held at ground by a bias resistor, R_b.

The current will be I_{DSS} and the transconductance is at its highest value when power is initially applied to the circuit. If the capacitor values and other components are correctly chosen, oscillation will begin. The gate potential will vary as amplitude builds. The diode from the gate-to-ground will prevent positive voltage excursions greater than 0.6 V. A slight diode current will flow. This current then charges the capacitors in the circuit. The dc gate potential will decrease with this mechanism, slowly driving the device toward pinch-off. Amplitude stabilization occurs when the combination of agc action of the diode–resistor–capacitor assembly and nonlinear FET conduction provides a transconductance that is proper for unity loop gain.

Sometimes a resistor is contained in series with the choke in the source lead. This serves a dual purpose. Not only does it place a lesser burden for stabilization upon the agc circuitry, but it will decrease the Q of the choke at lower frequencies, preventing possible spurious oscillations.

FET oscillators with agc are very common and have several virtues. Generally, field effect transistors are not well specified; there are large variations in both I_{DSS} and V_p, even for a given FET type. With the large variations, the transconductance is not as predictable as would be desired. This makes it difficult to build oscillators like those shown in Fig. 7.12 on a production basis. A worst-case analysis will ensure

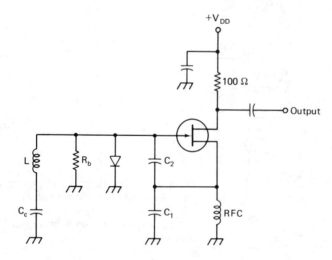

Figure 7.13 A practical version of a JFET Clapp oscillator. The diode will conduct over a small portion of the cycle, charging the circuit capacitances to an average negative dc voltage. This provides some automatic gain control for the oscillator, allowing the limiting to be well defined by the quadratic nature of the FET. The agc scheme is easily applied to other FET oscillators and allows such circuits to operate over very large frequency ranges.

proper oscillation over the frequency range of interest if agc is used. The dc voltages will change with changes in gain that result from device characteristic differences or changes as the circuit is tuned over a frequency range. Operating level is not as easily defined with the FET oscillator as it is with the bipolar.

A circuit of equal popularity is the Hartley. It is simple and generally similar to the Colpitts. A small-signal version of the Hartley oscillator is shown in Fig. 7.14a.

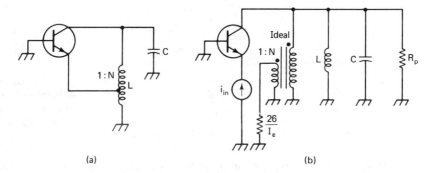

Figure 7.14 The basic Hartley oscillator is shown at (a) while the broken loop used for analysis is presented in (b).

The Hartley oscillator has the virtue of having fewer components in the resonator. A tapped inductor is used to provide the impedance transformation required to achieve the gain for oscillation.

The circuit of Fig. 7.14b is used for analysis. The inductor and capacitor are modeled as an ideal resonator with a loss resistance, $Q_u \omega L$. This is paralleled with an ideal transformer with a turns ratio of N. The smaller number of turns drives the emitter. The transistor model used is identical with the earlier one, an input resistor looking into the emitter and a noninverting current gain of unity. The transformer model works well with toroidal inductors. It does not do as well in predicting gain with a solenoidal inductor, for the impedance transformation properties do not then follow a turns-squared law. The gain calculations are similar to those with the Colpitts and will not be covered here.

Figure 7.15 shows the circuit of a practical Hartley oscillator. Collector bias is supplied to the transistor through the inductor. A blocking capacitor is included in the feedback path. A series resistor decreases the starting gain.

The circuit has some interesting and sometimes detrimental properties. Like the Colpitts, the gain calculations are in good agreement with measured data. Similarly, restrictions on resonator impedance and emitter current used earlier describe the operating level. However, the circuit is prone to spurious oscillations. This is a result of the broadband feedback provided by the transformer. In contrast, the Colpitts feedback path has a high pass filtering action.

One form of spurious oscillation is termed "squegging." This is a low frequency relaxation-type oscillation that modulates the desired sinusoidal rf variations. It can

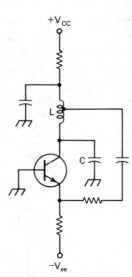

Figure 7.15 A Hartley oscillator with emitter degeneration. Unless built for very small gain at starting, this circuit is subject to "squegging" as described in the text.

occur when the starting gain is high and the series capacitor in the feedback loop is of relatively large value. If the capacitor is, for example, 0.1 μF for a 10-MHz desired operating frequency, the squegging may be so severe that oscillation will completely cease for short periods, only to build up again and again. A reduction in the value will decrease the magnitude of the squegging and sometimes stop it. An excess phase shift is introduced in the loop if the capacitor becomes too small. The better cure is to decrease starting gain by biasing or increasing the size of the series resistor.

Another form of spurious oscillation that can occur is at vhf. If the emitter current is high enough to allow voltage limiting to occur, the resonator Q decreases dramatically. Feedback is still present over a wide bandwidth. This allows the circuit to oscillate at higher frequencies. The output is a sinusoid at the desired output frequency with a "ragged" trace when observed on a high frequency oscilloscope. The multiple frequency output is readily observed with a spectrum analyzer.

The spurious oscillations mentioned are not restricted to the Hartley oscillator. However, they are certainly more predominant.

Figure 7.16 shows a Hartley oscillator using a JFET. This circuit is generally less prone to spurious oscillation problems than the bipolar one. This results from agc to partially determine operating level. The gain control voltage, the result of small current flow through the diode, is effective in reducing the gain at low frequencies as well as the operating frequency. The feedback is then effective only at the operating frequency. The capacitor in series with the gate and bias detector, 2.7 pF in the example, should be kept small. Similarly, the resistor should be large. This will ensure that the energy taken from the resonator for control purposes is small enough to not alter the loaded Q.

All of the analysis and discussion on oscillators has been devoted to questions of gain (enough to ensure oscillation) and operating level. Other factors are also of critical importance, especially in communications and rf instrumentation equipment. Noise will be covered in a later section. An equally important factor is temperature stability.

Most components have parameters that vary with temperature. The variations are often related to thermal expansion. For example, an inductor changes value with temperature from thermal expansion of the wire. Similarly, a capacitor will change in value as the plates change size with changes in temperature. Other effects are more subtle. The permeability of a powdered iron core used for an inductor will be a function of temperature as will the permittivity of a dielectric used in a capacitor. All of these changes will alter the frequency of an oscillator. Active devices will also change characteristics with temperature.

None of the temperature dependent parameters would be of significance if all components were in a constant temperature environment. This is strictly impossible although ovens are used to approximate the condition. When a circuit is first turned on, a previously cold transistor will begin to dissipate heat. Transistor characteristics will change. It will also radiate thermal energy to adjacent components, changing their temperature.

High frequency current flows in resonator components. Because they are not perfect, they will have some losses. The energy expended in these losses will cause temperatures to increase.

The selfheating effects will lead to a "warm-up" drift in frequency. The direction and magnitude of the change will depend upon the characteristics of the devices. The time to stabilize depends primarily upon the physical characteristics of the components. The more sizeable, the longer the stabilization time becomes.

Most components are specified for temperature on the basis of a fractional rate of change. For example, a high Q toroid inductor might have a temperature coefficient of inductance of $+50$ parts per million (ppm) per °C. Using the traditional resonance equation, $\omega^2 = (LC)^{-1/2}$, differentiation will lead to the relationship

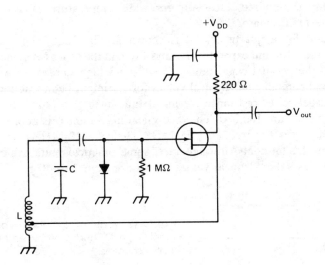

Figure 7.16 A practical JFET Hartley oscillator using a diode for agc. The series capacitor driving the gate should be very small, having a reactance of about 10 kΩ at the operating frequency. A typical inductor will have a reactance of 100 to 300 Ω at the operating frequency and is tapped at about 25% of the total number of turns from the ground end.

$\Delta\omega/\omega = -\frac{1}{2}\Delta L/L$. A 1 °C change in the temperature of a 1-μH inductor produces a change in inductance of $+50 \times 10^{-6}$ μH and a -25×10^{-6} fractional change in frequency. If the operating frequency is 10 MHz, the frequency will drift downward by 250 Hz.

Capacitors are available with fairly well specified temperature coefficients (TC), usually negative. The TC of most inductors is positive, indicating that the dominant physical phenomenon is a thermal expansion. Oscillators may be temperature compensated by using a small capacitor of the proper TC as a small part of the resonator. For example, if an oscillator used perfect capacitors (TC = 0) and an inductor like that specified earlier with a +50 ppm/°C TC, it would be compensated exactly if the net capacitance had a TC of −50 ppm/°C. Capacitors are available with a TC of −750 ppm/°C. One of these units could be used if it was a small part of the total capacitance. Specifically, the temperature compensating capacitor should be 1/15 of the total capacitance.

Temperature compensation is a difficult task and should be used only as a means of improving the stability of an already reasonable oscillator. The difficulty arises from a number of factors. First, a component TC is not constant with temperature. A component with a TC of −750 ppm/°C at room temperature might have a significantly different value at +80 °C. Moreover, warm-up drift must still be expected. Even if the components are exactly compensated, even over a wide temperature range, the thermal time constants will most likely be different. The best design philosophy uses components with low TC's; only then should the final drift be further decreased with compensation.

7.4 CRYSTAL CONTROLLED OSCILLATORS

The stability of an *LC* oscillator is limited by the component quality. Whenever significantly better stability is required, especially for wide temperature ranges, a quartz crystal replaces the *LC* resonator.

The equivalent circuit for a quartz crystal is shown in Fig. 7.17. The vital elements are the motional inductor and capacitor, L_m and C_m, and the series resistance, R_s. Also of importance is the parallel capacitance, C_p. A series high Q resonance is defined by the two motional components. Typical values for L_m might be in fractions of a millihenry while a usual series resistance for an hf fundamental crystal would be 10 to 20 Ω. Typical Q values are 100,000. The theoretical maximum Q is approximated by $1.5 \times 10^7/f$, where f is the frequency in MHz. Hence, a 10-MHz crystal could have a Q as high as 1.5 million! Although rare, some practical units exhibit performance approaching this value. Typical values for C_p are from 2 to 10 pF.

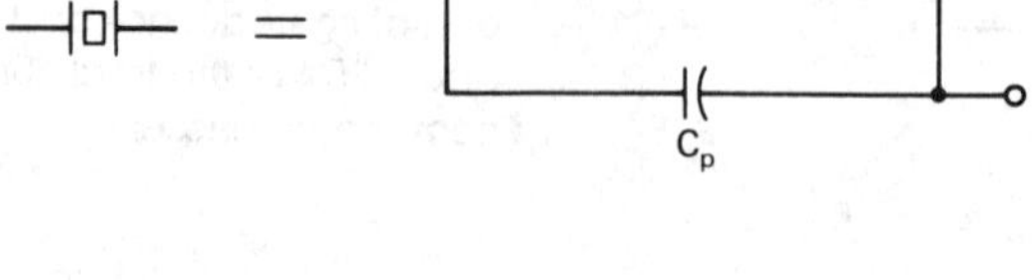

Figure 7.17 A quartz crystal offers extremely high Q_u and is often used in stable oscillators.

Sec. 7.4 Crystal Controlled Oscillators

The characteristic that makes the quartz crystal appealing for oscillators, besides the high unloaded Q, is the extreme mechanical stability of quartz. It is one of the more stable of all crystalline structures and certainly at the top of the list of materials which display a piezoelectric property.

Perhaps the simplest crystal oscillator is shown in Fig. 7.18a. If the equivalent circuit from Fig. 7.17 is substituted into the oscillator schematic, the result is a Clapp circuit. The circuit operates with a composite capacitance consisting of the motional one in the crystal plus the Colpitts capacitors. The crystal parallel capacitance will also alter the resonant frequency. The series resonant frequency defined by the motional elements differs from the operating frequency. The circuit is not an optimum one because of this difference. The better oscillators are those operating at the series resonant point.

The circuit of Fig. 7.18a is a practical one, operated in the "grounded collector" mode. Output is extracted from a current sampling resistor in the collector lead. If this circuit is redrawn as a common emitter configuration and put into a small-signal form, the circuit of Fig. 7.18b results. This is often called the Pierce circuit.

The circuits of Fig. 7.18 are both suitable for casual applications. They will function well on the fundamental modes of the crystal. The low pass nature of the amplifier with shunt capacitances tends to decrease the gain for crystal overtones. While these circuits are not optimum, they are still much more stable than an LC oscillator.

In some applications, it is desirable to tune a crystal oscillator over a narrow range. Investigation of the equivalent circuit, Fig. 7.17, reveals that this is possible with a series reactance. If a series capacitor is used, the frequency may be varied

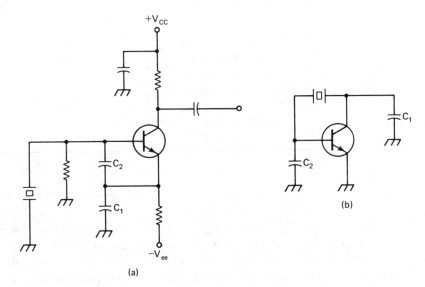

Figure 7.18 Two versions of the Colpitts oscillator using a crystal as the inductor. Oscillation is displaced from the series resonance of the crystal.

from the series resonant frequency up to the parallel, or antiresonant point. This may be enough range to be useful in many applications. Operation below the resonant frequency is possible if a series inductor is added. A circuit with extra elements added to the crystal to move the frequency is called a VXO or variable crystal oscillator. The X (rather than C) is historical; quartz crystal is often abbreviated as a *xtal*.

The effects of adding series (or parallel) elements with a crystal may be quite complex. Not only is the resonant frequency altered, but the resistance is scaled. This is exactly analogous to the transformations like the Clapp version of the *LC* Colpitts oscillator. Losses in the added components will alter the Q of the composite resonator. These effects vary as the circuit is tuned. The usual application of VXO circuits is to adjust a crystal oscillator exactly to a desired frequency. This compensates for inaccuracies in resonant frequency from the manufacturing process. It may also be required to compensate for aging effects.

Figure 7.19 shows an improved crystal oscillator circuit. It is essentially a Colpitts *LC* oscillator with a crystal in series with the feedback to the emitter. The design method is relatively straightforward. The crystal appears as a resistor of value R_s (Fig. 7.17) at the series resonant frequency. Hence, the circuit is designed to start and operate with this resistance in the feedback loop. The crystal is substituted for the resistor. It is experimentally useful to actually check operation with a resistor of the appropriate value.

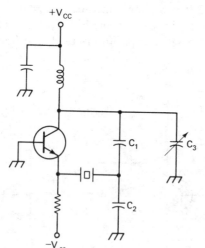

Figure 7.19 A Colpitts oscillator with a crystal inserted in the feedback path. This circuit will allow oscillation at the series resonant point of the crystal.

The crystal parallel capacitance can be of significance in this circuit. Some workers effectively cancel it by paralleling the crystal with an inductor of a value that will be resonant with the C_p. The circuit of Fig. 7.19 functions well with overtone crystals as well as with fundamental types. If overtone operation is used, the resonator should be loaded to a relatively low Q, ensuring only enough selectivity to prevent oscillation at the fundamental or the other overtones. A Q of about 10 is satisfactory.

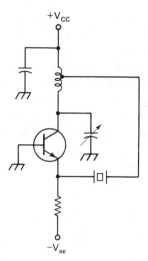

Figure 7.20 A Hartley oscillator with a crystal in the feedback path.

A similar variation can be built from a Hartley oscillator, shown in Fig. 7.20. The operation is virtually identical to the earlier Hartley with a series resistance equaling the crystal R_s. The crystal controlled version does not suffer from the problems outlined for the *LC* Hartley, for feedback is restricted to a well-defined spectrum.

Figure 7.21 shows a useful variation of this circuit. While still a Hartley crystal oscillator, there are two differences. First, the grounded position is shifted from the base to the tap. This places one end of the crystal at ground potential, a useful feature if several crystals are to be switched. It also works well in VXO applications. The second departure is the use of a broadband ferrite transformer, which allows

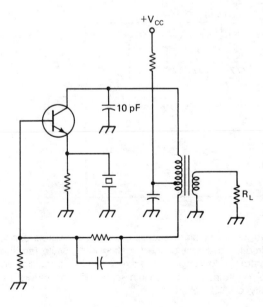

Figure 7.21 A variation of the Hartley crystal oscillator of Fig. 7.20. A broadband transformer provides the required impedance transformation. The 10-pF collector capacitor causes the circuit to have a decreasing gain with frequency, ensuring that operation is on the fundamental crystal mode rather than at an overtone. The ground point of the oscillator has been moved to the Hartley tap, allowing one end of the crystal to be at ground potential. This circuit and others using crystals may require a "swamping" resistance to load the output if the external load is not well defined.

the oscillator to operate with no circuit changes for frequencies over a very wide range. In one version built by the writer, the circuit worked well from 1.8 to 20 MHz. The 10-pF collector capacitor causes the gain to slowly decrease with frequency. This ensures fundamental rather than overtone mode operation.

The output of this circuit is obtained from a link attached to a low impedance load. The load should be present whenever the circuit is operated. Without the load, the impedance presented to the collector is too high for current limited operation.

VXO circuits are interesting and often useful in communications and rf applica-

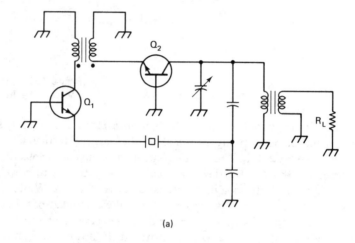

(a)

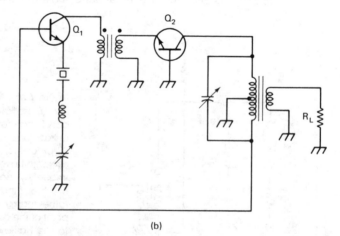

(b)

Figure 7.22 Small-signal equivalent circuits of crystal oscillators using two transistors. These circuits have well-defined limiting in Q_2 while terminating both ends of the crystal in a low impedance to ensure high loaded resonator Q. The circuit at (b) contains components in series with the crystal to pull the frequency slightly, a VXO action.

tions. They are, however, complicated to build. A VXO is generally best operated at a fundamental crystal mode. Then, the shift expected is between 0.1 and 0.2% of the crystal frequency. The fractional shift drops proportionally with the order of the overtone used.

Another problem encountered with VXO circuits is variation in resonator loss with frequency shift. This can lead to circuits that will not start reliably. More oscillator gain may be provided with a two-stage amplifier. This is illustrated by the small-signal circuits of Fig. 7.22.

The first (Fig. 7.22a) is a variation on a Colpitts, Fig. 7.19. The feedback path through the crystal is not directly to the emitter, but to another transistor, Q_1. The coupling between Q_1 and Q_2 is through a broadband transformer. The output transformer is tuned.

The second circuit, Fig. 7.22b, is a version of the crystal controlled Hartley oscillator. The ground point is moved, placing the tap at ground. The crystal is then placed in the emitter return to ground of Q_1. Components for VXO action are included in the second circuit.

Both oscillators have two useful features, each resulting from the biasing. Q_1 in each is biased for a relatively large current, while Q_2 runs with a low emitter current. Limiting occurs in Q_2. The output transformer ensures that the crystal is driven from a low impedance. The low input impedance of Q_1 with the lack of limiting in that stage terminates the other end of the crystal in a low impedance. This places the crystal in an environment to maintain the highest possible loaded Q. Limiting is still well defined and predictable.

Quartz crystals must operate with a limited amount of drive energy. One milliwatt is a typical upper limit. This is essentially the maximum power to be dissipated in the series resistance, R_s. A crystal with a maximum power of 1 mW and a 30 Ω series resistance then has a crystal current limit of 5.8 mA.

Crystal current may be calculated if the limiting characteristics of the oscillator are known. For example, the Colpitts LC oscillator with a crystal inserted in the feedback loop has a crystal current well approximated by the standing emitter bias current. Crystal current may also be measured. A current transformer is used as shown in the circuit of Fig. 7.23. The transformer is wound with a high permeability ferrite toroid core. N turns are used on the secondary connecting to the 50-Ω load.

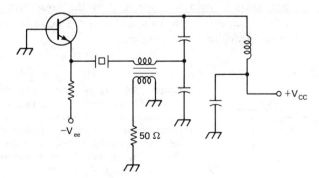

Figure 7.23 The current in a crystal may be measured with a simple current transformer. Crystal current is often well approximated by the standing bias current in the transistor emitter of the Colpitts circuit shown.

The primary is a single turn through the toroid. The impedance reflected back through the transformer is $50/N^2$. This would be 0.5 Ω if $N = 10$, a negligible amount. The current flowing in the 50-Ω load is that in the crystal diminished by N.

The 1-mW power level used in the discussion is an absolute maximum. Larger amounts may damage the crystal. Smaller levels are often recommended. Optimum thermal stability occurs for 50 μW or less.

There are many other suitable circuits for crystal controlled oscillators. Only a few have been presented here. Others are found in the literature (4). The stability of crystal controlled oscillators is generally excellent. Even a casually designed oscillator will deliver an output frequency which is within one part in 10^6 of the specified frequency. With temperature compensation and regulation with an oven, stabilities of one part in 10^8 and better may be achieved. These numbers reflect not only the thermal effects, but the crystal aging.

7.5 NOISE IN OSCILLATORS

Noise is usually not considered in the analysis of an oscillator except during the starting process. It should be, for it is of great importance. The desired noise performance will often dictate the methods used during oscillator design. The noise accompanying an oscillator output will be the factor which limits system performance in many communications and rf instrumentation systems.

Noise is the only output of an oscillator at the instant of initial power application. The noise is amplified and routed through a resonator where it is filtered and, perhaps, transformed in impedance level. It is then reapplied to the amplifier input. The noise builds in amplitude until a well-defined output signal exists. Although the amplifier gain is reduced by the limiting process, whatever it might be, there is still gain. Noise is still generated and will accompany the output signal.

Figure 7.24 shows a block diagram of an operating oscillator. It is labeled with the noise powers. The available noise power from the resonator at virtually any frequency is kTB where k is Boltzmann's constant, T is the absolute temperature, and B is the bandwidth in which the noise is observed. Fundamental noise concepts, primarily in amplifiers, were presented in Chap. 6. The input noise is increased by the amplifier. Also, there is excess noise contributed, specified by the noise factor. The net noise output is then $kTBGF$ where G is the gain and F is the noise factor, the algebraic equivalent of the usual logarithmic noise figure.

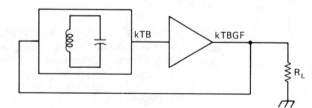

Figure 7.24 An oscillator block diagram at the instant power is applied. The available noise power from the resonator is kTB over any frequency segment. The output noise is increased over kTB by the noise factor, F, and the gain at starting, G.

Sec. 7.5   Noise in Oscillators

Only amplitude noise is usually considered when working with amplifiers. It is this noise which is measured with a traditional detector. In an oscillator, however, the noise is segmented into two parts, amplitude and phase.

The absolute level of the noise output is not as important as is the ratio of the noise to the signal level. The noise is regarded as a modulation of the dominant signal output.

The two kinds of noise are generally regarded as being equal. The rationale for this assumption is illustrated in Fig. 7.25. The signal is represented as a phasor, a rotating vector. The length is the amplitude while the angle represents the instantaneous phase. The angular velocity represents the frequency. At any instant, there will be a noise component that is added to the existing signal, represented by a second vector. The vector sum represents the instantaneous total signal.

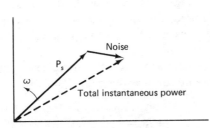

Figure 7.25 The output of an oscillator is represented as a rotating vector, or phasor, P_s. This output will have a small added noise component. The noise is random in magnitude and angle. The resultant vector is the dotted one. Because the noise is completely random, half of the noise power is parallel to the signal vector and half is normal to it. Half of the broadband noise output from an oscillator is then phase noise.

The noise is completely random. It may thus be resolved into two components. The part parallel to the signal vector represents amplitude noise. That at right angles to the signal is phase noise. Because of the randomness, half of the noise power is amplitude noise while the other half is phase noise. Note that the very concept of phase noise loses definition if there is no reference signal.

The more important component is phase noise in oscillator analysis. The reason for this is that amplitude noise is often reduced significantly by limiting. For example, if the oscillator output is used as the driving input of a switching mode mixer, the gain of the mixer is insensitive to the amplitude. The mixer limiting action removes the amplitude noise. In addition, the limiting nature of many oscillators during operation will limit some of the amplitude variation. Limiting will not alter the phase noise.

The absence of amplitude noise is an approximation that is often applied to oscillators. While limiting may reduce the presence of amplitude noise, it still can have impact. An amplifier following an oscillator may contribute amplitude noise. In addition, amplitude noise may be converted to phase noise. One example of such a conversion process was presented in the discussion of the Colpitts oscillator where an amplitude variation represented an instantaneous variation in gain. That was then converted to a change in the associated angle of the oscillator gain. Another example would be an amplifier with a reverse biased junction. The diode, such as the collector-base junction of an amplifier, will vary in capacitance value as the voltage varies.

The capacitance variation will then modulate the phase response of the amplifier. Amplitude-to-phase-noise conversion is a common phenomenon with amplifiers. Additional information is presented in the literature (5).

Returning to the oscillator block diagram, Fig. 7.24, the phase noise power may be evaluated and compared with the signal power. The gain is usually small in an operating oscillator. Strictly speaking, the loop gain around the complete circuit is unity. However, there is a loss in the resonator which must be compensated by the amplifier gain. The broadband phase noise output of the amplifier is thus $P_n = \frac{1}{2}kTBGF$. The factor of $\frac{1}{2}$ results from the segmenting of phase and amplitude noise.

The carrier or signal power in the loop is related to the energy in the resonator and its loaded Q. Q is defined as the ratio of energy stored to that lost in a cycle of oscillation, specifically, $Q = \omega U/P_s$ where U is the stored energy and P_s is the signal power flowing through the resonator. The energy stored may be related to either the current or the voltage associated with the resonator, $U = \frac{1}{2}CV_p^2$ or $U = \frac{1}{2}LI_p^2$. Peak values are used, for they occur at the instant when all energy is associated with the variable chosen. That is, the peak voltage occurs when all of the resonator energy is stored in the capacitor.

The usual practice in oscillator analysis is to normalize the results to a bandwidth of 1 Hz. Using this, the phase-noise-to-carrier ratio (NCR) is

$$\text{NCR} = \frac{kTFG}{2P_s} \qquad (7.5\text{-}1)$$

Usually, the approximation is made that the amplifier gain is unity, resulting in

$$\text{NCR} = \frac{kTF}{2P_s} \qquad (7.5\text{-}2)$$

The noise described by these equations is of a very wideband nature. The noise is present at virtually all frequencies. kTF is often termed the broadband noise "floor" of the oscillator output.

The resonator has a significant impact upon noise output. A segment of noise at a frequency far removed from the resonant frequency will be returned to the resonator input. However, in passing through the resonator, it is attenuated. Also, it is shifted in phase by more than 90 degrees. The resulting noise component presented to the amplifier is out of phase with the original output.

To the contrary, a noise component at a frequency very close to the center of the resonator response appears at the resonator input and passes through with little attenuation and little phase shift. The resulting component appears at the amplifier input where it may be amplified again. The component continues to go around the loop. It is shifted in phase and diminished in amplitude with each pass. Only the component exactly at resonance, the "signal," will continue unattenuated. A noise component separated from the carrier by half of the bandwidth is shifted in phase

Sec. 7.5 Noise in Oscillators

by 45 degrees with the first pass through the resonator. With a second pass, it is attenuated far enough that it is merely a contribution to the broadband noise floor. The noise within the resonator bandwidth is additive, causing the noise power per hertz of bandwidth to increase as the carrier is approached. It will grow at a rate of 6 dB per octave where the frequency related to the growth is f_m, the so-called modulation frequency. f_m is the separation between the carrier and the frequency where the noise is measured. For f_m greater than the half bandwidth, $f_0/2Q$, the noise is the broadband value.

A 6 dB per octave rate with respect to f_m corresponds to $1/f_m^2$ behavior. This leads to the traditional formula for close-in phase-noise-to-power ratio (6)

$$\text{NCR} = \frac{kTF}{2P_s}\left(\frac{f_0}{2Q}\right)^2\left(\frac{1}{f_m^2}\right) \tag{7.5-3}$$

where Q is the loaded value. This expression does not contain the gain of the amplifier. A sketch of the spectrum is presented in Fig. 7.26.

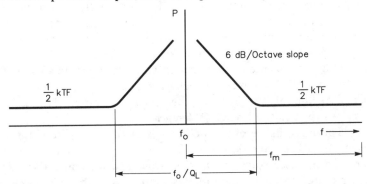

Figure 7.26 The phase noise output of an oscillator. The broadband phase noise appears far away from the carrier with a magnitude ½ kTF. The noise grows as the carrier is approached. The slope is 6 dB per octave with respect to the Fourier frequency, the separation from the carrier. The close-in noise growth is confined to the loaded resonator bandwidth.

The noise power behavior with f_m is modified further at very close spacings. There, low frequency transistor noise grows, the so-called excess or $1/f$ noise. This will further modulate the carrier, causing the noise to grow at a 9 dB per octave rate sufficiently close to the carrier. This usually occurs at f_m values of less than 5 to 10 kHz with bipolar transistors. An equation similar to Eq. 7.5-3 can then be written emphasizing a $1/f_m^3$ behavior.

Consider an example, a 10-MHz oscillator with $Q_L = 100$. Assume the inductor is 2 μH and that the peak voltage across it is 10 V. The net capacitance for resonance is then 127 pF. Using ½ CV_p^2 to evaluate the energy, p_s is 4 mW, or +6 dBm. Using Eq. 7.5-2, the broadband phase noise to carrier ratio is −173 dB/Hz. A 10-dB noise figure is assumed. This is the noise at 50 kHz, the value of f_m equaling the half bandwidth. At $f_m = 30$ kHz, the ratio is 168.5 dB/Hz. At 10 kHz it has degraded to 159 dB/Hz. These results come from application of Eq. 7.5-3 with no excess or $1/f$ noise.

Figure 7.26 and Eqs. 7.5-2 and 7.5-3 suggest that the only effect of Q is to determine the *breakout* frequency where additive noise begins to contribute. This is not accurate. Q is related to the power level and the stored energy by $Q = \omega U/P_s$. This was used to evaluate the broadband noise floor of the preceding example. However, Q will have a further effect, one that is often neglected.

Assume that the resonator in the example is doubly terminated. The insertion loss is then related to the Q by $IL = 20 \log(1 - Q_L/Q_u)$ dB. The loss is 4.4 dB for the example considered if $Q_u = 250$. The loss must be factored into the results, for the operating gain is now specified. Using Eq. 7.5-1, the predicted broadband phase noise to carrier ratio is 168.6 dB/Hz. The ratio would degrade to 156 dB/Hz if the loaded Q was increased to 200. The effect of changing Q is much more dramatic when the effect upon operating gain is considered.

It is not valid to assume that an oscillator resonator will be doubly terminated in equal and stable immittances. Still, it is not totally reasonable to neglect the variations in gain resulting from loading. Oscillator design is always a compromise. The best broadband noise performance results from a low value of Q_L/Q_u. The breakout frequency is then further removed from the carrier.

Other factors are also significant. The noise figure in an amplifier or an oscillator is a function of the source impedance. A noise match rarely coincides with a power match. As limiting is approached, the impedance presented to the input of the amplifier is altered, just as the loaded Q changes.

The noise figure of an oscillator is often found, from Eq. 7.5-2, to be higher than that of the same device operated at the same average current as an amplifier. Part of this discrepancy may come from not including the gain in the equation, gain required to compensate for resonator loss. It has also been suggested that the increase results from nonlinear operation of the oscillating transistor. This produces device currents at many harmonics. They mix to produce excess outputs at the fundamental, but of complicated phases. The net result is a degradation of noise figure.

With the uncertainties, there will be some difficulty in exactly predicting oscillator noise spectra. The traditional equations, those without the gain term included, are generally useful as a starting point. They provide an optimistic view of the results, the best that can be done. Generally, a high noise figure is assumed. If a transistor noise figure in an amplifier application is known, the value is increased by 6 to 10 dB for oscillator analysis.

In spite of the inaccuracy, the analysis presented may be utilized to guide the designer in building an oscillator for low noise. If the circuit is to have a large carrier to phase noise ratio, the carrier power in the loop must be as high as possible. This explains the utility of designs that allow this through appropriate impedance transformations such as the Clapp and Seiler variations of the fundamental Colpitts oscillator. Along with the high energy flowing in the loop, the loaded Q must be maintained reasonably high. If this is not done, the close-in additive noise will grow beyond the broadband value at frequencies further removed from the carrier. This suggests that the limiting mechanism should be one which does not degrade the resonator Q. Voltage limiting should be avoided. Current limited operation where transconductance at the operating frequency decreases with increasing loop power is vastly preferred. As was shown in the discussion of the Colpitts, the loaded Q actually increased over the value during starting when current limiting is employed.

Transistor type will also be of significance. The transistor should have a reasonably low noise figure, even at high current levels, for this will occur over a short portion of each cycle. The mechanism proposed to account for degraded noise figures in oscillator applications would suggest that some form of negative feedback be used in the oscillator to limit starting gain. This will ensure that the amount of harmonic distortion required of the active device is minimal when limiting occurs, aiding in preserving a low noise figure. Also, a transistor should be chosen for good $1/f$ noise characteristics.

The junction field effect transistor is appealing for many oscillator applications. It has very good $1/f$ noise properties when compared with the usual rf bipolar transistor. In addition, the limiting occurs only from a square-law nonlinearity. High order harmonic currents are small, especially if excursions into the pinch-off region are minimized.

The analysis points out a fundamental limitation of a quartz crystal controlled oscillator. It was recommended in the previous section that a maximum crystal power of about 50 μW be used. If this is followed, and a 10-dB noise figure is assumed, the broadband carrier to phase noise ratio is only 154 dB/Hz. This level is much worse than that calculated for an LC oscillator. It results from the limited power flowing through the crystal. The half bandwidth would be 50 Hz if the oscillator was operating at 10 MHz with a loaded Q of 100,000. The broadband noise would be maintained to very close spacings.

Modulation theory may be invoked to predict the carrier to phase noise ratio when the oscillator output is presented to other networks. Of special interest is the frequency multiplier or divider.

A frequency multiplier is a circuit driven or biased so that significant harmonic distortion occurs. A filter at the output will then select one of the harmonics while suppressing the others as well as the driving input. If the input signal was regarded as a carrier with phase modulation applied at a frequency corresponding to some spacing, f_m, small modulation-index fm theory may be applied (7). This results in noise on the multiplied output at the same spacing, f_m, growing at a rate of 20 log N dB where N is the order of the multiplication. The phase noise will degrade by 40 dB if an oscillator output is multiplied by $N = 100$. Clearly, multiplication is to be avoided when possible.

To the contrary, frequency division will improve noise performance. Frequency division is usually performed with digital-type integrated circuits. If an oscillator is divided by 2, the phase noise sidebands will drop by 6 dB. There is a limit to how far this "clean-up" can be applied. The digital IC's used have a propagation delay, the time difference between the instant an input signal is applied and the instant that the result occurs at the output. Divider noise will produce a fluctuation in the propagation delay, a phase modulation at the output. This accumulates with a cascade of many dividers, yielding degraded phase noise.

Circuit output was obtained from the amplifier output in the block diagram of Fig. 7.24. Hence, the output noise of the amplifier is included. A useful signal may, instead, be extracted from the input of the amplifier. The power available will be less by the insertion loss of the resonator. However, the broadband noise floor may be lower. This results from the added filtering action of the resonator. The noise then continues to drop at a 6 dB per octave rate with respect to f_m. This method is especially useful with crystal oscillators where the loop power is limited.

Measurement of oscillator carrier to noise ratio is not trivial. The calculations show ratios greater than 170 dB/Hz. If this noise was observed in an instrument with a 1-kHz bandwidth, the ratio would still be 140 dB. This far exceeds the dynamic range of most rf instrumentation.

Figure 7.27a shows a method where the oscillator output may be studied with a spectrum analyzer. An oscillator to be studied is set to a frequency, f_1. The output is applied to a filter with a center frequency, f_2. The separation between the frequencies is f_m, the spacing from the carrier where the noise is to be measured. The filter insertion loss, bandwidth, and stopband attenuation are known. The observed output seen in the spectrum analyzer is sketched in Fig. 7.27b.

Assume the output of the oscillator is +10 dBm and that the filter uses quartz crystals to produce a bandwidth of 3 kHz. For simplicity, assume that the filter has no insertion loss. The analyzer is set to a bandwidth of 300 Hz. We will assume that this is also the noise bandwidth of the analyzer, although this is not generally true. A slight correction may be applied (8).

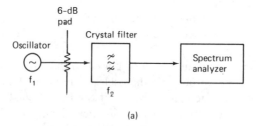

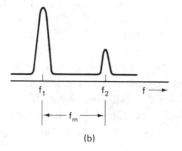

Figure 7.27 A simple method for measurement of oscillator noise. A crystal filter at a frequency slightly different than the oscillator provides suppression of the carrier. Without the filter the spectrum analyzer would be severely overloaded. The 6-dB pad ensures a proper termination for the oscillator and the crystal filter. A typical result is sketched at (b).

Initially, the oscillator is set at $f_1 = f_2$, producing an output of +10 dBm on the screen of the analyzer. Then, f_1 is moved to a different frequency. The gain in the analyzer is adjusted so that a display like that of Fig. 7.27b is seen. The value of the observed power at f_1 does not really matter so long as it is on the screen of the display, indicating that it is not overloading the analyzer. The power observed at f_2 is noted. Assume that it is −125 dBm. The carrier to noise ratio in a 300-Hz bandwidth is then 135 dB. This becomes 159.8 dB/Hz when normalized to a 1-Hz bandwidth.

The extra filter provides suppression of the carrier while the noise is being measured. The filter should have a bandwidth larger than that used in the analyzer. The filter shape will determine how close to the carrier the noise may be measured.

Spectrum analyzers use peak detectors and logarithmic amplifiers. Both of these circuits respond differently to random noise than they do to a sine wave. Suitable corrections, usually of the order of 3 dB total, must be applied (9).

Another method for noise evaluation is frequency multiplication. The output of the oscillator to be studied is applied to a suitable multiplier and then passed through a filter that passes only the desired output. The carrier to noise ratio may be observed directly with a spectrum analyzer if the multiplication factor is large enough.

Figure 7.28 shows an instrument that may be built for evaluation of oscillators. A balanced mixer drives a crystal filter, Z_1. This is followed by a high gain amplifier chain and a second crystal filter, Z_2, and, finally, a detector. The second filter serves the purpose of removing wide band noise that might be generated in the high gain i.f. amplifier. The bandwidth of Z_2 is the same as Z_1.

Two oscillator inputs are required at the mixer, the frequencies being f_1 and f_2. The nominal frequency difference is the i.f., that of the crystal filters. Assume initially that f_2 is the local oscillator (lo) for the mixer, the larger output. We assume that the lo has perfect spectral purity with a virtually infinite carrier to noise ratio. This is, of course, not valid but provides a starting point. The oscillator to be evaluated is set to $f_1 = f_2 \pm f_{i.f.}$. The attenuator is adjusted to produce a measurable output at the detector. Considerable attenuation may be required. It may also be useful to decrease the i.f. gain by a known amount. The detector response is noted.

The frequency of the oscillator being studied is now moved by a desired incre-

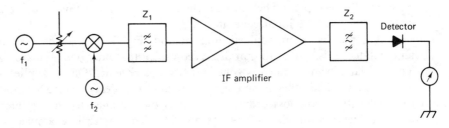

Figure 7.28 A specialized spectrum analyzer for the measurement of phase noise in oscillators.

ment, f_m. The detected response should disappear. The i.f. amplifier gain is now increased and the attenuator value decreased until the same response is noted. The oscillator under evaluation (f_1) is temporarily turned off, or more attenuation is momentarily added to ensure that the detected output is not the noise of the i.f. amplifier. The total change in gain and attenuation required to produce the response is then the carrier to noise ratio for the bandwidth of the filters used. The evaluation is performed over the spectrum of f_m of interest, within the limitations of the filter stopband attenuation. Care should be taken to avoid overdriving the mixer. Once completed, the data is normalized to the usual 1-Hz bandwidth.

The system of Fig. 7.28 is essentially a committed spectrum analyzer. It differs from the usual one only in the inclusion of a narrow filter at an early point in the system. It is, in this vein, much more like a modern communications receiver. A refined version of this system forms the basis of a measurement scheme used by the National Bureau of Standards for oscillator noise evaluation (10).

The assumption was made that the lo for the mixer, f_2, was spectrally very pure. This may be realized in practice by using a stable oscillator that is filtered by a separate crystal filter. This is not typical in either a receiver or a spectrum analyzer. In those applications, f_2 must be variable to allow the instrument to tune a wide range of input frequencies. When the lo driving the mixer is less than spectrally perfect, its noise may contribute to the measured output.

A phenomenon called *reciprocal mixing* occurs when noise appears in the i.f. amplifier from an lo accompanied by noise sidebands. Assume that two oscillators are adjusted for evaluation of noise at some offset frequency, f_m. Even if the input signal, f_1, was perfectly clean and noisefree, it would mix with the noise sidebands on the lo at f_2 to produce an output at the i.f. Reciprocal mixing is the undesired effect that is often the major limitation of the useable dynamic range of communications equipment and instrumentation.

Reciprocal mixing is based upon the multiplier characteristic of the usual mixer. The output will be the product of the input signals so long as the mixer is being switched off and on (assuming a switching-mode mixer). A receiver or rf instrument evaluation of reciprocal mixing may be complicated by noise sidebands on the input generator. For example, if the system of Fig. 7.28 was a communications receiver with an lo operating at f_2, the carrier to noise ratio of the lo could be evaluated by application of a suitably strong signal at the input, f_1. However, that signal should be spectrally very pure. The most practical way to achieve this is to use a cascaded crystal filter with the input generator at f_1.

A somewhat more sophisticated method, Fig. 7.29, may be used to evaluate an oscillator. It is similar to that discussed except that only one oscillator is required. The oscillator being studied is split into two components. One is applied to a mixer as a "signal." The other component is applied to a delay line, often nothing more than a length of low loss coaxial cable. It is then applied to an amplifier to obtain a well-defined amplitude to serve as an "lo." The mixer is examined with a low frequency spectrum analyzer or wave analyzer.

If there was no delay line in this system, the output would be nothing but a

Sec. 7.6   Negative Resistance Oscillators

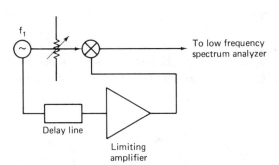

Figure 7.29 A method for measuring oscillator noise. A doubly balanced mixer, typically a diode ring, operates as a phase detector if driven at both inputs with the same frequency. The addition of a delay line in one path will then allow the system to operate as a discriminator. The delay line length is adjusted for 0 Vdc output. The ac output is observed with a low frequency spectrum analyzer, usually preceded with a low noise preamplifier.

dc component resulting from slight phase differences between the two outputs. Because of the delay line, the oscillator signal at one point in time serves as the lo to mix the signal originating at a slightly later point in time. The output observed with the spectrum analyzer is the noise spectra with respect to f_m, the modulation or sideband frequency. The analyzer must be capable of operating at frequencies close to dc to observe the noise at low f_m values. Also, the time delay in the line will determine how close to the carrier the measurements may be performed. Additional information on this method is found in the literature (11, 12, 13).

7.6 NEGATIVE RESISTANCE OSCILLATORS

The oscillator designs considered have used the simplest of models. Still, analysis has not been trivial. Oscillators are nonlinear circuits; the nonlinear behavior must be present to define the operating level and preserve amplitude stability.

A simple model, like that used, is suitable for oscillators at lower frequencies, perhaps up to about 50 or 100 MHz. Eventually, it becomes necessary to use more sophisticated models, or measured data. The early work may be extended through the application of two-port methods as described in Chap. 5. The opened loop analysis of the Colpitts oscillator used a ladder topology, one that is easily analyzed with two-port parameters to represent active devices.

Most microwave oscillator design is performed with scattering parameters. This set is the most measurable of all two-port representations. Other sets are just as viable, but calculations are not as readily compared with measured data. The essential detail which emerges from an S parameter viewpoint is that the active device is characterized by a reflection coefficient greater than unity, corresponding to an immittance with a negative real part. This approach will be the subject of this section.

The example to be considered is an oscillator studied in detail earlier, the 10-MHz Colpitts using a bipolar transistor. The microwave designer might question the use of two-port methods to describe a circuit that is so easily analyzed with more rudimentary techniques. However, approaching a familiar circuit with a different viewpoint will allow us to transfer some of the intuition gained from earlier work to the microwave methods to be used later.

The circuit to be analyzed is shown in Fig. 7.30. A bipolar transistor operates as a common base amplifier. Feedback is introduced through paralleling by C_1 while the input is terminated in a capacitor, C_2. A resonator is attached to the output.

A simple transistor model is chosen, a beta generator with an emitter resistance, $r_e = 26/I_e$ (mA, dc). The admittance parameters for the common base transistor are derived from earlier work

$$Y_b = \begin{pmatrix} \dfrac{1}{r_e} & 0 \\ \dfrac{-\beta}{r_e(\beta+1)} & 0 \end{pmatrix} \tag{7.6-1}$$

The admittance matrix for the feedback capacitor, C_1, is written easily and added directly to that of the transistor to produce the two-port representation of the composite circuit. The input is terminated in a "source" admittance, $j\omega C_2$, and the output admittance is calculated. The composite y parameters are

$$Y = \begin{pmatrix} \dfrac{1}{r_e} + j\omega C_1 & -j\omega C_1 \\ \dfrac{-\beta}{r_e(\beta+1)} - j\omega C_1 & j\omega C_1 \end{pmatrix} \tag{7.6-2}$$

while the output admittance is given by the relationship derived in Chap. 5

$$Y_{\text{out}} = y_{22} - \frac{y_{12} y_{21}}{y_{11} + y_s} \tag{7.6-3}$$

Substitution yields

$$Y_{\text{out}} = j\omega C_1 - \frac{\left[\dfrac{j\omega C_1 \beta}{r_e(\beta+1)} - \omega^2 C_1^2\right]}{\dfrac{1}{r_e} + j\omega(C_1 + C_2)} \simeq \frac{-\omega^2 C_1 C_2}{\dfrac{1}{r_e} + j\omega(C_1 + C_2)} \tag{7.6-4}$$

The approximate form represents the condition of very high beta, the assumption used in the earlier analysis.

The "resonator" admittance may also be evaluated. The final oscillator operates at 10 MHz. The active device contains capacitance which will be part of the composite resonator. The net device capacitance is calculated without the transistor, $(1/C_1 + 1/C_2)^{-1} = 90.91$ pF. We find that C_3 must be 415.7 pF using the inductor to evaluate resonance at 10 MHz. The "resonator" portion of Fig. 7.30 alone will have a parallel resonance at a higher frequency, 11.04 MHz. It will appear inductive at 10 MHz.

An equivalent circuit may be drawn if Eq. 7.6-4 is evaluated for a selected set of parameters. A dc emitter current of 2 mA is chosen. The equivalent circuit is

Sec. 7.6 Negative Resistance Oscillators

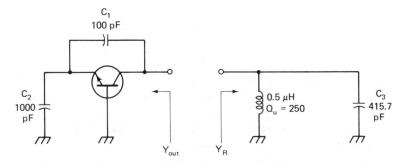

Figure 7.30 A 10-MHz Colpitts oscillator viewed as the interconnection of two single-port networks.

shown in Fig. 7.31. The active device equivalent resistance is the reciprocal of the real part of Eq. 7.6-4 while the parallel capacitance is the reactive part divided by the angular frequency.

Two vital details immediately appear. First, the equivalent resistance is negative. Second, the capacitance is not that expected, but a smaller value.

The resonator may also be evaluated. The unloaded Q is assumed to be 250, resulting in a parallel resistance at 10 MHz of 7854 Ω. This resistance will be a weak function of frequency.

The equivalent resonator inductance is combined with the active device output C to predict frequency. The net parallel resistance is calculated by adding conductances. The net parallel resistance is -382.8 Ω. Because it is negative, oscillation will start.

An ideal resonator paralleled with a negative resistance is an oversimplified circuit. The power in the resonator would continue to grow forever, without bound. As was found earlier, something must occur to limit the amplitude. Specifically, the magnitude of the negative resistance at the operating frequency must increase as the power level grows.

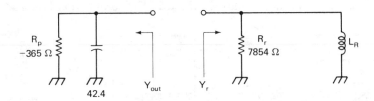

Figure 7.31 Equivalent representations of the circuit of Fig. 7.30. The real part of the output admittance of the "transistor" is negative. The total admittance when the two networks are connected still has a negative real part, indicating that oscillation will start. The magnitude of the real part of the total admittance is zero for an operating oscillator. The R_p value will increase to equal that of the resonator if current limiting defines the operating level. Resonator equivalent resistance, R_r, will decrease with voltage limiting.

Other bias conditions may be evaluated. If the transistor is biased at 0.1 mA, the active device parallel resistance becomes $-3272\ \Omega$ while the equivalent capacitance is 90.64 pF. The net resistance of the complete circuit is still negative; oscillation will still occur. The capacitance is now much closer to that without the transistor, indicating an operating frequency close to the desired 10 MHz. If an even smaller bias current is used, say 0.01 mA, the device resistance becomes $-32.6\ \text{k}\Omega$. The net resistance is now positive. Oscillation will not occur.

Operating level is not predicted with this modeling, for no nonlinear elements are included. However, the analysis may be used to estimate something about operating conditions. The active device negative resistance magnitude equals that of the resonator at a current of 0.04 mA. Hence, in stable oscillation, the transistor must be operating at a gain commensurate with 0.04 mA. This will be the gain at the operating frequency. Significant harmonic currents will flow if the bias is much greater. As expected, the results are in exact agreement with earlier analysis.

The conditions for oscillation with a stable amplitude are summarized using Fig. 7.31. The susceptances must cancel to produce resonance. The parallel conductances, one positive representing the resonator, and the other negative representing the active device, must add to zero. This will correspond with a net parallel impedance which is infinite.

The analysis has used parallel admittances, but could also have been done with series impedances. The series reactances would then cancel to produce resonance while the two resistances would add to zero.

Most higher frequency design is performed using scattering parameters. Figure 7.32 shows an S parameter representation of a single port oscillator. The resonator is characterized by a reflection coefficient which is a function of frequency. The "output" of the active device is also characterized by a reflection coefficient. The frequency dependence is usually small compared with that of the resonator. The transistor is shown as a single port device.

Noise from the resonator results in a voltage wave toward the active device when power is initially applied to the oscillator. If the active device were nothing but a resistor of value Z_0, the characteristic impedance used for definition of reflection coefficients, nothing would be reflected. The termination is not Z_0, so some of the

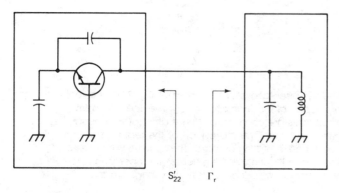

Figure 7.32 An S parameter representation of a negative resistance oscillator. Oscillation is possible only when S'_{22} has a magnitude exceeding unity. An operating oscillator has $S'_{22}\Gamma_r = 1$.

resonator energy is reflected. When the active device has a reflection coefficient greater than unity, the voltage wave from the resonator reappears with a magnitude greater than the original. The wave, now incident to the resonator, will be reflected again. It will lose some magnitude, for there is loss in the resonator. However, stable amplitude oscillations will occur if the loss is small enough that it may be compensated by the "reflection gain" of the active device. The condition for oscillation is

$$S'_{22} \Gamma_r = 1 \tag{7.6-5}$$

This equation is applied to the previous oscillator by converting the Y_{out} values to a reflection coefficient, S'_{22}. The conversion uses the standard formula

$$\Gamma = \frac{Z - Z_0}{Z + Z_0} = \frac{Y_0 - Y}{Y_0 + Y} \tag{7.6-6}$$

The resonator has a 10-MHz reflection coefficient of $\Gamma_r = 0.98829$ at $+31.9$ degrees. A 50-Ω characteristic impedance is assumed. At $I_e = 2$ mA, the active device has $S'_{22} = 1.31$ at -15 degrees. Evaluation of the product in Eq. 7.6-5 yields 1.29 at 16.5 degrees. Oscillation is clearly possible, for the magnitude of the product is greater than unity. However, it is clear that limiting is yet to occur.

If I_e is 0.05 mA, $S'_{22} = 1.014$ at -31.9 degrees. The product is now 1.002 at 0 degrees. Equation 7.6-5 is close to being satisfied. If I_e is 0.01 mA, the product reduces to 0.991 at 0 degrees. Oscillation is now not possible.

It is always desirable to use the Smith chart when working with S parameters. This is not directly possible with a standard chart, for reflection coefficients greater than unity are involved. They will lie outside of the unit circle perimeter of the chart. There are two alternatives. One is the use of a compressed Smith chart. This is a chart compressed in scale with the unit circle occupying only a small fraction of the chart. The compressed Smith chart is especially handy for the oscillator designer, or for the amplifier designer concerned about plotting stability contours.

The other alternative is to plot a reciprocal. The rationale for this is seen by examining Eq. 7.6-5. Solving for the resonator reflection coefficient

$$\Gamma_r = \frac{1}{S'_{22}} \tag{7.6-7}$$

If the magnitude of the resonator reflection coefficient is less than unity, as it must be, the reciprocal of the active device reflection coefficient must lie within the perimeter of the standard Smith chart for allowed oscillation.

The impedance of a resonator is easily plotted as a function of frequency on a Smith chart. It is usually a circle close to the perimeter of the chart. It will occupy 360 degrees as the frequency varies from zero to very large values if it is a simple resonator with one inductor and one capacitor. If the resonator is more complicated, such as that used in the Seiler oscillator, the "circle" may actually go around the chart for more than one complete rotation.

Once the resonator characteristic has been plotted for the desired center frequency, $1/S'_{22}$ is plotted. If this lies within the resonator circle, oscillation can occur.

Our example Colpitts oscillator is graphed in Fig. 7.33. The characteristic impedance chosen was 1000 Ω rather than the usual 50-Ω value. The resonator has a Γ value at 10 MHz near the perimeter at an angle of +160 degrees. Parallel resonance occurs at 11.04 MHz, the point where the impedance is high and purely resistive. $1/S'_{22}$ is plotted at 10 MHz but for a variety of bias current values ranging from 0.1 to 10 mA. The complete plot of S'_{22} lies within the resonator circle; hence, oscillation can occur for any of the bias conditions shown. The curve is extended, although not labeled for current, to higher I_e values. Eventually, at I_e approaching 100 mA, $1/S'_{22}$ goes outside the resonator circle.

Note that Smith chart positions for the graphed example are much different than the values found at a 50-Ω Z_0. A given impedance will move considerably on the chart as the characteristic impedance used for reflection coefficient definition is changed. Points on the outer perimeter will remain on the perimeter. Also, points on the line of pure resistance will remain on that line. Angular positions can change dramatically though. The conclusions regarding oscillation will not be altered by a change in Z_0 even though the appearance of the plot may be much different. It is not necessary to design an oscillator such that the $1/S'_{22}$ plot lies close to the real axis. Further, it is not necessary to bias the oscillator for a starting condition with $1/S'_{22}$ of high gain, well removed from the resonator circle. Indeed, just the opposite is desired.

The example presented has been viewed as a two terminal, or one-port device. There are several two terminal devices that function as oscillators including tunnel and IMPATT diodes. They must be biased for a negative resistance characteristic. The oscillator design is then performed as outlined. It is not, however, complete. Only the small-signal details have been considered, showing the conditions for starting. Once the oscillator has started, the characteristics of the device will change until an operating point is established where the device $1/S'_{22}$ curve exactly intersects the resonator plot. Equation 7.6-5 is then satisfied exactly.

In the Colpitts example with a reasonable starting current, oscillation will grow. Current limiting forces movement generally along the $1/S'_{22}$ plot until it enters the resonator plot. Alternatively, if voltage limiting occurs such that the resonator Q is degraded, the resonator circle will decrease in size.

The active device data presented in Fig. 7.33 is a result of the simple model chosen. Variation of emitter bias current allows the S parameters to vary in a simple and predictable way. As we found earlier, evaluations at a variety of bias conditions will allow us to estimate the nature of the limiting, at least in a qualitative way. The simple models are not so readily applied at higher frequencies. The calculations will be performed using existing S parameters, resulting in a single point instead of a curve representing different operating conditions.

In many practical designs, a curve of $1/S'_{22}$ is plotted as a function of frequency.

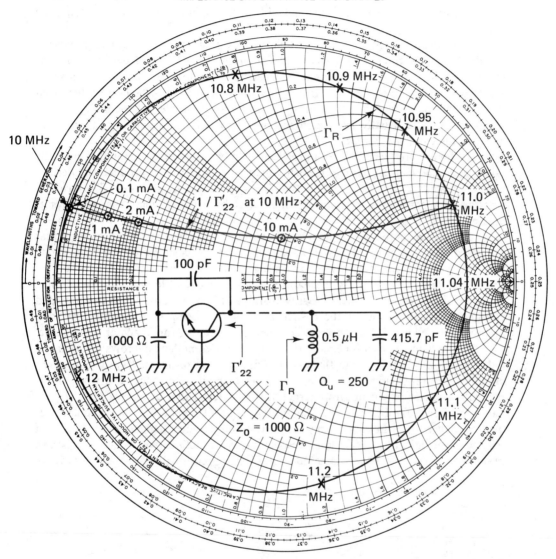

Figure 7.33 Use of the Smith chart for oscillator design. The active device will have an "output" reflection coefficient exceeding unity and is, hence, outside the standard chart. The reciprocal is plotted in its place. The resonator is represented by a circle. The active device is described by plotting $1/\Gamma'_{22}$ for various bias current values. Oscillation may start so long as the point representing the active device is within the resonator circle. Current limiting of the transistor will force the point representing the device to move from the starting current to one commensurate with a lower value (reduced gain at the operating frequency). Voltage limiting will load the resonator, causing that circle to shrink. Operation will occur at an intersection of the two curves.

A family of resonator responses are also plotted on the same chart, each representing different tuning frequencies. It is then possible to ascertain oscillation over a frequency band.

The Colpitts oscillator has been viewed as a single port negative resistance circuit. However, the active device was actually a three terminal device, a bipolar transistor. The input was terminated in a specific impedance, a capacitor, to produce a suitable negative resistance at the output. The negative resistance resulted from the introduction of feedback through C_1 in an otherwise stable, nonoscillating device. The oscillation is usually associated (erroneously) with an instability at only one port, in this case, the output.

As higher frequencies are approached, a more generalized approach must be taken. Series as well as parallel feedback may be considered to produce the desired negative resistances. Intrinsic device instability may be used to advantage. The terminations at both ports must be considered. It must be decided where the resonator is best placed.

Up until recently, the approach used by oscillator designers was primarily that presented. Some high level power measurements have been performed to evaluate the effective S parameters that occur during operation (rather than at starting) to better understand the limiting mechanism. Conceptually, little more has been done. In 1979 a paper was published by Basawapatna and Stancliff (14) which has provided considerable intuition about the design of oscillators using three terminal devices. Some of their work will be summarized here owing to its fundamental significance.

Figure 7.34 shows a schematic that is typical of a modern microwave oscillator. Other topologies are used, but this one is common. The resonator has the equivalent circuit of a parallel tuned circuit in series with a small inductance. The resonator is typically a sphere of yitrium-iron garnet (YIG) material that displays a high Q resonance at microwave frequencies. The virtue of the YIG sphere is that resonant frequency may be moved by altering the surrounding static magnetic field. Hence, it may be electronically tuned. Some resonators will function over nearly a 4-octave range. Moreover, the elements in the equivalent circuit will both change (15). A

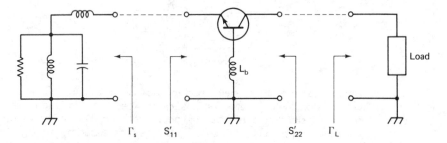

Figure 7.34 A typical microwave oscillator. The active device is a transistor (or FET) operated in the common base mode. Feedback is introduced with some base inductance, allowing oscillation to start. The resonator is a sphere of yitrium-iron-garnet (YIG), modeled as a parallel resonator. Stable operation occurs when $\Gamma_s S'_{11} = 1$ and $S'_{22} \Gamma_L = 1$.

typical hf or vhf resonator has a variable capacitance or inductance, but not both. This restricts tuning range and causes resonator impedance to vary significantly with frequency. The YIG resonator is more nearly ideal. The series inductor represents a small coupling wire that communicates energy to and from the sphere.

The active device shown is a bipolar transistor. It may be replaced with a suitable GaAs FET device. The S parameters are first investigated for stability. Unlike the amplifier, the oscillator is designed to prohibit stability. Positive feedback is introduced with an inductor in series with the common base lead. The S parameters of the composite circuit will often show reflection coefficients greater than unity at both the input and the output. Very little base inductance is required for application above 2 GHz. Recall that a small wire has an inductance of around 1 nH per mm. Hence, the lead length of a packaged transistor may ensure oscillation.

The output of the network is terminated in a suitable load, usually the input of a buffer amplifier. This is the point where power is extracted from the oscillator.

Viewing this circuit as a single port negative resistance oscillator, as discussed above, we would surmise that oscillation occurs between the input port and the resonator while the output port is passive. This is not true. Indeed, this subtlety is the significant detail presented by Basawapatna and Stancliff.

Relationships using scattering parameters were presented in Chap. 5 that expressed input and output reflection coefficients as functions of the S matrix and the loads at the opposite port. Specifically

$$S'_{11} = S_{11} + \frac{S_{12}S_{21}\Gamma_L}{1 - S_{22}\Gamma_L} \qquad (7.6\text{-}8)$$

and

$$S'_{22} = S_{22} + \frac{S_{12}S_{21}\Gamma_s}{1 - S_{11}\Gamma_s} \qquad (7.6\text{-}9)$$

where the primes indicate the input or output parameter measured with a specific termination, Γ_L or Γ_s.

We are concerned now with the conditions present at the output port, assuming a stable oscillating condition at the input. Noting the earlier significance of the reciprocal of the input reflection coefficient, Eq. 7.6-9 is solved for $1/S'_{22}$

$$\frac{1}{S'_{22}} = \frac{1 - S_{11}\Gamma_s}{S_{22} - S_{11}S_{22}\Gamma_s + S_{12}S_{21}\Gamma_s} \qquad (7.6\text{-}10)$$

Equation 7.6-8 is solved for S_{11}

$$S_{11} = S'_{11} - \frac{S_{12}S_{21}\Gamma_L}{1 - S_{22}\Gamma_L} \qquad (7.6\text{-}11)$$

Substitution into Eq. 7.6-10 yields

$$\frac{1}{S'_{22}} = \frac{1 - S'_{11}\Gamma_s + \dfrac{S_{12}S_{21}\Gamma_s\Gamma_L}{1 - S_{22}\Gamma_L}}{S_{22} - S_{22}\Gamma_s\left(S'_{11} - \dfrac{S_{12}S_{21}\Gamma_L}{1 - S_{22}\Gamma_L}\right) + S_{12}S_{21}\Gamma_s} \qquad (7.6\text{-}12)$$

which reduces to

$$\frac{1}{S'_{22}} = \frac{(1 - S'_{11}\Gamma_s) + \dfrac{S_{12}S_{21}\Gamma_s\Gamma_L}{1 - S_{22}\Gamma_L}}{S_{22}(1 - \Gamma_s S'_{11}) + \dfrac{S_{12}S_{21}\Gamma_s}{1 - S_{22}\Gamma_L}} \qquad (7.6\text{-}13)$$

Assume that the circuit is oscillating. Hence $S'_{11}\Gamma_s = 1$ and Eq. 7.6-13 reduces to $S'_{22}\Gamma_L = 1$. We conclude that if oscillation at a stable amplitude is taking place, it occurs at both ports simultaneously. The oscillation may not be isolated to just one port.

Consider an example: A device is chosen and a feedback inductor is picked to yield an input reflection coefficient, S'_{11}, greater than unity, usually over a band of frequencies. Examination of device curves superimposed on those of the resonator shows that oscillation is possible over the desired band. Curves like Fig. 7.33 are sufficient to predict starting. They provide little information about operating conditions though.

When the device is examined for starting, as outlined, the output may or may not be capable of oscillation. That is, nothing specific is said about S'_{22} during starting merely because S'_{11} is greater than unity in magnitude. However, the previous theorem ensures that once oscillation has commenced, both ports will have reflection coefficients capable of sustaining oscillation. Their magnitude and phase may be determined by examination of the terminations. This gives the designer a tool to use in choosing terminations. Output power may be optimized as can the flatness, the constancy of power over a tuning range. Conditions may be picked to ensure that oscillation is confined to the desired oscillator transistor and does not occur in the buffer amplifier that usually follows.

There is considerable subtlety still involved, especially in designing oscillators which must tune over a wide frequency range. Many examples are found in the current literature, especially that specializing in microwave circuits. Much of the design problem is involved in synthesis of the network between an oscillating stage and the buffer amplifier. The Basawapatna and Stancliff paper is recommended reading.

The topology of Fig. 7.34 is popular for good reason. The feedback shown usually produces an S'_{11} which is inductive over a wide frequency spectrum. A resonance would exist with the series inductor representing the coupling loop used to excite the YIG sphere if the input was capacitive. Spurious oscillations would then be possible at undesired and uncontrolled frequencies.

7.7 METHODS OF FREQUENCY SYNTHESIS

The performance of free running oscillators is limited. The frequency stability of *LC* oscillators is limited by thermal drift effects, especially at higher frequencies. While the crystal controlled oscillator is better in this regard, it is not variable in frequency.

The problems are overcome largely through the application of frequency synthesis. The subject is broad. Frequency synthesis may be defined in the most general sense as being a system where circuitry operates on one or a few oscillators to produce an output that is different than the original frequencies. While frequency synthesis has many virtues, it can introduce severe problems. The greatest of these is poor noise performance in all but the best designed systems.

A complete treatment of synthesizers will not be attempted. There are complete texts on the subject (16). The presentation of this section and those following is aimed at presenting the fundamental concepts with an emphasis on the problems introduced as well as the virtues. Some simple systems, including the popular phase-locked loop (PLL) types will be covered in more detail.

Frequency synthesizers are generally divided into two types: direct and indirect synthesizers. A direct synthesizer is one using frequency mixing and multiplication with appropriate filtering to achieve the desired output. Indirect synthesizers utilize phase-locked loops with voltage controlled oscillators. There is overlap between the two types. For example, synthesizers for microwave application may use a PLL system operating at vhf which is then multiplied to microwave frequencies.

Figure 7.35 shows the simplest of direct synthesizers, often called a "mixmaster." Two crystal oscillators are applied to a mixer. The resulting output is then filtered to produce the desired frequency. Each oscillator has a number of crystals which are closely spaced in frequency. Assume for example that oscillator A has 10 crystals

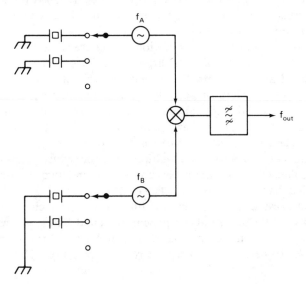

Figure 7.35 The simplest of frequency synthesizers, often termed a "mixmaster."

in the range of 20.000 to 20.450 MHz, each frequency differing from the other by 50 kHz. The other oscillator, P, has 10 crystals at 5-kHz spacings in the range of 25.000 to 25.045 MHz. By choosing appropriate combinations, 100 output frequencies are available in the range of 45.000 to 45.495 MHz, each separated by 5 kHz. The bandpass filter selects only the sum frequency. It should offer good suppression of the 5-MHz image as well as the harmonics of the two oscillators.

The concepts of the simple direct synthesizer may be expanded. Other conversions with additional filter and oscillator groups may be used to expand the frequency coverage. The total number of output frequencies will be a product. In the example shown, 20 crystals are used, 10 in each oscillator. There are 10×10 possible outputs. If the synthesizer is expanded to include one more bank of 10 crystals, there will be 1000 possible output frequencies. Through multiplication of some outputs and division of others with suitable filtering, the number may be expanded indefinitely. The circuitry becomes very complex, owing predominantly to the high performance filters required. Other variations of this system will allow all of the frequencies to be derived from a single crystal oscillator. Combinations of multiplication, mixing, and division are all used.

Direct synthesizers offer good spectral purity, approaching that of the crystal oscillator(s) used. The close-in noise will be dominated by the oscillator designs while the spurious responses will be dominated by the filter designs. Direct-type synthesizers are very frequency "agile," meaning that the output frequency can be changed by either a small increment or a large one in a very short time. A typical switching speed is around 1 μs. Switching is usually done with PIN diodes. The largest problem with direct synthesizers is their high cost and complexity. They are generally limited to laboratory applications.

Most basic building blocks of a direct synthesizer have already been covered. The mixers should operate at high levels. Broadband noise performance may be compromised with low level mixing. The crystal oscillators should be designed for good phase noise performance. Other resonator types are attracting considerable attention. These include surface acoustic wave resonators.

Little has been said about frequency multiplication. There are several methods that may be used. Bipolar transistors work well through the lower uhf spectrum. The circuit resembles a tuned amplifier with the output resonant at a multiple of the input frequency. The input power must be well beyond the usual small-signal limits, allowing the circuit to work with high harmonic distortion. The output current will contain the fundamental input signal in large quantities as well as harmonics other than the one desired. Hence, filtering must be carefully done.

The filtering requirements at the output of a multiplier may be reduced considerably through the use of balance in the circuit. Figure 7.36 shows two multiplier circuits, both frequency doublers, that employ balance. The circuit in Fig. 7.36a uses two diodes. The circuit is essentially a traditional full-wave rectifier circuit with an rf choke at the output to short the dc output component to ground. The suppression of the fundamental driving input is limited by the diode matching and transformer symmetry, with values of 40 dB being realized with careful construction. The usual

Sec. 7.7 Methods of Frequency Synthesis 313

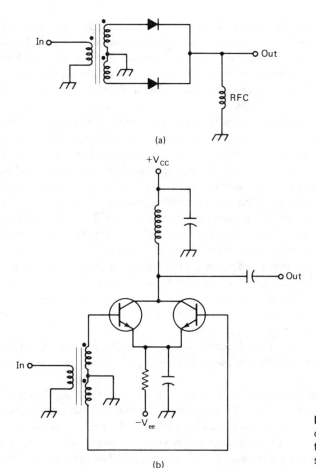

Figure 7.36 (a) A simple frequency doubler using diodes. (b) A push-push frequency doubler using bipolar transistors.

circuits employ hot-carrier diodes. A virtue of this circuit is the relatively broadband response. It is useful in some applications with no output filtering if the presence of higher order even harmonics is not objectionable. Some filtering is often useful though. Even a single resonator will produce a surprisingly clean output. The circuit has a typical loss of 7 to 9 dB and requires an input drive power sufficient to overcome the diode turn-on voltages. +10 dBm is usually sufficient.

The circuit in Fig. 7.36b is similar except that bipolar transistors are used. The bases are driven in push-pull, but the outputs are tied together. This provides balance which suppresses the fundamental driving signal. This circuit will function with smaller input power because the emitter-base junctions are forward biased. However, the input power should still be large enough to ensure severe harmonic distortion. The output impedance is relatively high. The filter can contain impedance transforming properties. This circuit is called a *push-push* doubler and is capable of power gain. If a simple multiplier chain is desired, consisting of a cascade of several multiplier–

filter combinations, a suitable circuit may be built by cascading the two doublers of Fig. 7.36. The gain of the second will compensate for the loss of the diode doubler.

Frequency triplers or even quadruplers may be built with transistors. Single devices with suitable output filters may be used. Alternatively, balance may be employed to suppress some of the output components. Multiplication beyond a factor of 4 or 5 per stage is generally not recommended with transistors.

High order frequency multiplication is often performed with varactor diodes, or step-recovery diodes. The step-recovery diode is a junction device with relatively large amounts of charge storage. The fundamental driving signal will forward bias the diode over part of the operating cycle. As the input polarity reverses, the charge stored in the diode does not disappear immediately. When the driving polarity reverses, current continues to flow just as if the device were a resistor. However, when the stored charge is finally "swept out" of the junction, current ceases. The sudden decrease leads to a large output voltage of short duration. The pulse is very rich in harmonics. The output voltage pulse results from driving the diode with an inductive source.

The output of a step-recovery diode multiplier may be filtered to achieve a desired frequency. The filter must generally be of relatively narrow bandwidth owing to the high order of multiplication employed. Multipliers of this type find frequent application in systems where the pulse is used directly, eliminating the need for a filter. Step recovery type multipliers are covered well by Manassewitsch (17).

Numerous other methods are available for frequency multiplication. All of them will degrade the phase noise characteristics of the source. This was covered in Sec. 7.5.

Frequency division is a means of achieving outputs at a frequency lower than the input. There are numerous methods for frequency division, but the most common ones use digital integrated circuits. An example, a type D flip-flop is shown in Fig. 7.37. This integrated circuit is an edge triggered device. Assume that it is designed to respond to a positive-going transition of an input signal. The type D, or "data" flip-flop will transfer the logic voltage at the D input to the Q output at the positive transition of the clock line. The $\overline{Q}$ output is opposite to that of Q and achieves the opposite state after a small propagation delay. A timing diagram is shown with the circuit of Fig. 7.37. The output appearing at Q_A is at half the frequency of the input clock. This is then a divide-by-2 circuit.

An arbitrary number of dividers may be cascaded to produce very low frequencies. The circuitry is generally straightforward. The input "clock" signal must be of the proper voltage level and must usually have a reasonably fast transition to ensure that the flip-flop is properly triggered.

In a cascade, a later flip-flop (ff) may be clocked by either the Q or $\overline{Q}$ output of the preceding stage. It does not matter which is used if the only outputs of interest are division by 2^n. The method of interconnection will determine if the total circuit counts "up" or "down" specifically, if the outputs of the cascade of ff's are assigned a binary value. The timing diagram of Fig. 7.37 shows the output of the flip-flop shown in the circuit, Q_a, and of a second one that is driven from the Q_a output. The resulting output, Q_b, is at one quarter the frequency of the original clock. Noting

Sec. 7.7 Methods of Frequency Synthesis   315

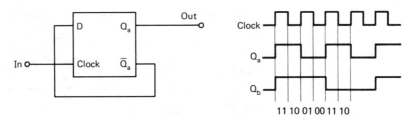

Figure 7.37 Frequency divider action with a type-D flip-flop.

that the input ff represents the least significant binary bit while the second ff is the most significant, the sequence shown is 11, 10, 01, and 00. The decimal equivalent is 3, 2, and 1, and 0; the circuit is thus a down counter. Note that after the 0 state the counter cycles back to the 3 state with the next clock pulse.

An upward counter is fabricated with similar ease. Each stage in the chain is clocked by the $\overline{Q}$ output of the preceding one.

The utilization of the full binary output of the cascade can be of great importance for it allows division by factors other than powers of 2. For example, the counter discussed so far can be forced to divide by 3 with the addition of a NOR gate. The gate has the characteristic (formally, truth table) that the output is zero if either or both inputs are high. A high output results if and only if both inputs are low.

Assume that the ff's also have a "preset" input, a line that overrides other input signals. A high signal on the preset input will force a 1 to appear at the corresponding Q output. The typical D ff will also have a reset that forces the Q to a logical zero stage.

With reference to the schematic and timing diagrams, Fig. 7.38, the action of the down counter is evaluated. Starting at a count of 3 (binary 11), the counter decreases until it reaches the 0 state. At that instant, both inputs to the NOR gate are low, producing a high at the gate output. This immediately presets the counter to the 3 state. The output at Q_b is now a rectangular wave with a frequency one third that of the input clock.

The preceding examples of frequency division have been presented to illustrate the concepts. They may be extended to very large division ratios. A complete discussion of the digital methods is away from the goal of this text.

One practical difficulty with frequency dividers is the presence of very narrow pulses. In the example of Fig. 7.38, the ff's were preset as soon as a binary 00 condition occurs at the ff outputs. While this will produce the desired NOR gate output, there exists the possibility that one of the ff's will be preset before the other. A one-shot multivibrator is often useful in such applications.

Another problem with circuits such as shown in Fig. 7.38 is that it is a ripple counter. That is, each ff is triggered (clocked) by the transition on the output of the previous one. Owing to propagation delay, this may produce some very brief invalid output states which could preset the circuit at times other than desired. The best solution to this problem is the utilization of a synchronous counter, one where

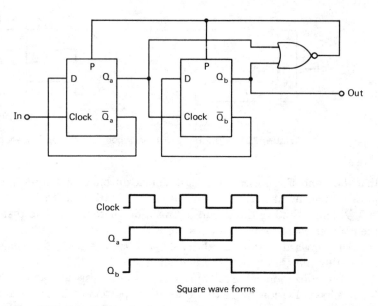

Figure 7.38 Two D-type flip-flops connected with a NOR gate to form a divide-by-3 circuit.

the transitions at all ff's occur at the same time. Examples are the 74193 or 74LS193, four-bit programmable synchronous up-down counters. When operated as down counters, all four Q outputs change in synchronism with the driving clock. This device is called programmable, for when a low signal is applied to the "load" input, the data on four input lines is immediately loaded into the counter. When in the down count mode, a "borrow" output goes low after all four ff's are in a zero state. The borrow output, usually with some buffering, controls the load line.

Another frequency division method that is often overlooked is with a shift register. A four-bit shift register built from four cascaded D ff's is shown in Fig. 7.39. With each clock pulse, the information at the D input of each ff is transferred to the corresponding Q output. Assume that a logical 1 is applied to the input and that initially all Q values are zero. When the first clock pulse arrives, Q_1 will be 1 while the rest remain zero. After the second clock pulse, Q_2 is 1 as well. After the

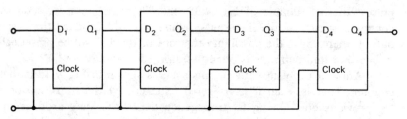

Figure 7.39 Use of a chain of D flip-flops as a shift register for frequency division.

fourth clock pulse, a logical 1 appears at the output. This may be used to trigger a one-shot multivibrator to reset all of the ff's. The result is a divide-by-4 circuit. The N number is the same as the number of stages in the register.

The propagation delay of several stages will accumulate in a ripple counter; uncertainty or jitter in the delay will lead to unwanted phase modulation of the output signal. This is eliminated through the application of a D ff at the output of a large divider chain. While there may be considerable variation in the signals rippling through the basic counter, the final D ff is being clocked by the same signal as is the input. Hence, outputs will be synchronous with the basic clock. The extra D ff will add 1 to the divisor, usually a minor problem.

There are many logic families. All are suitable for frequency division, although some are better than others. CMOS is generally limited to the lower frequencies. Although it will function at the lower part of the hf spectrum, the transitions are usually relatively slow when compared with faster logic types. CMOS is usually used only in the more casual applications.

Transistor-transistor logic (TTL) is very popular and usually performs well in divider applications throughout the hf spectrum. The low power Shottkey versions have reasonable power consumption while offering high speed and fast transitions. Some TTL devices will function at frequencies as high as 100 MHz.

Emitter-coupled logic (ECL) is the fastest general family available. Some ECL devices will function to above 1 GHz and are well suited to frequency division applications. Logic families are often mixed in synthesizers. The high speed devices are used where needed while the slower ones are used when possible to conserve power.

Frequency division may be applied in the design of a direct synthesizer. For example, a high frequency crystal oscillator may be divided by a programmable circuit to yield an output that is applied to a mixer. It is then heterodyned to a desired output. Appropriate filters are required at the output of the divider and of the mixer. The more common use for dividers is in the fabrication of PLL, or indirect synthesizers. The rest of our discussion of frequency synthesis will be confined to such designs.

Figure 7.40 shows a basic phase-locked loop. A reference from one oscillator is applied to the phase detector. Also applied is the output of a voltage controlled oscillator (vco). The phase is compared, resulting in a dc output voltage. This is filtered and perhaps amplified in a low pass filter, usually described as the $H(s)$ or loop filter. The resulting voltage is routed back to the vco. If the frequency of the vco equals that of the reference (or suitably close), the loop will control the voltage applied to the vco such that the phase of the two oscillators differs at most by a constant.

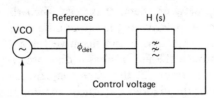

Figure 7.40 A simple phase-locked loop. The frequency of a voltage controlled oscillator (vco) is forced to equal that of a reference by comparison of the two phases. The $H(s)$ filter is a low pass type that removes ac components from the voltage control line.

The usual goal of a frequency synthesizer is the control of frequency. Phase is of little concern, at least in typical receiver and spectrum analyzer applications. However, a phase detector is still used. There are good reasons for this.

The output of the vco is generally a sinusoid, $V_{\text{out}} = \sin(\omega t + \phi_0) = \sin \theta$. The angle θ is the instantaneous phase, $\theta = \omega t + \phi_0$ where ϕ_0 is a phase difference with respect to the reference. Differentiating, the angular frequency is the rate of change of instantaneous phase. If the phase is held constant, that is without change, the only solution is that the frequency is zero. If the phase difference between the vco and a reference is held constant, the frequency must be a constant.

The PLL is a control system, a collection of elements with negative feedback applied. Because there are many elements in the loop, the feedback is more complicated than with a simple feedback amplifier like those studied earlier. The concepts are still identical. Feedback in an operational amplifier was never perfect. The gain of the amplifier was not infinite; hence there was always a small error voltage existing at the input. This is a general characteristic of any feedback loop, including the PLL. However, the PLL has the virtue that feedback is applied to phase, the integral of the frequency. While phase errors exist, there is essentially no frequency error.

The three basic elements of the PLL of Fig. 7.40 are the vco, phase detector, and filter. The voltage controlled oscillator is like those presented earlier except that the variable tuning capacitor is replaced with a varactor diode, a voltage variable capacitive element. More will be said later about the design and compromises of vco's.

The $H(s)$ filter is usually simple, using an operational amplifier, a capacitor, and a pair of resistors. Sometimes additional filtering is cascaded with the normal filter. The details of the filter design will be presented later.

There are numerous circuits that will function as a phase detector. Perhaps the simplest is that shown in Fig. 7.41 with two diodes and two transformers.

Consider a signal applied to input A. With the positive-going polarity at the input, D_1 will conduct, charging C_1. Capacitor C_2 will be charged during the negative polarity of signal A by conduction of diode D_2. The polarity of the two signals will be opposite, causing the sum as viewed at the output to be zero. All components are assumed to be matched. While the net output voltage is zero from signal A, this is only a result of balance.

The action from application of a signal only at input B is similar. During the positive polarity, both diodes conduct, placing charge on both capacitors. However, since the connection of T_2 is to the junction of the two diodes, no net voltage appears. No conduction occurs during the negative polarity of signal B.

When both signals are present, the voltage appearing at the output depends upon the phase relationship between them. It is zero when the signals differ by 90 degrees. The output changes in polarity as the phase difference changes. The output is positive when the signals are in phase and negative for a 180 degree difference. The details are analyzed through the use of phasor diagrams (18).

If the two input signals are not at the same frequency, the output will be an ac signal. The capacitors cause the difference frequency to dominate. Depending upon

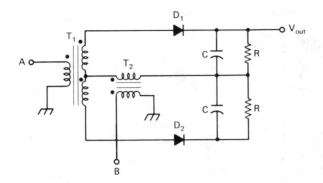

Figure 7.41 An analog phase detector using a pair of diodes.

the difference frequency and the capacitor reactances, the ac signal may or may not cause the vco to move in frequency. Often, if the two input frequencies are close, resulting in a low difference frequency, the vco will change in frequency, allowing the two frequencies to eventually be equal. The loop is then said to have "acquired" the signal (19).

It is more common in synthesis applications to force the frequencies to coincide through an external stimulus. A voltage is applied to the vco, causing it to sweep over the range of interest until the frequency is the same as that of the reference. Then, the resulting dc output from the phase detector is allowed to assume control, providing the required vco voltage.

There are many other phase detectors that operate similar to the one described. This includes the traditional diode-ring mixer. The dc is then extracted from the "i.f." port. Again, a zero output corresponds to a 90 degree phase difference. Detectors of this general type have the same frequency restrictions that the corresponding mixer would have.

Another phase detector using diodes is the sampling type. An example is shown in Fig. 7.42a. The reference input is in the form of a narrow pulse chain with a frequency much lower than the vco. When the positive pulse is applied, it causes all four diodes to conduct. The vco is attached momentarily to the sampling capacitor when the strobe is applied.

Figure 7.42b shows a representative waveform with the strobe pulse superimposed. The first pulse occurs at a vco zero crossing. Hence, no voltage will appear on the capacitor. The next strobe pulse occurs several vco cycles later and is at a position to cause charge transfer into the sampling capacitor. The voltage will increase owing to the difference in phase.

The voltage would be representative of the phase difference if the strobe pulse was to occur only at zero crossings. The output could be applied to an amplifier and $H(s)$ filter to achieve phase lock. Note that lock will occur for frequencies where the vco is a harmonic of the strobe. The gain will decrease as the harmonic number increases. A typical application of this type of phase detector is in microwave rf instruments where a vco operating at several GHz is locked to a harmonic of a strobe signal that is only a few MHz.

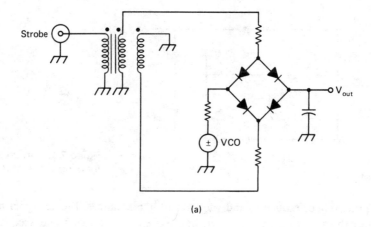

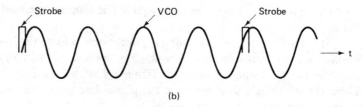

Figure 7.42 A sampling-type phase-gate. The strobe input is a low frequency pulse of very narrow width which periodically turns the bridge circuit on. The vco voltage is sampled during the strobe periods. The vco frequency will be a multiple of the strobe frequency if a feedback loop forces the dc output voltage to be constant.

Phase detection is often performed at lower frequencies with digital integrated circuits. There are numerous possible schemes. Perhaps the simplest is an Exclusive-OR gate, shown in Fig. 7.43a. The truth table for the Exclusive-OR gate is paraphrased by stating that the output is high if either input is high, but is zero if both are high or are low. The characteristics of the Exclusive-OR phase detector are evaluated with this information. If both signals are in phase, the output will always be zero. If they are exactly out of phase, a difference of 180 degrees, and the signals are square waves, the output will be a logical one. Figure 7.43b shows an example of the output with a 90 degree phase difference. The output is a square wave with a 50% duty cycle. Hence, the average voltage is 2.5 V, assuming a 5-V supply. The transfer characteristic is shown in Fig. 7.43c.

The transfer curve is a triangle wave. If the loop using this detector is adjusted for a lock with 2.5 V output (90 degrees) only one slope will function as desired. If the detector should end up operating on the wrong slope, the phase of the vco will be forced to move until it is operating on the stable one.

Although digital, the operation of the Exclusive-Or gate detector is similar to

Sec. 7.7   Methods of Frequency Synthesis   321

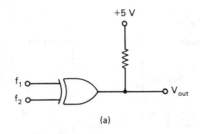

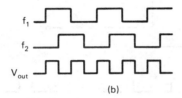

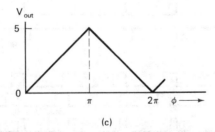

Figure 7.43 (a) shows a digital Exclusive-OR gate used as a phase detector. Timing diagrams are shown in (b) while the average dc output is shown in (c).

the diode circuit of Fig. 7.41. It may be necessary to sweep the vco to achieve a locked condition in a reasonable time period.

The most popular phase comparing circuit using digital IC's is the phase-frequency detector. An example using LS-TTL logic is shown in Fig. 7.44a. The circuit uses edge-triggered flip-flops. The D-type ff's shown have their data inputs permanently tied to a logic one voltage. They set the output at logic one when clocked by a positive-going waveform. The two Q outputs are monitored by a NAND gate. When both Q outputs are one, a logical 0 appears at the gate output. This is applied to the reset line of both ff's, setting both back to 0. One ff is clocked by the vco while the other is clocked by the reference signal.

The detector output is the sum of two voltages, those on Q_1 and on $\overline{Q}_2$. Usually, these points each drive a resistor that drives current into the summing node of an operational amplifier, part of the $H(s)$ filter. The filtering action removes the pulsing ac waveforms while retaining the average dc value.

The phase detection action is shown in the waveforms accompanying the circuit. Those shown in Fig. 7.44b show f_1 lagging behind f_2 by 45 degrees, where f_1 and f_2 are the input digital signals. Each ff is set by the positive-going edge of the corresponding clock. Both ff's reset as soon as Q_1 and Q_2 are both high. The average value of

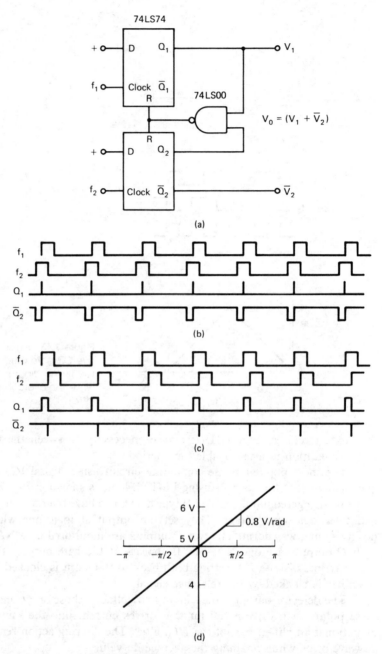

Figure 7.44 A digital phase-frequency detector using TTL logic. The schematic is shown in (a). Timing diagrams for f_1 lagging or leading f_2 are shown in (b) and (c). The average output voltage is shown in (d).

Q_1 is zero, for the narrow pulse has little energy in it. The average value of $\overline{Q_2}$ is slightly under 5 V. Hence, the sum output is under 5 V.

When both input signals are exactly in phase, they are triggered on simultaneously, only to be reset by the NAND gate. The average output voltage is 5 V. The opposite phase, f_2 lagging f_1 by 45 degrees, is shown in Fig. 7.44c. The average output is now slightly over 5 V.

If the exact dc values are calculated with numerical integration by inspection, a curve of output as a function of phase results. This is shown in Fig. 7.44d. Of vital interest is the slope, or gain of the detector, 0.8 V per radian.

The detector of Fig. 7.44 has a very useful property that is not obvious from the data presented. If the two inputs are at differing frequencies, the resulting outputs are dc values that, when averaged in the $H(s)$ filter, are in the proper direction to move the vco toward zero frequency difference. Hence, the system will achieve lock without the need for external sweeping circuitry. This characteristic may be demonstrated with timing diagrams with $f_1 \neq f_2$.

There is something of a dilemma with the timing diagrams shown in Fig. 7.44b and c. They are based upon the assumption that both ff's are reset at the left before arrival of the first clock pulse. This assumption may not be valid. The initial conditions could be different, leading to altered waveforms. However, the output voltage is still that shown in Fig. 7.44d.

The circuit shown (Fig. 7.44a) is a practical one. It should function well to at least 5 MHz using TTL logic. Even higher input frequencies are usable if ECL is substituted. The limit will depend upon the specific type of ECL chosen. The higher speed capability makes the ECL logic slightly more difficult to use, a problem for the logic designer, but certainly not for the rf engineer.

CMOS may also be used for the phase frequency detector. The upper working frequency is generally limited to somewhere near 1 MHz.

7.8 PHASE-LOCKED-LOOP FREQUENCY SYNTHESIS

Most modern frequency synthesizers use at least one phase-locked loop (PLL). This section is devoted to examination of the details of such a system. Many of the basic components and the underlying concept were presented in the previous section. The following section will emphasize the design of voltage controlled oscillators. Here we examine the loop as a control system and the implications with regard to spectral purity.

The previous PLL example, Fig. 7.40, had the vco operating at the same frequency as the reference. This system is practical, especially in applications such as frequency standards. However, it lacks flexibility—only one possible output frequency is available. A minor modification to this system is shown in Fig. 7.45. The vco output is heterodyned with a crystal oscillator to a lower frequency and then applied to a phase-frequency detector. The resulting error is filtered in the $H(s)$ filter and used for control of the vco. A typical system would use a reference frequency which

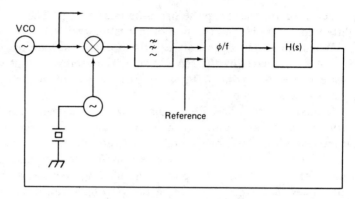

Figure 7.45 A simple application of the PLL, a tracking or one-on-one loop. The vco is the sum or difference of the crystal oscillator and the reference.

is variable. The vco will then move linearly with changes in the reference. This system is termed a tracking or one-on-one PLL.

The tracking loop could be replaced with a hetedrodyne system that did not utilize a PLL. The reference and crystal oscillator would be mixed with the sum or difference filtered and amplified. The results would be similar. However, the PLL offers simplicity. The only filter in the PLL system is a low pass. A narrow bandpass filter is required for the heterodyne replacement. PLL systems like that of Fig. 7.45 find wide application in both communications and rf instrumentation.

The most common type of synthesizer utilizing a PLL is that shown in the example of Fig. 7.46. A 10-MHz crystal oscillator serves as the standard. This is divided by a factor P, 1000 in the example, to produce a 10-kHz reference for the phase detector.

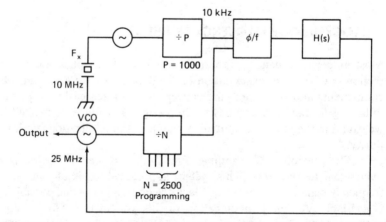

Figure 7.46 A simple, divide-by-N synthesizer. The vco frequency is divided by a programmable divide-by-N circuit to provide a drive for the phase-frequency detector. Vco frequency is given by $f_{vco} = f_x N/P$.

A vco operates at 25 MHz. This is applied to a second divider with a division ratio of $N = 2500$. The output is also 10 kHz. The two signals are compared in the phase detector, filtered, and used to control the vco frequency. The virtue of this system comes from the programming of N. If N is changed to 2501, an error appears at the phase-frequency detector. The signals produced force the vco to move to 25.01 MHz where a "locked" condition is again realized. The range of such a synthesizer is limited only by the available N numbers in the programmable divider and by the vco. The possible output frequencies are all multiples of the reference frequency at the phase detector.

Although common, the divide-by-N PLL synthesizer is not without problems. One is frequency multiplication. Any phase noise on the reference applied to the detector will be multiplied, causing a potential degradation of the phase noise close to the carrier. The response time of the system is limited. This is a result of the restricted bandwidth of the $H(s)$ filter. More will be said about this later. The loop can also degrade the output spectra of the vco by adding sidebands at a separation equal to the reference frequency. All of these problems must be considered in design of the overall system, and especially in the design of the $H(s)$ filter.

The PLL is a control system. As such, it must be analyzed for stability and effectiveness. A control system, in this case one with negative feedback, is analyzed in the same way that an oscillator is studied. The loop is broken at an arbitrary point and a signal is injected. The overall gain is evaluated for that signal to evaluate the performance of the closed loop.

The PLL, broken for evaluation, is shown in Fig. 7.47. This loop is much different than that of an oscillator. The desired gain has a negative sign, for it is a negative feedback control system. The "desired" signal is a dc voltage, the control required for the vco. However, the loop must be able to respond to changes, so input signals other than dc must also be considered. All frequencies where the loop has an overall voltage gain greater than unity must be examined. The phase of the gain at all of these frequencies must be evaluated as well. If the phase should change from the 180 degrees required for negative feedback at dc, the loop could oscillate. This is certainly not desirable. The input frequency for the voltage source of Fig. 7.47 is f_m, the modulation or Fourier frequency. It is not related to the frequency of the vco, f_v.

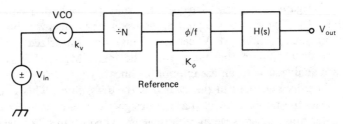

Figure 7.47 Analysis of a PLL. The loop is broken just as an oscillator was evaluated. The individual element gains will determine the response and stability of the loop.

Many negative (or positive) feedback systems may be opened experimentally as shown in Fig. 7.47. This is very difficult with the PLL and is rarely done. The scheme is still valid for analysis. Each system element must be evaluated for its gain at various input frequencies, f_m.

The first element in the loop is the vco. It has an operating frequency which is proportional to voltage, ideally with a linear characteristic

$$f_v = f_0 + GV_{in} \qquad (7.8\text{-}1)$$

where f_0 is a lower limit of the control range and G is a constant of proportionality in Hz per V. Applying a sinusoidal input voltage at a frequency f_m

$$f_v = f_0 + GV_m \sin \omega_m t \qquad (7.8\text{-}2)$$

where ω_m is the angular frequency, $2\pi f_m$, and V_m is the input amplitude.

The vco output will eventually be applied to a phase-sensitive detector. Our concern is primarily with phase rather than frequency. But for a sinusoidal oscillator, the frequency is merely the time rate of change of instantaneous phase. Hence, phase is the integral of the time dependent frequency

$$\phi = \int f_v \, dt = f_0 t - \frac{V_m K_v}{\omega_m} \cos \omega_m t \qquad (7.8\text{-}3)$$

where $K_v = 2\pi G$, the sensitivity of the oscillator in radians per second per volt. This equation is vital to understanding the PLL. The vco operates as an integrator so far as f_m signals are concerned. Dividing by the input magnitude, the frequency domain voltage transfer function of the vco is

$$H_{vco}(s) = \frac{K_v}{s} \qquad (7.8\text{-}4)$$

Any phase variation in the vco will be diminished by the programmable divider. That gain is then $1/N$.

The gain of a phase detector, K_ϕ, may be evaluated either experimentally or analytically. The gain of a digital phase-frequency detector using TTL was calculated in a previous section, 0.8 V per rad. This is a "linearized" value representing the average dc output of the pulses from the logic. Most PLL systems are nonlinear, but are analyzed with an assumption of linearity.

The final element in the system is the $H(s)$ filter. This circuit is critical, for it represents the major control that the designer can exercise over the system response. The filter must satisfy several requirements. The output of a phase-frequency detector is a chain of pulses. The $H(s)$ filter must have a low pass response that will extract the dc component, but virtually eliminate ac components. The strongest component to be attenuated is the reference frequency. Inadequate filtering will allow some reference energy to reach the vco control line, leading to phase modulation of the output.

Although the $H(s)$ or loop filter must be a low pass circuit, it must have a wide enough bandwidth to allow the PLL to respond to changes such as might occur when the N number in the synthesizer is altered. Finally, the filter must not introduce too much phase shift at frequencies where the loop has a net gain greater than unity. Excess phase will change the nominally negative feedback to an unstable loop oscillation.

A typical circuit for the $H(s)$ filter is shown in Fig. 7.48. Some initial low pass filtering is supplied by the split input resistors and C'. The dominant filtering is achieved by the integrator action of the input resistors, R_1, and the feedback capacitor, C. The input connections are the outputs of the digital phase-frequency detector, Fig. 7.44. Neglecting the effects of the input capacitors, C', the frequency domain response of the filter is

$$H(s) = \frac{1 + sCR_2}{sCR_1} \tag{7.8-5}$$

An ideal operational amplifier with infinite gain is assumed in the analysis. The response has a pole at the origin and a zero in the left half plane. The zero is vital in preserving system stability and in controlling the step response.

The transfer function of the PLL is the product of the individual functions. The steady state response is then

$$H(j\omega) = \frac{-K_v K_\phi}{N} \frac{1 + j\omega_m R_2 C}{\omega_m^2 R_1 C} \tag{7.8-6}$$

Note that ω_m^2 occurs in the denominator indicating two poles at the origin, one resulting from the integrator action of the vco and the other from the $H(s)$ filter.

Consider as an example a simple synthesizer operating at 5 MHz. The reference frequency is picked to be 5 kHz, allowing the vco to take on frequencies with a 5-

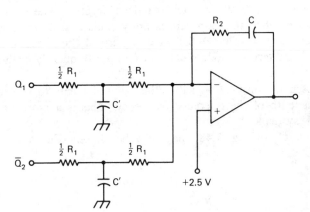

Figure 7.48 A typical $H(s)$ filter to be used with the phase-frequency detector of Fig. 7.44.

kHz spacing. Assume that the vco tunes from 5 to 5.15 MHz with control voltages from 5 to 10 V. Then, $G = 30$ kHz per V and K_v is $2\pi G = 1.88 \times 10^5$. A digital phase-frequency detector is used, $K_\phi = 0.8$. The N chosen for evaluation will be 1000, yielding

$$\frac{K_v K_\phi}{N} = 150.8 \qquad (7.8\text{-}7)$$

Values are now chosen for R_1 and C: 10 kΩ and 0.01 μF. If R_2 is temporarily assumed zero, the loop gain may be evaluated at $\omega_m = 1$ radian

$$\begin{aligned}\text{Loop gain} &= \frac{K_v K_\phi}{N}\frac{1}{R_1 C} \\ &= 1.508 \times 10^6 \\ &= 123.6 \text{ dB}\end{aligned} \qquad (7.8\text{-}8)$$

The gain will decrease at a 12 dB per octave rate, since R_2 has not yet been picked. The loop, potentially unstable at this point in the design, will have unity gain (0 dB) at a frequency of 195 Hz.

The value of R_2 may now be chosen. The zero frequency is given as

$$\omega_z = \frac{1}{R_2 C} \qquad (7.8\text{-}9)$$

The zero frequency should be set at or below the frequency of gain crossover of the unstable loop. $R_2 = 200$ kΩ yields f_z of about 80 Hz. The loop is now designed. A complete response of the system is shown in Fig. 7.49. Both the amplitude and phase are presented in that figure.

The gain at $f_m = 1$ Hz is high at 92 dB. It decreases at a 12 dB per octave rate, but eventually decreases in slope, crossing unity gain at 485 Hz at a slope just over 6 dB per octave. The loop should be quite stable. The phase response at 1 Hz is 180 degrees, showing that the overall feedback is indeed negative. The phase increases, reaching 260 degrees when the loop gain is unity. This is still 100 degrees away from the unstable 360 degree value. 100 degrees is termed the phase margin. Plots like that of Fig. 7.49 are termed Bode plots (20) and are frequently evaluated for any control system.

R_1 and C were chosen in conjunction with other gain elements to provide a gain crossover at about a tenth of the reference frequency. This is a more or less standard practice for PLL systems.

The input capacitors, C' of Fig. 7.48, may now be chosen. The associated pole frequency is

$$\omega'_p = \frac{4}{R_1 C'} \qquad (7.8\text{-}10)$$

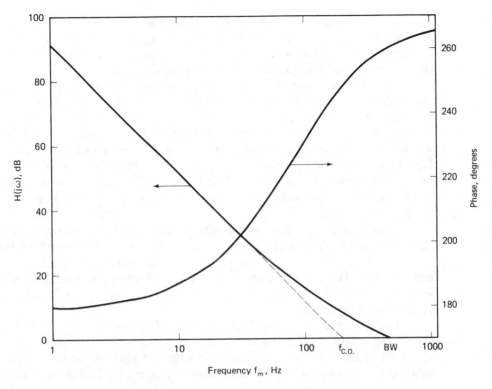

Figure 7.49 A Bode plot, or frequency and phase response of the sample PLL described in the text.

This value should be large compared with the gain crossover frequency. Picking $C' = 0.01$ μF yields f_p of about 6 kHz. This is high enough that the addition of C' will introduce a minimum of phase shift at lower frequencies below the gain crossover.

Note that the circuit described does not really have a bandwidth in the traditional sense. Gain increases forever as dc is approached, a result of using an operational amplifier with infinite gain for the analysis. The gain crossover frequency, 485 Hz in the example, is usually termed the loop bandwidth for a PLL. It is labeled as the BW in Fig. 7.49. The corresponding value for the loop prior to insertion of R_2 is also shown, $f_{c.o.}$. Note that the change in slope occurs at about 80 Hz, the zero frequency.

The loop has over unity gain for frequencies within the loop bandwidth. If the vco is modulated by some extraneous source, that noise will be removed if it is at a frequency within the loop bandwidth. An example might be 60- or 120-Hz power supply noise. On the contrary, noise at a higher frequency is not suppressed by the loop filter. An example for the synthesizer described might be a mechanical vibration that was at a higher frequency. Such a "microphonic" effect would not be eliminated so long as it was at a high frequency, beyond the loop bandwidth.

The same arguments apply as far as oscillator phase noise is concerned. Noise on the reference will be multiplied and will appear on the vco output if it is within the loop bandwidth. Reference noise at higher frequency will not be multiplied. Then, the vco noise output will be that of the free running vco.

The PLL has a gain less than unity at f_m outside the loop bandwidth. Reference sidebands, at 5 kHz in the example, are outside the loop BW. The gain is decreasing at a single pole rate (neglecting the effects of C') for f_m above the loop BW. Hence, reference sidebands are reduced by only 20 dB in the example. The addition of C' will increase this attenuation. More filtering may also be added. This may be in the form of an active circuit, either a low pass or bandstop filter. Alternatively, an LC filter could be used. In either case, the response of the additional filtering should be investigated to ensure that a minimum of additional phase shift is created within the loop bandwidth.

Phase-locked loops are often analyzed with regard to "natural frequency and damping factor." These are parameters akin to those used to describe a second order bandpass filter, a single resonator. The differential equation for the PLL is also of second order. The different parameters merely represent the domain in which the system is viewed. As with any filter, the response may be viewed in either the time or the frequency domain.

Analysis of a PLL with the assumption of approximation as a second order loop is represented by Eq. 7.8-6. Addition of C' changes the loop to a third order system. The further addition of more filtering will also add to the complexity. A subtle effect in wideband loops comes from the operational amplifier which usually has a gain which decreases at a 6 dB per octave rate, a single-pole response. This can be a severe problem if not accounted for in the design. A frequency domain analysis, usually through the application of the Bode plot concept, will provide the information for design of the loop and analysis of the results.

7.9 VOLTAGE CONTROLLED OSCILLATORS AND IMPROVED SYNTHESIZERS

The vco is perhaps the most critical element in a phase-locked loop synthesizer. If it is spectrally clean, having very low phase noise, it may be controlled with a narrow bandwidth PLL with no compromise in overall noise performance. The major limitation will then be response time.

A noisy vco may be "cleaned-up" with a PLL. If controlled by a low division ratio loop, it will be stabilized and the spectral purity of the reference will be transferred directly to the output for spacings within the loop bandwidth. Like so many engineering problems, synthesizer design is a problem of compromises.

The most common means for electronically tuning an oscillator is with a varicap, or varactor diode. This diode is operated with a reverse bias on the junction. Capacitance is a function of the reverse bias. A typical oscillator using a varactor diode is shown in Fig. 7.50. A Colpitts circuit is used with the major portion of the resonator

Sec. 7.9 Voltage Controlled Oscillators and Improved Synthesizers

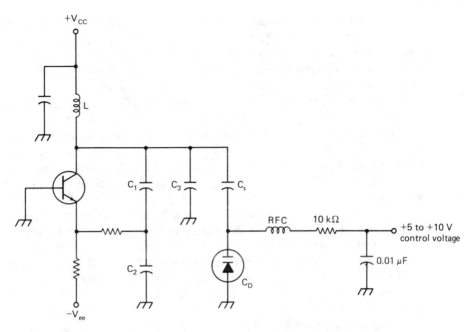

Figure 7.50 A Colpitts oscillator tuned with a varactor diode, C_D. The tuning range is restricted by the diode parameters and the series capacitor, C_s.

being L and the C_1-C_2-C_3 combination. The circuit is tuned with the varactor diode, C_d. The capacitance variation presented to the resonator is determined by the diode and by the series capacitor, C_s.

A typical abrupt junction varactor diode might have a capacitance of 21 pF with a bias of 5 V. The capacitance would decrease to 5 pF with a bias of 30 V, the breakdown value. The value at 10 V might be 12 pF. If the control voltage is restricted to the range from 5 to 10 V, the variation available is 9 pF. If C_s is large with respect to the diode capacitance, 9 pF is the variation available at the resonator. Decreasing C_s allows smaller variations at the resonator.

The capacitance variation available may be used to determine the inductance required for the oscillator. In the example cited, diode capacitance varies from 12 to 21 pF. The inductance would be evaluated at the upper edge of the frequency band to be tuned using the minimum capacitance. The lower frequency is then determined by the larger diode capacitance. The tuning range is further restricted in the Colpitts oscillator shown, for some capacitance will be required for the Colpitts feedback elements, C_1 and C_2.

The frequency range to be tuned is often small with respect to the operating frequency. The diode is then not a severe restriction.

The diode described was one with an abrupt junction, a sharp doping profile within the junction. The other popular diode type is the hyperabrupt junction. The ratio of maximum capacitance at a given bias level such as 4 V to that at a higher value is greater. Some hyperabrupt diodes have a range of up to 10, allowing vco's to be built with a 3-to-1 frequency variation. The hyperabrupt diodes have the added virtue that oscillator frequency versus voltage characteristics are more linear, especially if the tuning range is severely restricted. Hyperabrupt diodes typically have a lower Q and lower breakdown voltage than the abrupt junction devices.

Generally, varactor diode Q is inversely proportional to frequency. Hence, loss is often well modeled with nothing more than a series resistor. At 50 MHz, a typical Q might be several hundred for an abrupt junction device, while the hyperabrupt version would have a much lower value. Diode losses must be taken into account during the oscillator design, for they may often be a dominant element. In the more traditional hf oscillator design, the inductor is so dominant that capacitor losses are ignored.

Earlier we found it desirable to have a high voltage across the resonator, for this determined the carrier-to-broadband noise performance. The varactor diode must never be forward biased by the rf voltage. If this occurs, the oscillator limiting mechanism will be altered and resonator Q will be degraded. This presents a severe restriction on the design and construction of voltage controlled oscillators. Not only must a compromise be accepted in broadband noise performance, but the oscillator design must allow the peak operating voltages to be predicted, especially if the circuit is to be manufactured repetitively.

The diode capacitance, C_d, and the series capacitor, C_s, form a divider for the rf voltage. The rf voltage across the diode is smaller if C_s is small with respect to C_d. Resonator energy may then be increased while always maintaining sufficient diode bias. However, the tuning range is then restricted. The smaller the tuning range of a vco, the better the possible noise performance may be. Also, the diode loss is of less significance in lowering resonator Q with a restricted tuning range.

Linear tuning characteristics are important. One of the key gain elements in a PLL analysis is the sensitivity of the vco, $G_v = df/dV$, with $K_v = 2\pi G_v$. If the tuning is not linear with voltage, the loop gain will vary.

The voltage controlled nature of the oscillator presents a possible source for additional phase noise. Consider an oscillator with a sensitivity of 100 kHz per volt. Assume that the control voltage has a superimposed noise component of 10 μV. This noise will cause a frequency modulation of 1 Hz. While this is a small amount, it is enough to degrade vco noise performance. The exact nature of the degradation depends upon the spectrum of the superimposed noise. Small deviation fm theory may be applied for calculations (21, 22).

Most vco's are used in synthesizers. As such, thermal drift characteristics are not of great concern, for the vco is within a control loop. Sometimes a vco is used as a free-running oscillator. An example is the swept oscillator in a spectrum analyzer. Thermal stability then becomes of greater concern. Varactor diodes usually have poor temperature coefficients. However, the drift arises from a relatively simple mechanism; the offset voltage of the diode varies as a function of temperature. The varactor may be temperature compensated with the addition of a temperature sampling diode. The stability improvement is dramatic (23).

Sec. 7.9 Voltage Controlled Oscillators and Improved Synthesizers

Tuning linearity can be very important in free-running vco's. Linear tuning allows simple frequency calibration of a receiver. If used in a spectrum analyzer, the tuning linearity determines the frequency accuracy of the display. Varactor tuning, especially with abrupt junction diodes, can be quite nonlinear. This is often altered with appropriate shaping circuitry. This usually consists of a network of diodes and resistors connected to produce a nonlinear transfer function. The characteristic is designed to complement the nonlinear nature of the diode to produce a linear frequency versus voltage. It is not unusual to find networks with over 10 diodes. The shaping circuitry must be designed with care to ensure that excess noise is not introduced.

With all of the problems outlined, one might conclude that it is not possible to build a voltage controlled oscillator which has good spectral purity. While this is not true, special care must be used if reasonable performance is to be maintained. It is not surprising that many simple synthesizers used in some communications equipment have horrible phase noise problems!

A PLL is often used to improve vco spectral purity. The vco is heterodyned to a lower frequency and compared with a clean reference in a wideband loop. Specifically, the spectral purity of the reference is transferred to the vco output for spacings within the loop bandwidth. The loop is effective only close to the carrier. It will do nothing for the broadband noise output.

The best vco designs are those with the smallest tuning sensitivities. This, of course, leads to a restricted tuning range. The overall range is then expanded by switching reactances into the circuit. An example is shown in Fig. 7.51 where capacitors are switched with diodes. Three control lines are shown. One is the normal line supplying the varactor diode. The associated tuning range is restricted. The other two inputs, A and B, drive the diode switches. With no voltage applied to point A, the associated capacitor will charge to a dc voltage, with the required rectification coming from the diode. The dc current will be small. As long as the diode is a low capacitance type, the capacitance added to the circuit is also small. Current will flow in the diode if a positive voltage is applied to point A. This effectively places C_A in parallel with the resonator, shifting the frequency downward. The amount of shift is adjusted to be slightly less than the total range realized with the varactor tuning.

The action of the second capacitor, C_B, is identical. The frequency shift is twice that associated with C_A. More capacitors may be added, each representing a frequency shift twice that of the previous one. The switch lines are then controlled with digital circuitry.

If the oscillator of Fig. 7.51 was part of a synthesizer, the output of the $H(s)$ filter would be applied to the varactor. The control voltage would also be sampled and applied to a set of differential comparators, operational amplifiers operated without feedback. When the control voltage exceeds a set limit, a pulse would be generated to cause an up-down binary counter to increment by one step. The counter outputs would control the switching lines.

Normal silicon switching diodes may be used in the circuit. The bias current should be large compared with the ac current flowing in the capacitors. It should

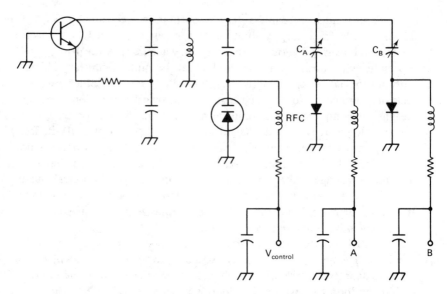

Figure 7.51 A refined Colpitts oscillator. A varactor diode provides tuning over a small frequency range. The frequency is stepped by turning on one or both of the diodes, placing C_A and/or C_B into the circuit. Only two switched states are shown, but more could be used. This scheme will offer significantly improved noise performance over the simple system of Fig. 7.50.

also be large enough that the dynamic resistance of the diode, when on, is small enough to preserve system Q. The better choice for diodes would be PIN types. They should have low capacitance when off and should have an external source of reverse bias. Bipolar transistors may also be used for the switching.

Figure 7.52 shows the resonator section of a Seiler oscillator where inductors are switched with PIN diodes. L_0 is the main inductor. Other inductors are switched in with removal of bias from the appropriate lines, A through D. This scheme is generally better suited to vhf vco's than is capacitor switching. The circuit of Fig. 7.52 is sequential rather than binary as that of Fig. 7.51 (24).

There are system methods which may be used in designing a synthesizer to improve the performance over that expected with the simple divide-by-N techniques described so far. One variation is to perform the synthesis at much higher frequency than the desired output. Then division is used to obtain the required signal. Assume that a synthesizer is needed for a communications receiver. It must tune the range of 5 to 5.5 MHz in 0.1-kHz steps. An improved way to build such a synthesizer would be to use a 50- to 55-MHz vco with a programmable divider providing a 1-kHz output. The 50-MHz signal is then divided by 5 and then by 2 to provide the desired output at 5 MHz. The final division factor of 2 is chosen to provide a square wave output, providing the proper waveform to drive a switching mode mixer. VCO operation at 50 MHz with subsequent division would provide a 20-dB improvement in phase noise. This will help to overcome the vco design problem related to

Sec. 7.9 Voltage Controlled Oscillators and Improved Synthesizers

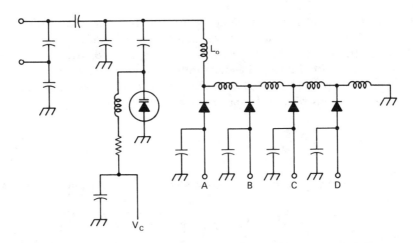

Figure 7.52 The resonator portion of a Seiler oscillator. A varactor diode provides tuning over a small range. Diode switches will alter the total inductance, changing the frequency in larger steps.

using limited resonator power. It will not change the effects of superimposed noise on the control line, for the gain will be 20 dB higher in the loop using the vhf vco. The time response is improved owing to the larger allowed loop bandwidth.

Figure 7.53 shows a single loop synthesizer. A crystal oscillator is divided by N to produce the desired reference. The vco is divided by M, producing the same reference. The vco frequency is then

$$f_v = f_x \frac{M}{N} \qquad (7.9\text{-}1)$$

where f_x is the frequency of the crystal oscillator. N has been a fixed number in the designs considered. This is, however, an unnecessary restriction. The output frequencies possible are expanded greatly if both N and M are allowed to vary.

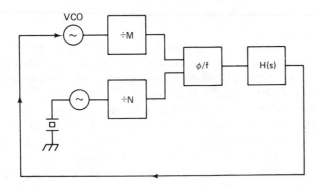

Figure 7.53 A block diagram of a single loop, divide-by-M synthesizer. Most loops like this use a fixed value for N, leading to evenly spaced vco frequencies. Considerable flexibility is provided if both N and M are allowed to vary although the step increments between vco frequencies may not be equal.

One particularly simple application of this generalization is to allow M to be related to N through a simple relationship. Specifically, let $M = N + K$ where K is a smaller integer. The dividers are then programmed together. The $N + K$ division is realized with a divider identical to the divide-by-N circuit except that K extra D-type flip-flops are cascaded with the output in a shift-register–type circuit. See Fig. 7.39. With this restriction applied to M, Eq. 7.9-1 becomes

$$f_v = f_x \left(1 + \frac{K}{N}\right) \quad (7.9\text{-}2)$$

The change in frequency associated with a change in N is evaluated by

$$\begin{aligned}\Delta f_v &= f_x\left(1 + \frac{K}{N+1}\right) - f_x\left(1 + \frac{K}{N}\right) \\ &= f_x\left(\frac{-K}{N(N+1)}\right)\end{aligned} \quad (7.9\text{-}3)$$

As N becomes even moderately large, the step size is well approximated by $f_v = f_x K/N^2$.

Consider an example, a synthesizer for the 5-MHz region for receiver applications. Allow N to vary over the range of 64 to 128, relatively small numbers in comparison with the earlier communications examples. K is set to 1. The crystal frequency is 4961 kHz. With N set at 128, the output frequency will be 5000 kHz, the step size will be 0.3 kHz, and the reference frequency will be 38.8 kHz. When N is decreased to 64, the corresponding numbers are 5038.8 kHz, 1.21 kHz, and 77.5 kHz. Clearly, the loop filter can be of much greater bandwidth than would normally be required for such close frequency steps. The response time will be correspondingly faster.

This scheme has its problems, the main one being the uneven, continually changing step size. Also, the frequency range is restricted for a 2 to 1 variation in N. A reasonable and practical synthesizer could be built by using a second loop to generate frequencies at a spacing of about 30 kHz to replace the crystal oscillator. This would expand the range quickly to whatever might be desired. A detail that must be taken into account is the 2 to 1 variation in N, resulting in a change in loop gain as N is altered.

This example is of greatest interest as an illustration. However, the writer has used the scheme to build synthesizers for simple communications applications where the uneven step size is not a problem. The more general and versatile synthesizer is one which places no severe restriction on the relationship of M and N of Fig. 7.53. Consider, for example, a synthesizer where both N and M are allowed to vary from 256 to 512. The total number of combinations possible is then $(256)^2 = 65{,}536$. Each combination will not yield a unique frequency. Still, a very large number are possible. Tuning such a synthesizer could be done with a committed microcomputer to generate the required M and N numbers. Alternatively, a large read-only-memory (ROM) could be used to store the M and N values corresponding to the "channels" desired. This is entirely practical considering the decreasing price of semiconductor memory.

The simple divide-by-N synthesizers described have used but a single loop. In the simple cases, the step size equals the reference frequency. The methods presented to achieve smaller increments are more complicated and produce an uneven step. Most synthesizer designs demand that the frequencies generated be at a constant spacing.

Figure 7.54 shows part of a synthesizer that will yield arbitrary resolution with a high reference frequency. Assume that an lo is required for a communications receiver. The lo should tune the relatively narrow range of 5 to 5.1 MHz with a 10-Hz step. The reference sidebands should be well suppressed.

The loop of Fig. 7.54 generates a signal at 50 to 51 MHz. This is divided by 10 to achieve the desired output. The vco output is also applied to a mixer and heterodyned with a 5-MHz signal. This could be from a crystal oscillator. The lower sideband is selected with a bandpass filter, producing an output at 45 to 46 MHz. This is divided in a programmable divider to produce a 0.1-MHz output which is compared with a reference, filtered in an $H(s)$ filter, and applied to the vco. The vco will have outputs separated by 100 kHz. Once divided to the desired 5- to 5.1-MHz output, the resolution will be 10 kHz.

The mixer input, $f_2 = 5$ MHz, may be tuned over a 100-kHz range to "fill in" the gaps in the output spectrum. The loop of Fig. 7.54 is called an offset loop.

Note that the offset frequency, f_2, has the same range as the final output. Hence, rather than using a variable frequency oscillator for f_2, another loop exactly like that shown may be used. This loop will produce outputs in the original separated by 1 kHz. A third loop will produce outputs with 100 Hz spacing while a fourth will yield the desired 10-Hz resolution. Each loop operates with a 100-kHz reference frequency. Reference sideband suppression should present no problem.

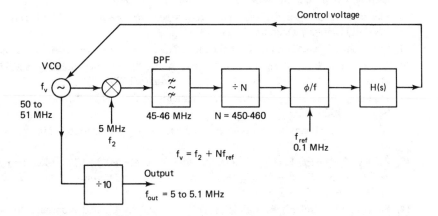

Figure 7.54 An example of an offset-divide-by-N PLL. A large number of these loops may be used to provide a synthesizer of arbitrary resolution while using a reference frequency much higher than the step size.

The synthesizer described uses a multiplicity of PLL's, all operating in a similar frequency range. To be practical, such a system would need to be constructed with great care. The shielding should be excellent between sections. Power supply decoupling should be extensive. The mixers should display good balance and the signals should be well buffered. Still, this approach to synthesizer design is popular where high resolution is required.

The phase noise performance of this synthesizer should be excellent. Note that the results of each added block will be divided prior to use as an offset. The extensive division allows the loops determining the highest resolution to be relatively noisy without adverse effects in the final system performance.

REFERENCES

1. Oliver, Bernard M., "The Effect of μ-Circuit Non-Linearity on the Amplitude Stability of RC Oscillators," *Hewlett-Packard Journal,* **11,** April–June 1960.
2. Clarke, Kenneth K. and Hess, Donald T., *Communication Circuits: Analysis and Design,* Addison-Wesley, Reading, Mass., 1971.
3. Clarke, Kenneth K., "Design of Self-Limiting Transistor Sine-Wave Oscillators," *IEEE Trans. on Circuit Theory,* **CT-13,** *1,* pp. 58–63, March 1966.
4. Frerking, Marvin E., *Crystal Oscillator Design and Temperature Compensation,* Van Nostrand Reinhold Co., New York, 1978.
5. Manassewitsch, Vadim, *Frequency Synthesizers—Theory and Design,* John Wiley & Sons, New York, 1976.
6. See reference 5, p. 102.
7. Krauss, Herbert L., Bostian, Charles W., and Raab, Frederick H., *Solid State Radio Engineering,* Chap. 8, John Wiley & Sons, New York, 1980.
8. Engelson, Morris and Telewski, Fred, *Spectrum Analyzer Theory and Applications,* ARTECH House, Dedham, Mass., 1974.
9. Engelson, Morris, Tektronix Applications Note AX-3259, Beaverton, Oregon.
10. Shoaf, John H., Halford, D., and Risley, A. S., "Frequency Stability Specifications and Measurement," NBS Technical Note 632, January 1973.
11. Tykulsky, Alexander, "Spectral Measurements of Oscillators," *Proc. IEEE,* **54,** *2,* p. 306, February 1966.
12. Scherer, D., "Today's Lesson—Learn About Low Noise Design," *Microwaves,* April and May 1979.
13. Ashley, J. Robert, Barley, Thomas A., and Tast, Gustaf J., "The Measurement of Noise in Microwave Transmitters," *IEEE Transactions on Microwave Theory and Techniques,* **MTT-25,** *4,* pp. 294–318, April 1977.
14. Basawapatna, Ganesh R. and Stancliff, Roger B., "A Unified Approach to the Design of Wide-Band Microwave Solid-State Oscillators," *IEEE Trans. on Microwave Theory and Techniques,* **MTT-27,** *5,* pp. 379–385, May 1979.

15. Ollivier, Pierre M., "Microwave YIG-Tuned Transistor Oscillator Amplifier Design: Application to C Band," *IEEE Journal of Solid-State Circuits,* **SC-7,** pp. 54–60, February 1972.
16. See reference 5.
17. See reference 5, pp. 342–350.
18. See reference 5, p. 398.
19. Blanchard, Alain, *Phase-Locked Loops—Application to Coherent Receiver Design,* John Wiley & Sons, New York, 1976.
20. Bode, H. W., *Network Analysis and Feedback Amplifier Design,* D. Van Nostrand, Co., New York, 1945.
21. See reference 8.
22. Bales, Robert, "Suppress Noise Output of YIG-Tuned Sources," *Microwaves,* pp. 87–91, May 1980.
23. Johnson, Doug and Hejhall, Roy, "Tuning Diode Design Techniques," Motorola Application Note AN-551.
24. See reference 12.

SUGGESTED ADDITIONAL READINGS

1. Hamilton, Steve, "FM and AM Noise in Microwave Oscillators," *Microwave Journal,* pp. 105–109, June 1978.
2. Kurokawa, K., "Noise in Synchronized Oscillators," *IEEE Trans. on Microwave Theory and Techniques,* **MTT-16,** *4,* pp. 234–240, April 1968.
3. Baghdady, E. J., Lincoln, R. N. and Nelin, B. D., "Short-Term Frequency Stability: Characterization, Theory and Measurement," *Proc. IEEE,* **53,** *7,* pp. 704–722, July 1965.
4. Cutler, L. S. and Searle, C. L., "Some Aspects of the Theory and Measurement of Frequency Fluctuations in Frequency Standards," *Proc. IEEE,* **54,** *2,* pp. 136–154, February 1966. (This complete issue of the *Proceedings* is devoted to the problem of frequency stability and contains many vital papers.)
5. Barnes, James A., et.al., "Characterization of Frequency Stability," *IEEE Transactions on Instrumentation and Measurement,* **IM-20,** *2,* pp. 105–120, May 1971.
6. Leeson, D. B., "A Simple Model of Feedback Oscillator Noise Spectrum," *Proc. IEEE (Letters),* **54,** *2,* February 1966. (This is THE classic paper on oscillator noise.)
7. Cote, A. J., Jr., "Matrix Analysis of Oscillators and Transistor Applications," *IRE Trans. on Circuit Theory,* **CT-5,** *3,* pp. 181–188. September 1958.
8. Payne, John. B., "Synthesizer Designs Depend On Satcom Uses," and Winchell, Doug, "Single-Loop Synthesizer Upgrades Satcom Links," both papers in *Microwaves,* March 1980.
9. Gardner, F. M., *Phaselock Techniques,* John Wiley & Sons, New York, 1966.
10. Clarke, K. K., "Transistor Sine Wave Oscillators—Squegging and Collector Saturation," *IEEE Trans. on Circuit Theory,* **CT-13,** *4,* pp. 424–428, December 1966.
11. Strid, Gene, "*S*-Parameters Simplify Accurate VCO Design," *Microwaves,* pp. 34–40, May 1975.

12. Ruttan, Tom, "GaAs FETs Rival Gunns In YIG-Tuned Oscillators," *Microwaves*, pp. 42–48, July 1977.
13. Wagner, Walter, "Oscillator Design by Device Line Measurement," *Microwave Journal*, **22**, *2*, pp. 43–48, February 1979.
14. Johnson, Kenneth M., "Large Signal GaAs MESFET Oscillator Design," *IEEE Trans. Microwave Theory and Techniques*, **MTT-27**, *3*, pp. 217–227, March 1979.
15. Herbert, Charles N. and Chernega, John, "Broadband Varactor Tuning of Transistor Oscillators," *Microwaves*, **6**, *3*, pp. 28–32, March 1967.
16. Clapp, J. K., "An Inductance-Capacitance Oscillator of Unusual Frequency Stability," *Proc. IRE*, **34**, *3*, pp. 356–358, March 1948.
17. Egan, William F., *Frequency Synthesis by Phase Lock*, John Wiley & Sons, New York, 1981. (An outstanding treatment of the methods used in modern synthesizers.)

8

The Receiver:
An RF System

Receivers have been of interest to engineers since the very inception of radio near the beginning of the twentieth century. A receiver is an instrument intended to obtain rf signals from an antenna and amplify them, producing an output that replicates the original information transmitted. The antenna is an integral part of the receiving system, the transducer that converts a propagating electromagnetic wave to a voltage. The receiver is the instrument attached to the antenna, usually through a transmission line, that does the processing required to extract the information.

There are numerous receiver types. They may be classified according to the type of information to be received, a factor that also determines the bandwidth. For example, a television receiver must have a bandwidth of several MHz, for that is the extent of the transmitted signal. A receiver intended for the standard AM broadcast band must have a bandwidth of only a few kHz.

We will severely restrict the receivers discussed in this chapter. The intent is not to survey the many types. Instead, it is to present a system example to illustrate application of the circuit concepts presented in earlier chapters. Our discussion will be confined to two types.

The first is the communications receiver for the reception of relatively narrow bandwidth information. The usual emissions received are single sideband (ssb) voice and continuous wave (cw). Among the latter types are cw signals where a carrier is turned on and off in a predetermined pattern, usually International Morse Code. Also included would be frequency shift keyed signals used for standard radio teletype or ASCII, an encoding scheme used for communications with computers.

The second type is the instrumentation receiver, a system like the communications receiver except that it is calibrated with regard to amplitude. It is usually intended for measurement applications. The better general purpose communications receivers are well suited to instrumentation use. Rarely are instrumentations receivers ideal for communications though. The ultimate instrumentation receiver is the spectrum analyzer.

The scope of the discussion is also limited. The goal is to present receivers as an example of a radio frequency system. No attempt is made at covering all of the subtleties of the subject.

8.1 RECEIVING SYSTEMS

The superheterodyne concept is the basis of virtually all modern communications receivers. Figure 8.1 shows the block diagram for such a system, a single conversion design. The incoming signal is converted only one time, reaching an intermediate frequency (i.f.) where selective filtering occurs.

The system shown is one that might be used to tune a relatively limited frequency range. A typical application might be for amateur ("ham") communications. Although such receivers are designed for consumer application and must, consequently, be of relatively simple and economical design, the performance requirements are among the most severe.

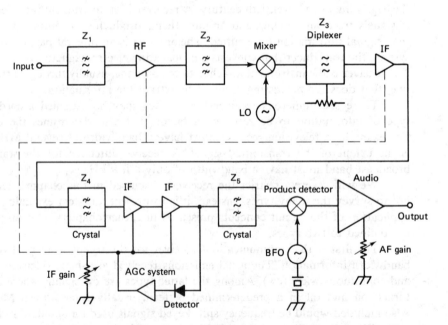

Figure 8.1 Block diagram for a modern single conversion superhetrodyne receiver. This design is intended primarily for the reception of cw and ssb signals.

The first element in the block diagram (Fig. 8.1) is a preselector filter, Z_1. The bandwidth is usually wide to keep insertion loss as low as possible. Any insertion loss detracts directly from the system noise figure. The next stage is an amplifier operating in the radio frequency spectrum being received. The rf amplifier must have a reasonably low noise figure and only enough gain to dominate the system noise. Excess gain will not significantly improve the noise figure, but may severely increase the signal levels presented to the mixer. The result will be a degradation in system input intercept. Chapter 6 presented methods for calculating the noise figure and intercept of cascaded stages. More is said about these details later in this chapter.

The filter following the rf amplifier, Z_2, serves two functions. It provides additional preselection, enough to ensure whatever level of image rejection needed. It must have excellent stopband rejection for this function. The second role is as a noise filter. It must reject noise at the image frequency which might otherwise appear at the mixer input. It is because of the noise "stripping" action and input noise figure constraints that the major selectivity ahead of the mixers appears in the second filter rather than that at the system input.

The output of the image stripping filter is applied to the mixer which may have some gain, although this is not necessary. The mixer is usually the dominant element in determining intermodulation distortion performance of the overall receiver.

The mixer is driven by a suitable local oscillator (lo). Depending upon the frequency range to be covered, this may be a free-running *LC* oscillator or it may be a synthesizer. The performance of the lo is extremely critical. It will determine system frequency stability. Although not shown, it is often supplemented with a digital counter to provide frequency readout. Mechanical dials are sometimes used in low cost equipment. However, a digital readout is easily achieved with modern circuitry and is so inexpensive that it is easily justified, even as a substitute for the traditional dial.

The local oscillator system must have low phase noise; lo noise sidebands will mix with incoming signals separated from the receiver frequency to produce an output at the i.f. This *reciprocal mixing* is often a more severe problem than is intermodulation distortion. An otherwise excellent receiver may be rendered useless by a poorly designed (noisy) synthesizer serving the lo function.

The mixer is followed by a diplexer, Z_3. This combination of filters, discussed in Chap. 4, may or may not be needed, depending upon characteristics of the mixer and the amplifier that follows.

The next stage is an i.f. amplifier. This overcomes the loss of the next filter, Z_4, usually a crystal type, and that of the mixer if a diode type is used. The i.f. amplifier must have both low noise figure and good imd performance. A poorly designed postmixer i.f. amplifier may degrade the system intercept. Excess noise will degrade overall noise figure, for a carefully designed system will have relatively low gain preceding the amplifier. Preservation of both noise figure and imd performance while keeping reciprocal mixing low involves a set of compromises. This tradeoff is the essence of designing the front-end of a superhet. The "front-end" is that section which precedes the most selective element in the receiver.

The next element in the block (Fig. 8.1) is Z_4, the most selective filter in the system, usually a crystal ladder or lattice type. It has a bandwidth compatible with the information being processed. A typical ssb bandwidth is 2 to 3 kHz while the bandwidth may be from 100 to 500 Hz for cw reception. This filter should have a stopband attenuation exceeding 100 dB. The shape factor is also of significance. Shape factor is the ratio of the bandwidth at two different attenuation levels, usually 6 and 60 dB. Filters designed for ssb with 8 to 10 resonators often display a shape factor of 1.5 or better. Similar filters with a 500-Hz bandwidth may have a shape factor of 2. Shape factor is often a more significant measure of selectivity, the ability to reject an adjacent channel, than is bandwidth. Receivers built for instrumentation applications, including spectrum analyzers, often have a larger shape factor, sometimes 10 or more.

The shape factor of an i.f. filter has a significant effect upon lo spectral purity requirements. Consider two filters, each with a 1-kHz bandwidth. One has a 6- to 100-dB shape factor of 20 while the other is a very selective one with a shape factor of 2. A strong, but absolutely pure signal is applied to the receiver input. The input signal has no noise sidebands whatsoever, a condition approximated by a quality signal generator followed by a crystal filter. Assume that the receiver lo has noise sidebands which are 140 dB/Hz down at a spacing of 3 kHz. If the receiver is tuned 3 kHz away from the input signal while using the 20-shape factor filter, a response will be noted merely from the signal. Noise sidebands may produce some output as well, but it is masked by the signal. To the contrary, there will be no carrier within the passband if the filter with a shape factor of 2 is used. However, lo noise sidebands will reciprocally mix with the input carrier to produce an output within the i.f. filter. The noise sidebands at 3-kHz spacing will be down by 110 dB in a 1-kHz bandwidth. The results of reciprocal mixing are evident in the receiver output if the incoming signal is strong enough. While filters with a low shape factor (steep skirt response) are preferred for communications, they represent a worst case for observation of reciprocal mixing. Again, a compromise is involved in design.

An often neglected filter characteristic is transient response. The information bandwidth may be evaluated if incoming signals are analyzed theoretically. A cw signal at a typical speed (20 baud or 24 words per minute) will have an information bandwidth of about 20 Hz. Yet, a 20-Hz wide filter is virtually never used. The step response of such a filter would exhibit severe ringing (see Chap. 2). Incoming antenna noise is not the thermal noise considered in earlier discussions and used for most theoretical analysis. Instead, it is a train of more or less random pulses, usually resulting from electrical storms or manufactured equipment. When this signal is applied to a 20-Hz wide filter, the ringing will be intolerable; the receiver output will sound like a continuous sine wave. Results are similar even with thermal noise.

Filters of very narrow bandwidth are not uncommon in instrumentation equipment. For communications, they are implemented with special circuitry following the receiver. The filter has a response that is "matched" to the information received, not only in the bandwidth sense, but in the time domain. This is termed a *matched filter*. The human listening to the receiver serves the same function in a traditional communications circuit. The processing ability of the human brain cannot be ignored when attempting to analyze a communications system!

Ringing must be avoided. A trained operator listening to the receiver output or a computer performing the functions of a matched filter are of little value if incoming noise pulses are converted to something resembling a sinusoidal output. The usual design uses filters wider than the incoming information bandwidth, especially when narrow bandwidth information is encountered (cw). Narrower filters are used if designed for a suitable response. Specifically, filters designed for a Gaussian or an equal-ripple phase response are usable at bandwidths down to perhaps 100 Hz where an equivalent Chebyshev filter would be completely useless. Group delay should be investigated in detail when designing filters. This information is presented in Zverev (1). Generally, filters with good transient response characteristics have poor (large) shape factors.

Following the first i.f. filter, Z_4, is the main i.f. amplifier. The gain of this section may be quite high, depending upon the signal to be received. Gain is not critical if the only signals are cw and ssb. However, if standard am signals are to be detected, the i.f. amplifier output must be high enough to drive a diode detector. Levels as high as 0 dBm are common for am receivers. The maximum level needed in an ssb receiver is around -20 to -30 dBm.

A major function of the i.f. amplifier is to provide a convenient means for automatic gain control (agc). The only stages controlled may be those following Z_4. Sometimes the gain of earlier stages is also varied. More will be said about agc systems.

Following the main i.f. amplifier is another filter, Z_5. This filter is missing in many receivers, but can be of great importance. It restricts the effective noise bandwidth to be equal to that of the main filter, Z_4. If Z_5 is not included, broadband noise generated within the i.f. amplifier will reach the detector and will produce an audio output. The total noise power may be large. This results in a significant reduction in the effective receiver noise figure, for the gain ahead of the main filter, Z_4, is small. High front-end gain would cause the amplifiers there to dominate the overall noise figure, but would then be a potential source of imd. Z_5 is usually called a "noise filter." Z_5 can be especially important in cw bandwidths (500 Hz and less) but is less significant for ssb reception, for the bandwidth of the later audio amplifier is confined.

The detector follows the noise filter. The detector is actually a mixer in the receiver shown in Fig. 8.1. It is driven with a crystal controlled lo, or beat frequency oscillator (bfo). This mixer is termed a "product detector" in this application. The function is to mix the i.f. signals down to an audio baseband output.

The product detector should be balanced. This restricts the bfo energy that finds its way back into the i.f. amplifier. The bfo should be well shielded from the i.f. amplifier, especially if i.f. gain is large. A common problem, and one that is subtle and difficult to diagnose, is a result of bfo energy in the i.f. amplifier. While this energy produces no direct audio output, the noise sidebands are still present. They will be amplified and possibly phase modulated by additional noise in the i.f.

amplifier. The noise sidebands are then detected and present in the audio output. The result is a receiver that sounds "mushy," lacking the crispness predicted from a simple analysis.

The audio section of a communications receiver is generally straightforward. Some distortion is allowed, for it will generally be masked by i.f. amplifier distortion. This is particularly true of harmonic and intermodulation distortion. To the contrary, crossover distortion in an audio output amplifier can be very annoying and should be avoided. The crossover distortion is found with push-pull circuits that operate predominantly in class B, with one of a pair of transistors conducting on each half cycle. Distortion may not be noticeable when the amplifier operates with large output power. It becomes a large fraction of the output, though, when the signal levels are small, as would be found when headphones are used. Because it is so easy, there is no reason not to incorporate reasonable audio fidelity in a communications receiver. An audio gain control is usually incorporated somewhere in the audio amplifier chain.

The receiver of Fig. 8.1 is an example used to illustrate the concepts. It may not be optimum, but it is practical.

Generally, a better hf receiver is one with no rf amplifier. The mixer is then subjected to the smallest possible signals, keeping distortion effects at a minimum. If a diode-ring mixer is used, the usual choice for most modern receivers, an additional virtue is realized. The mixer is a broadband device, offering a gain (conversion loss) that is quite independent of frequency. A receiver with a mixer input then has a very flat amplitude response. This is of vital concern for instrumentation applications.

The noise figure of a receiver with a mixer input is limited. The conversion loss of the typical doubly balanced ring mixer, and hence, a good approximation to its noise figure, is 6.5 dB. A typical amplifier following the mixer might have a noise figure of 4 dB. A contribution of another 2 dB or more from loss in the following crystal filter and the subsequent i.f. amplifiers is also common. The lowest possible noise figure is then 12.5 dB plus the insertion loss of whatever preselector filter might be used. The preselector must be a sharply tuned filter if the i.f. is moderately close in frequency to the input signal. This usually entails insertion loss. Hence, a typical system noise figure might be from 13 to 17 dB. This is adequate for many applications, especially at the lower end of the hf spectrum. It is not good enough for use at frequencies above about 20 or 30 MHz. One receiver manufacturer has used a simple, but viable compromise. An rf amplifier is included, but it is switched into the system only when needed.

Spurious responses can be a serious problem if lo and intermediate frequencies are not chosen with care. For example, a receiver built by the author for the 14-MHz region used a 9-MHz i.f. The lo was set at 5 MHz to ensure good close-in phase noise performance and thermal stability. A low loss preselector was initially used with no rf amplifier. The filter was a combination of a ninth order peaked high pass filter cascaded with a seventh order low pass. The insertion loss was under 1 dB, yielding a noise figure of 11 dB. However, the attenuation at 16 MHz was only about 30 dB, resulting in spurious responses. The fifth harmonic of the lo was mixing with 16-MHz signals to produce a 9-MHz i.f. output.

The preselector was changed to a third order bandpass filter with a cascaded low pass. The insertion loss climbed to 3.5 dB, producing a corresponding degradation in noise figure, making the addition of an rf amplifier mandatory. However, the 16-MHz spurious responses disappeared.

Figure 8.2 is the block diagram of a modern general coverage communications receiver tuning up to 30 MHz. The input is a preselector filter, followed by a mixer. The output is at vhf, 70 MHz in the example shown. A postmixer amplifier provides some front-end gain to drive the first filter, Z_1. In an optimum system, the bandwidth of this filter would equal that of the arriving information. For a general-purpose receiver, the filter would be switched as the system bandwidth is altered. This is not typical. Not only is it expensive, but there are limits on the performance of the vhf crystal filters that may be built. A more common configuration uses a filter equal to the widest bandwidth desired from the receiver, perhaps 10 kHz.

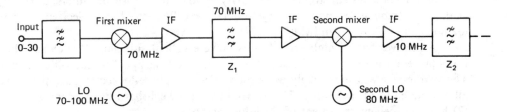

Figure 8.2 Block diagram for a modern general coverage receiver tuning the range from very low frequencies up to 30 MHz. The input filter is a low pass type. The local oscillator (lo) would probably be a frequency synthesizer.

Following the first filter is another amplifier and a second mixer. The receiver is thus a dual conversion design. The i.f. following the second mixer is 10 MHz in the example shown. It could, however, be any convenient frequency. The major receiver selectivity is provided by Z_2. Usually, this filter is switched to provide a variety of bandwidths.

A dual conversion receiver of the type shown in Fig. 8.2 has a number of virtues. The image rejection is excellent, for the image lies well into the vhf spectrum. The preselector is a simple low pass filter. It is relatively easy to obtain over 100 dB of image rejection with such a filter. Moreover, the high frequency first lo produces a minimum of possible frequencies for harmonic mixing, a frequent cause of spurious responses. The low pass preselector filter also has the virtue of offering very low loss. As such, receiver noise figure is maintained at a usable level throughout the hf spectrum.

The first intermediate frequency must be chosen with care. Ideally, it should be greater than twice the highest input frequency anticipated. This reduces the possibility of spurious responses from second order intermodulation distortion. Some designs are even more restrictive, placing the first i.f. at greater than three times the highest input. Others, primarily designed for amateur application, use a first i.f. as low as 40 MHz.

Owing to the frequency stability requirements usually found with hf communications receivers, the only suitable lo is a frequency synthesizer. The performance must be outstanding. The phase noise should be very low to minimize the problems from reciprocal mixing. In those receivers using a moderately wide first i.f. filter, Z_1, there will be some performance compromise. Strong signals may lie within the passband of the first filter, but outside of that of the second. This is generally a viable tradeoff, although it places severe design requirements on the i.f. amplifiers between the two mixers and the lo system.

Another virtue of a scheme like that of Fig. 8.2 is the use of a second i.f. This allows gain to be distributed more evenly. Having too much gain at one frequency, especially a high one, can lead to potential instability. Multiple i.f.'s decrease the likelihood of bfo energy appearing within the i.f. passband. To this end, some designs use a third conversion prior to product detection. Several i.f.'s also ease the problem of obtaining good system stopband selectivity. Two or three cascaded filters, especially at different frequencies, make it easy to obtain excellent stopband selectivity and to maintain the i.f. noise bandwidth.

A receiver with a high first i.f. must have excellent shielding. If such shielding is not present, the receiver will respond to the many strong signals in the vhf spectrum. The synthesizer will most likely use a multiplicity of phase-locked loops which must all be isolated from each other. The book on frequency synthesizers by Manassewitsch (2) has an excellent treatment of shielding methods.

Most spectrum analyzers use a block diagram much like that of Fig. 8.2. However, the frequencies may be much different owing to the much wider tuning range to be covered. A typical first i.f. is 2 GHz. The filter at that frequency is usually relatively wide, perhaps 15 to 20 MHz. The signal is then heterodyned to a second i.f. around 100 MHz and, eventually, to a third at 10 MHz. A fourth i.f. may be used in some designs primarily to obtain very narrow bandwidths.

The two block diagrams presented both represent high performance receivers. Both may also be simplified. There will, of course, be a compromise in performance. For example, a simple hf receiver may be constructed using dual conversion. Essentially, a single conversion receiver is constructed to tune a restricted range. Then, that receiver is preceded with a "converter," a system comprised of preselector filters, amplifiers, a mixer, and crystal controlled oscillators. This converts a desired input band to the input of the main receiver, serving now as a tuneable i.f. system. The largest deficiency is that signal levels may become large prior to selective filtering, leading to distortion problems. Dual conversion systems of this simple kind are rarely used today, although if carefully done, they can provide satisfactory performance.

Perhaps the simplest of receivers suitable for communications is the direct conversion type, shown in Fig. 8.3 (3, 4). This receiver is suitable for reception of cw and ssb signals. A direct conversion receiver uses no intermediate frequency; instead, incoming signals from the antenna are preselected and applied to a product detector. The output is at audio. Virtually all of the system gain is obtained in the audio amplifier, usually over 100 dB. Owing to the ease of constructing high gain audio systems, such a receiver is especially simple and is attractive for use by the home

Sec. 8.2 Receiver Evaluation and Measurements

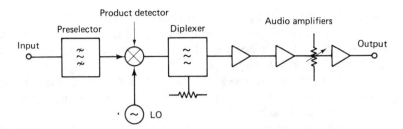

Figure 8.3 The simplest receiver type suitable for the reception of cw and ssb signals, a direct conversion design. Signals from the antenna are applied directly to a mixer with output at audio. Selectivity is typically obtained with *RC* active filters in the audio amplifier chain.

experimenter. *RC* active audio filtering is usually integrated into the amplifier to provide system selectivity.

The largest deficiency of the direct conversion receiver is the presence of an audio image. This can be eliminated through application of the phasing method of ssb reception (5, 6). Sideband suppression is usually only 40 to 50 dB. Another problem is the possibility of leakage of lo energy through the antenna port, for the lo differs from that of the incoming signal only by an audio difference frequency.

8.2 RECEIVER EVALUATION AND MEASUREMENTS

Receiver performance is best evaluated with a suitable combination of test equipment, much of it specialized and found only in the better equipped laboratories. However, surprisingly good measurements may also be performed with relatively simple equipment that may be constructed by the user. The frequency range will not be as great though (7). The usual equipment includes two well calibrated and shielded signal generators, a step attenuator with a range of at least 60 dB in 1-dB steps, a hybrid combiner, and an audio voltmeter. Some measurements require a crystal filter within the input spectrum of the receiver. An oscilloscope is also useful.

The first measurements performed relate to determining sensitivity or noise figure. The audio voltmeter is attached to the receiver output while a signal generator is applied to the input. Signal generators are usually calibrated in terms of the power available to a 50-Ω termination. Because the usual communications receiver is designed for use with a similar source impedance, no conversions are necessary in interpreting the levels. The receiver agc, if any, is defeated and the audio output is noted with no signal applied from the generator. The generator is then turned on and adjusted to produce a variety of outputs above the background noise. Many manufacturers specify available power (or generator voltage) required to achieve a 10-, 15-, or 20-dB increase in output above the thermal noise. The increase is the ratio of the signal plus the noise to the noise, or $(S + N)/N$.

A similar measurement is the minimum detectable signal, or mds. This is the power required from the generator to produce a 3-dB (S + N)/N, that response where the input signal equals the background noise. This measurement is usually done with the manual i.f./rf gain control set at maximum and the audio gain set for what might be a typical listening level.

It is informative to repeat the (S + N)/N measurements, whichever one might be employed, at a number of gain settings. This will show the way the S/N ratio might be altered as gain is reduced.

All (S + N)/N measurements are repeated for each available bandwidth. The evaluations are performed at several frequencies throughout the tuning range.

The bandwidth is measured while the signal generator is tuned to the input frequency. This requires careful plotting on a point-by-point basis of the filter response. Either receiver tuning or the signal generator frequency may be altered for this measurement. The noise bandwidth is calculated using the methods outlined earlier (Chap. 6). The filter response curves should be extended as far as they can with the instrumentation available. The usual data of interest include not only the 3- and 6-dB bandwidths, but the 60-dB figure used to calculate shape factor. The stopband or ultimate attenuation is also of great interest. Often, it may not be as good as the individual filters are capable of delivering, owing to stray coupling paths around the filters. This measurement may be difficult owing to reciprocal mixing from signal generator noise sidebands if the stopband attenuation is over about 110 to 120 dB. If available, a crystal filter is inserted in cascade with the generator output. The results may still be complicated by noise sidebands from the receiver local oscillator.

Measurement of reciprocal mixing may be integrated with the sensitivity measurements. First, a measurement is performed of (S + N)/N, usually at either the 3- or 10-dB level. The signal generator is then set to the frequency of the external crystal filter and the receiver is tuned away from the incoming signal. Generator output is increased until the response in the audio voltmeter is identical to that seen during the sensitivity evaluation. The power ratio is noted and normalized to a 1-Hz bandwidth. The result is then the carrier-to-noise ratio of the receiver lo or synthesizer.

Consider a receiver with an mds of −140 dBm and a noise bandwidth of 500 Hz. This is measured directly. The receiver is then tuned 10 kHz away from the frequency of the generator–crystal filter combination. The output is increased until the response in the meter is identical to that seen when measuring the mds. The increase is 127 dB. This is normalized to a 1-Hz bandwidth by adding 27 dB, 10 log bw$_n$, to obtain a carrier-to-noise ratio of 154 dB/Hz at a 10-kHz spacing. The reason for performing this measurement with the mds rather than, for example, a 20-dB (S + N)/N input is to allow for less than perfect stopband attenuation. Also, this will keep the input signal levels low enough that gain compression is not a problem, at least not in a modern, well designed receiver. Gain compression evaluation is described later.

Reciprocal mixing measurements may be extended to higher input levels within the limitations of the filter stopband attenuation and system gain compression. The noise output should increase in direct proportion to the signal generator output setting, assuming the lo noise is dominant over the background noise of the receiver. These measurements are difficult if the receiver lo performance is good. In some synthesized receivers, the measurements are altogether too easy though!

A parameter related to the sensitivity is the noise figure. It may be measured directly in much the same way it would be with an individual amplifier. A noise source with a known excess noise ratio is attached to the receiver input. With the noise source off, the output voltage is noted. The noise source is then turned on and the new level is noted. The ratio is the Y factor, described in Chap. 6. The Y factor is used to calculate the noise figure.

The minimum detectable signal is related directly to the noise figure, allowing noise figure to be inferred from an mds measurement. If the noise bandwidth has been measured, mds and NF are related by

$$\text{mds (dBm)} = -174 \text{ dBm} + 10 \log BW_n + \text{NF} \qquad (8.2\text{-}1)$$

where the NF is in dB. If the measurements, including that of noise BW, are done with care, mds results agree well with a direct NF measurement.

The virtue of a direct noise figure measurement is that it is bandwidth invariant. This results from the receiver measuring noise at two different levels. The response of the i.f. system is identical for both except for the total power output. The noise output will increase when the bandwidth is increased. However, the ratio seen when the noise source is turned off and on should be the same.

While the noise figure should not vary as the bandwidth is changed, it is wise to perform the measurement at differing bandwidths. A difference may result from i.f. system noise contribution. The errors are most pronounced in the direction of increased noise figure when the narrower bandwidths are used. Rarely will the difference exceed 2 or 3 dB with differing bandwidth.

Image rejection and i.f. feedthrough are measured with a single signal generator attached. The generator is set to a high output level and tuned to the image frequency. The receiver is tuned to produce an output and the level on the audio voltmeter is noted. The signal generator is then returned to the desired input frequency. The output level is decreased until the audio output is identical. The power ratio is then the image rejection. The measurement of i.f. rejection is virtually identical.

This procedure is also extended to measure other spurious responses. Frequencies to be checked include those resulting from harmonic mixing with the inputs. Ideally, a search should be conducted, scanning the generator over a wide spectrum with a variety of receiver frequencies. This can be rather difficult and tedious, although it is important. The search should be extended to frequencies well above the receiver input range.

The agc response may be evaluated with just one signal generator attached. Starting at the mds, the receiver audio output is plotted as a function of generator power. This is done for both the agc activated and off. An agc threshold may be defined as the point where overall gain is reduced by 3 dB by having the agc activated. A typical threshold for a communications receiver is around -110 dBm. An instrumentation receiver would have enough gain in the agc system to be activated even by background noise. Most instrumentation units will not incorporate an agc; instead, a logarithmic amplifier is employed.

The extent of the agc control range should be evaluated. It should be well in excess of 100 dB, producing perhaps a 6-dB variation in audio input.

It is often useful to evaluate the dynamic characteristics of the agc system. This is done with the signal generator applied to a doubly balanced modulator with the "i.f." port driven by a function generator set to deliver a very low frequency square wave. The output of the balanced modulator should be investigated with an oscilloscope terminated with a 50-Ω resistor to confirm that the rf envelope has a suitably fast rise and fall characteristic, essentially following the shape of the applied modulation. The modulated output is applied to the receiver input. The audio output envelope is monitored with the oscilloscope. The attack rate may then be monitored directly. The decay rate may be evaluated by monitoring the signal on the agc line in the receiver.

If the receiver contains a signal strength indicating meter, its accuracy may be confirmed during evaluation of the agc system. The accuracy should be good in a receiver designed for measurement applications. It may have little meaning in a communications receiver.

The next set of measurements uses two signal generators and is aimed at evaluating the response to strong signals away from the frequency to which the receiver is tuned. The first measurment is an evaluation of the intermodulation distortion. The test setup is shown in Fig. 8.4. Two signal generators are set to produce equal outputs (tones) at f_1 and f_2. The spacing is usually close, 20-kHz difference being a common starting point. The outputs are added in a hybrid combiner, described in Chap. 4 as a return loss bridge. The output is then applied to a fixed attenuator, usually of at least 10 dB. This attenuator will ensure that the hybrid is terminated to provide the desired isolation between generators. The fixed attenuator is applied to a step attenuator and then to the receiver. The output is monitored with an audio voltmeter.

The first two-tone measurement is an evaluation of the third order imd input intercept. This measurement is virtually the same as with an amplifier, described in Chap. 6. The receiver (agc off) is tuned to one of the third order intermodulation distortion frequencies, $(2f_1 - f_2)$ or $(2f_2 - f_1)$. These frequencies are above or below

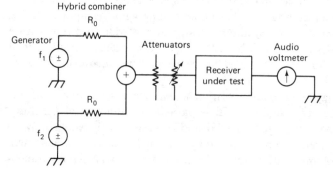

Figure 8.4 A system for the testing of receivers. Two signal generators are used to measure two-tone intermodulation distortion, gain compression, minimum detectable signal, and cross modulation.

the two tones by an amount equal to the spacing. For example, if the generators are set at 20 and 20.02 MHz, the imd frequencies will be at 19.98 and 20.04 MHz. The output at the imd frequencies is noted. The receiver is then tuned to one of the input frequencies and the step attenuator is adjusted until an identical audio voltmeter response is noted. The amount of attenuation inserted to produce this response is the imd ratio.

Consider an example. The power available from each generator at the receiver input terminal is -20 dBm. The response is noted at the imd frequencies and the receiver is then tuned to one of the input tones. 60 dB of attenuation is inserted to produce an identical output. The imd ratio is then $R = 60$ dB. Using the third order imd relationships developed in Chap. 6, the input intercept, IP_i, may be calculated

$$IP_i = P_i + \tfrac{1}{2} R \qquad (8.2\text{-}2)$$

where P_i is the input power per tone in dBm and R is the imd ratio in dB. For this example, $P_i = -20$ dBm and $R = 60$ dB, yielding $IP_i = +10$ dBm.

While this measurement is generally straightforward, there are subtleties that must be taken into account. First, the receiver may or may not be well-behaved. We found in Chap. 6 that imd power varies in proportion to the cube of the input tones. In the example, the equivalent input power for the imd was -80 dBm. If the input tones were decreased by 10 dB to -30 dBm, the imd responses should drop by 30 dB to an equivalent level of -110 dBm. The imd ratio should increase by 20 dB. The receiver is well-behaved if this occurs. In many cases, especially where high level diode mixers are used, the response may not be well-behaved. With a 10-dB decrease in input tones, the imd ratio may increase by as little as 10 dB instead of the predicted 20 dB. The predicted input intercept from Eq. 8.2-2 would then be -30 dBm $+ \tfrac{1}{2}(70$ dB$) = +5$ dBm. If the receiver is not reasonably well-behaved, it is not viable to specify an intercept, for it has no utility.

Another subtlety to be investigated when evaluating two-tone imd is the equality of the imd responses. They should be, and generally are, equal. They may be different, usually the result of distortion of a frequency dependent nature. An example would be distortion in a crystal filter.

Input intercept is a useful parameter, for, like noise figure, it is bandwidth invariant. It is not generally a strong function of frequency spacing as long as the tones are both within the passband of the preselector. Intercept is a measure of the strength of unwanted input signals which may be tolerated at the input.

Intercept alone is not a very meaningful measure of a receiver's performance. If an attenuator is placed ahead of the receiver, the intercept is increased by that same amount. The receiver mentioned with an input intercept of $+10$ dBm would have a $+40$-dBm input intercept with the addition of a 30-dB pad. On the other hand, the sensitivity would be reduced by 30 dB. Such an instrument is hardly useful for communications applications, even at the low end of the hf spectrum. Intercept is only a viable specification when compared with a sensitivity parameter.

One comparative parameter is the so-called two-tone dynamic range (8). The measurement may be performed directly. First measure mds. Then, two tones are applied as shown in Fig. 8.4. The tones are adjusted so that they produce an output imd response equal to that seen from the mds. The two-tone dynamic range, D, is then the ratio of the power in one of the two tones to the mds.

The dynamic range may be related directly to input intercept if the third order imd performance is well-behaved. Assume a system with known mds and IP_i. The equivalent distortion power, P_d, is then evaluated from the definition of intercept

$$P_d = IP_i - 3(IP_i - P_i)$$
$$= -2IP_i + 3P_i \qquad (8.2\text{-}3)$$

where P_i is one of the two arbitrary input tones. For the special case where the input distortion causing tones are adjusted to give a distortion response equal to the mds, the input is $P_i = \text{mds} + D$. Substitution into Eq. 8.2-3 yields the dynamic range

$$D = \tfrac{2}{3}(IP_i - \text{mds}) \qquad (8.2\text{-}4)$$

Consider as an example a receiver constructed for amateur communications by the author. The measured input intercept was +10.5 dBm while the mds was −142 dBm. The noise bandwidth was 500 Hz, indicating a noise figure of 5 dB from Eq. 8.2-1. The calculated two-tone dynamic range from Eq. 8.2-4 is then 101.7 dB. Direct measurement produced a value of 102 dB, well within the errors associated with the measurement of noisy signals.

The two-tone dynamic range parameter is especially useful for instrumentation applications. It will indicate the signal range that may be applied to the input of, for example, a spectrum analyzer, while still not creating internally generated third order imd.

The parameter has problems though. First, it is difficult to measure, especially in a narrow bandwidth receiver. The statistical variations in receiver output are large when presented only with noise, causing large variations in the audio voltmeter readings. Variations in system gain stability can also introduce errors. Many receiver manufacturers prefer not to specify either mds or two-tone dynamic range owing to the difficulties in measurement.

A second problem with D as a figure-of-merit is that it is a function of bandwidth. It is evident from Eq. 8.2-4 that D will vary as the $-\tfrac{2}{3}$ power of mds which is directly proportional to noise bandwidth. The mds will increase to −136 dBm if the bandwidth of the example receiver described above is increased to 2 kHz. The dynamic range will then be 97.7 dB.

A third problem is that dynamic range cannot always be realized in practice. This is a result of antenna input noise that is much higher than the thermal noise from a measurement input resistor and internally generated noise. The practical dynamic range is increased by the insertion of input attenuation until the input noise is just barely dominant.

Two-tone dynamic range does have one significant feature. It is a worst-case specification. Even if a less than well-behaved input mixer causes intercept to be an invalid parameter, D may still be measured directly. Usually, if a mixer is ill-behaved, it will be in such a direction that measurements at higher levels imply a larger value for D from Eq. 8.2-4.

While dynamic range is not a bandwidth invariant parameter, the significant contributing elements, noise figure and input intercept, are. They may be combined to define a bandwidth invariant *Receiver Factor*, $RF = Ip_i - NF$ (9). For the receiver described earlier with $IP_i = +10.5$ dBm and a 5-dB noise figure, the value is $RF = +5.5$ dBm. The receiver factor is an extension of a parameter, the amplifier factor, used to describe the dynamic range characteristics of amplifiers by Anzac Electronics (10). The two-tone dynamic range is calculated for any given bandwidth if the receiver factor is specified

$$D = \tfrac{2}{3}(RF + 174 - 10 \log BW_n) \text{ dB} \qquad (8.2\text{-}5)$$

Considerable emphasis is often placed upon the receiver imd performance. In practice, imd may not be as severe and common as the emphasis might imply; imd is measured with relative ease. Moreover, most efforts devoted to improving imd performance indicate improved linearity. Other nonlinear phenomena are also reduced.

A related measurement to imd evaluations is determination of the blocking, or gain compression. This is performed with a modification to the system shown in Fig. 8.4. The variable attenuator is moved to the line from the generator at f_2. The main generator at f_1 is set to a moderate signal, typically around -100 dBm, and the receiver is tuned to that frequency. The level of the f_2 generator is then increased until the output seen at f_1 decreases by a specified amount, usually 1 dB. The power available at the receiver input from the f_2 generator is then a measure of the signal strength required to compress the receiver gain, commonly known as blocking.

Blocking is easily measured with a poorly designed or older receiver. Typical performance might be for blocking to start at a level of about 120 dB above the mds. As such, the 120-dB ratio might be described as a "single-tone" dynamic range.

Blocking may be very difficult to measure if the receiver is of more modern design. First, the front-end stages are usually designed for linearity. Hence, they will handle very large signals before gain compression begins. The more common problem, though, is the observation of reciprocal mixing. Noise sidebands from either the receiver lo or the signal generator (or both) mix to produce an i.f. noise output. Blocking measurements are best done with a crystal filter cascaded with the stronger signal generator. In some cases, the blocking may not even be measureable owing to the reciprocal mixing. The output observed during the measurement will increase before it begins to decrease.

A related measurement is of cross-modulation. This is especially significant for receivers designed for amplitude modulated signals. The same test setup is used that measured blocking, except that the audio output is monitored with an oscilloscope instead of the audio voltmeter. The stronger signal generator at f_2 is amplitude modulated at a 30% level. The receiver is tuned to f_1, again set for a level of about −100 dBm. The level of the stronger, modulated generator is increased until a 1% amplitude modulation is observed on the weaker signal. A crystal filter in cascade with the stronger generator must have a wide enough bandwidth to pass the modulation sidebands without attenuation. A typical modulation frequency is either 400 or 1000 Hz. If this measurement is performed with a narrow bandwidth receiver, the receiver itself may be tuned away from the weaker carrier to observe the modulation sidebands.

Little has been said about the frequency separation used for the two-tone measurements described. Usually, separations of 15 or 20 kHz are utilized. Ideally, the measurements should be performed with a variety of separations. As the test tones are moved further apart, at least one will be attenuated by the preselector filter, increasing the observed dynamic range. The oscillator noise sidebands become more apparent as the tones become very close together. Also, with close spacings at least one of the test tones may appear inside the passband of a wider i.f. filter that may precede the narrower ones. This would occur, for example, with the up-converted receiver with a vhf first i.f. where the first filter, Z_1, is wider than those in the second i.f. This effect is especially significant with traditional dual conversion receivers where the first i.f. is very wide.

At very close spacings, both test tones will appear within the narrow i.f. passband. The output can then contain imd components that are also "in-band." The receiver cannot supply the selectivity required to observe these distortion products; an audio spectrum analyzer must then replace the audio voltmeter. In-band distortion is rarely of great concern unless it is severe. All in-band information is detected and displayed in an instrumentation receiver. There is little need to pursue minor in-band distortion for communications receivers, for typical speakers or headphones used with such equipment also create distortion.

Many of the measurements described are detailed and complicated. They are not all performed when evaluating a receiver. However, all are of significance to the receiver builder during construction and design. In this vein, the more subjective details should not be ignored. One of the most valuable assets available to a communications designer is a knowledge of how a good receiver should sound.

8.3 INTERMEDIATE FREQUENCY AMPLIFIER SYSTEMS AND GAIN CONTROL

A modern superheterodyne receiver will have much of the available gain at one, or perhaps several intermediate frequencies. The requirements for this part of the receiver will depend upon the intended application. Here we present an overview of some of the design details.

A traditional i.f. system for a superheterodyne communications receiver is shown in Fig. 8.5. The example is a single conversion receiver. It could equally well be a multiple conversion design with additional mixers.

The first element is the selectivity determining filter, Z_1. The following amplifiers

Sec. 8.3 Intermediate Frequency Amplifier Systems and Gain Control

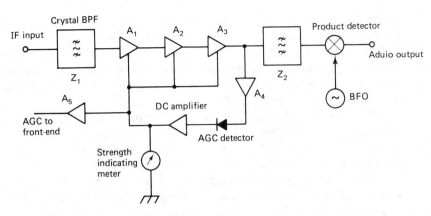

Figure 8.5 The intermediate frequency section of a communications receiver showing the system used for automatic gain control. Gain is reduced in the front-end only to allow for the reception of very strong signals. The amplifiers are arranged so that front end gain is not changed for "average" signal inputs.

are the main system gain elements. They provide a large fraction of the overall receiver gain. Integrated circuit amplifiers may be used in a typical communications system.

The net gain required will depend largely upon the information to be processed. A diode detector is usually employed if the receiver is intended for amplitude modulated signals. This will require a signal voltage that is large with respect to the diode offset. The signals arriving at the output of the main filter, Z_1, may be as weak as -100 to -120 dBm. A typical diode detector might require a minimum signal of -20 dBm. A total i.f. gain of up to 100 dB is then needed. Some of this may be obtained after the noise filter, Z_2. It might be advisable to employ a second conversion considering the large gain required.

The noise filter, Z_2, is a vital part of the overall i.f. system. The role has been described earlier. Consider a numerical example. Assume that the total front-end gain preceding Z_1 is 20 dB and that the noise figure of that section is 8 dB. The noise figure looking into the filter, Z_1, will actually depend upon the overall noise bandwidth of the i.f. system, determined primarily by the bandwidth of Z_2. Assume that the receiver is being driven from a 290 degree resistor. The noise power presented to the receiver input is then -174 dBm $+ 10 \log B_1$ where B_1 is the bandwidth of Z_1 (or of Z_2 if it should be less than Z_1). The power available at the input to Z_1 is then this amount increased by the front-end gain, 20 dB, and the front-end noise figure, 8 dB. For this example, assume that $B_1 = 1$ kHz. The input noise power at the receiver input is -144 dBm. The noise power at the filter input is then -116 dBm.

Assume that the loss of filter Z_1 is 5 dB, leaving a noise power at the first i.f. amplifier of -121 dBm. The total noise power available at the input of Z_2 is increased by the amplifier, assumed to have a gain of 80 dB. That noise originating at the system input and available at the input to Z_2 is then -41 dBm. Assume that the i.f. amplifier chain has a noise figure of 10 dB. The noise power resulting from noise generated within the i.f. amplifiers is then -174 dBm $+ 10 \log B_2 + NF_2 + G_{i.f.}$. If $B_2 = B_1$ at 1 kHz, i.f. generated noise is then -54 dBm. This is small compared with the -41 dBm resulting from front-end amplified noise and no system compromise is involved. The total noise power is the algebraic sum of the powers, or -40.8 dBm available at the input to Z_2.

Consider a less optimistic (and more typical) situation where Z_2 is nothing but a low Q tuned circuit with a bandwidth of perhaps 100 kHz. The i.f. generated noise is then 20 dB worse, -34 dBm. This is dominant over the front-end generated noise and will severely compromise system performance.

The numbers generated in this example would be those measured if the receiver were intended for instrumentation applications. The compromise would be severe. It would not be as bad in a communications application though. First, the audio amplifier that follows after detection would probably have a restricted bandwidth. If product detection is used, the system is still linear and the audio filtering will be effective. If a diode detector is employed, the audio fidelity has less significance. A second factor is the filtering capability of the listener. While considerable noise may be present, it is distributed over a wide spectrum. The ear and brain of the operator will respond primarily to that noise occupying the same spectrum as the information. The optimum design is still one where the filtering at Z_2 is matched to the first filter, Z_1.

The example has been simple with but one filter bandwidth. The methods may be extended to multiple conversion systems with a multiplicity of selectable filter bandwidths. Such calculations can be of great significance in the design of instrumentation such as a spectrum analyzer.

A second role of the i.f. amplifier, and often the more important one, is that of providing a convenient means for varying system gain. This is especially important in a communications receiver where the input signals may vary in strength by 120 dB or more. The receiver must have a means for agc that will encompass this variation.

Designing a receiver for good agc performance is not a trivial task. The controlled amplifiers should, ideally, have a logarithmic characteristic, a gain reduction in dB which is directly proportional to the control voltage. This is approximated in carefully designed amplifiers, but lacking in others. The overall system must be built for a desired transient response, both on application of a signal, the attack, and afterward, the decay. These characteristics should not depend upon the level of the signal. The following discussion will examine only the general concepts.

The signal at the noise filter input (Fig. 8.5) is sampled with an additional amplifier, A_4, and then applied to a detector. Extra gain is required to isolate the detector from the rest of the amplifier and to provide the added signal level required to drive a diode detector. This is more important in a receiver for cw or ssb where product detector signal levels are small.

A typical agc detector system is shown in Fig. 8.6a. The extra amplifier, A_4, uses a bipolar transistor with negative feedback, providing both stable gain and a low driving impedance for the diode detector, D_1. Positive-going peaks at the diode are rectified. A positive excursion on the anode of D_1 will charge the memory capacitor, C_2. The rate that the gain is reduced is a function of the overall system gain and of the attack time constant, primarily composed of C_2 and the driving impedance. The attack will be fast in the system shown, usually much faster than the risetime of the information appearing through the main i.f. filter, Z_1 of Fig. 8.5.

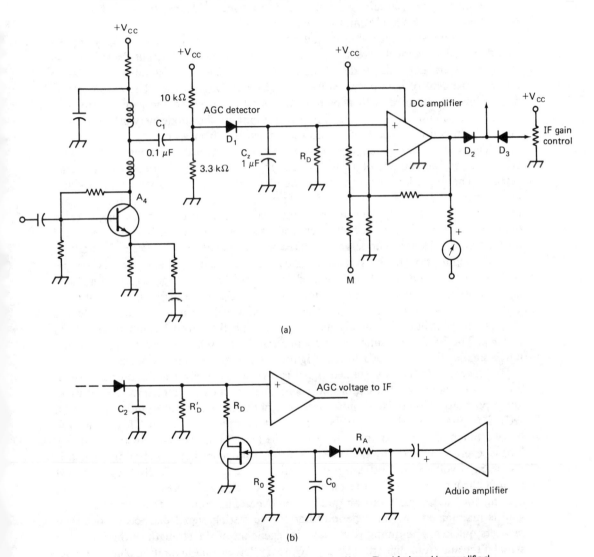

Figure 8.6 Details of a typical agc detector system. The i.f. signal is amplified and detected with the diode. The resulting dc appears across the "memory" capacitor, C_2. It is amplified and level-shifted with the operational amplifier in (a). The modification in (b) shows the method used to obtain a hang-type agc response.

The decay slope, the rate that gain is restored when the i.f. signal is removed, is determined primarily by the time constant $C_2 R_d$. An increase in R_d provides a slower decay.

The gain control characteristics of the amplifiers are of vital importance to the stability of the control loop and to the accuracy of the meter readout. Gain variable i.f. amplifier circuits were discussed in Chap. 6. Ideally, the agc loop should be studied with a Bode plot and treated as a control loop.

The sampled voltage appearing across the "memory" capacitor, C_2, is applied to a dc amplifier, shown here as an operational amplifier. The output drives a signal strength meter. The other side of the meter is returned to a suitable potential to allow adjustment of a zero response when no signals are present. The output of the dc amplifier is summed with a pair of diodes and applied to the gain control line going to the main i.f. amplifiers. The other input to the diode network is from a manual gain control, usually mounted on the receiver front panel.

The system shown has one feature that is not typical, a detector diode with positive bias. The cathode of D_1 is "pulled" toward ground potential with the decay resistor. The diode bias serves two functions. First, sensitivity is increased. Second, the approximately constant current flowing in the decay resistor, R_D, allows the gain to recover at a relatively constant rate. Full gain will be present when the detector diode again begins to conduct after removal of a detected signal. It is not necessary for the timing network to undergo a long recovery to establish a steady state condition.

A modification to the system described is the so-called "hang" agc. A sample is shown in Fig. 8.6b. Two loops are used for timing. The main agc detector functions as described above, charging the timing capacitor, C_2. If the arriving information is a short pulse, such as a burst of noise, the timing capacitor will charge, reducing the system gain, but will quickly discharge through the parallel combination of R_D and R'_b. The FET is assumed to be conducting. The discharge resistor, R_D, is made much smaller than that used in the original system, so recovery is quick.

Assume now that a sustained signal is present. A receiver audio output results. Some of the audio signal is sampled and amplified in an extra amplifier. The negative-going portions of that signal, now large in amplitude, are rectified and applied to a second timing network, C_0 and R_0. The voltage applied to the FET gate is well beyond pinch-off. FET conduction ceases. The attack time constant of this auxillary loop is severely restricted by R_A. When the signal disappears, the main timing capacitor, C_2, changes voltage very little, for R'_b is quite large. However, C_0 discharges toward ground through R_0. When the pinch-off is reached, conduction now begins again, causing the charge on C_2 to be quickly removed. The result is that the overall i.f. gain is maintained low for a period after the activating signal disappears, but then recovers quickly. The timing is virtually independent of the strength of the incoming signal if the audio amplifier driving the secondary detector has a limiting characteristic. Moreover, the agc system is not adversely disturbed by isolated noise pulses. Hang-type agc systems are described in the literature (11).

AGC is applied primarily to the amplifiers following the main filter, Z_1. However,

Sec. 8.3 Intermediate Frequency Amplifier Systems and Gain Control

some agc signal is applied to an additional dc amplifier, A_5, and then to gain reducing elements in the receiver front-end section. This might be a PIN-diode attenuator. Whatever system is used, it should be one with low distortion, for it will be subjected to large signal levels to which the receiver is not tuned. The agc signal which controls the front-end gain should be "delayed." That is, it should not be applied until a significant reduction has occurred in the main i.f. system gain, A_1 through A_3. System signal-to-noise ratio will be degraded from excess i.f. noise if front-end gain is reduced too soon.

The main i.f. system, Fig. 8.5, has no selective filter within the agc control loop itself. This is quite important if the system is to have a fast attack characteristic. A narrow filter will exhibit a time delay, a period between the instant of application of an input and the moment when a significant filter output occurs. If this filter is within a control loop, the transient characteristics will be severely compromised, usually by the addition of an overshoot. Conversely, if the bandwidth of the control loop is large compared with that of the preceding filter, the loop will virtually follow the risetime of the input signal.

The detection system shown in Fig. 8.6 presumes an i.f. amplifier with a forward-type agc characteristic, one where the gain decreases with increasing potential on the control line. This may not be the case. Minor changes in the operational amplifier circuitry or in diode polarities will allow the circuit to be adapted to available amplifier characteristics.

It is convenient in some simple receivers to derive the agc signal from the audio. Conceptually, this is no different than detecting at a higher frequency. There are serious practical problems though. First, the frequency response of the usual audio amplifier is purposefully confined. The filtering action causes a time delay that restricts attack time.

A second problem is the nature of the usual detector, a peak responding circuit. As such, the incoming waveform is sampled for only a short interval during each cycle of oscillation. This is not enough information. The fundamental sampling theorem states that an arbitrary signal may be reconstructed from a sampling rate twice that of the highest frequency present. The usual agc detector misses the theoretical minimum sampling rate by a factor of 2. Full wave rectification in the detector will help. Generally, audio derived agc is not recommended except in the simplest of receivers.

The role of an agc system is to adjust the gain to accommodate the incoming signals to which the receiver is tuned. Often, there are other signals that must also be used for gain reduction purposes, even though they contain no vital information to be processed by the receiver. One example of such a signal is a noise pulse. These are short duration, but very strong broadband pulses that might originate from automotive ignition or from an over-the-horizon radar. The narrow, but loud pulses will be converted by the front-end mixers, producing an input signal at the selective filters. The filter will then have an output much like the theoretical impulse response. It will be a longer duration signal that decays only after a suitable damping period.

The impulses are especially annoying in a communications receiver. Not only do they mask a weaker signal, but they activate the agc, causing system gain reduction. This is partially alleviated by restricting agc attack time.

The more effective way to deal with noise pulses is with a noise blanker circuit. A block diagram is shown in Fig. 8.7. Although operating at the intermediate frequency, the noise blanking circuitry precedes the dominant selective filter. As such, the noise blanker should be considered part of the receiver front-end.

The i.f. signal from the mixer is amplified and applied to a relatively wide filter, Z_b. It is not necessary or even desirable that this filter have a steep skirt response. Often it is nothing more than a second order monolithic crystal filter with a bandwidth of approximately 10 times that of the information processing filter, Z_1.

Some of the signal prior to the noise filter is extracted and processed. Another mixer is often used in conjunction with an auxiliary i.f. amplifier as shown in the figure. However, the processing could be a wide bandwidth amplifier operating at the same intermediate frequency as that of the main signal processing chain. After suitable amplification, the noise is detected, producing a narrow pulse. This is then applied to a gate diode within the main signal path. The gate diode is momentarily reverse biased, prohibiting the large noise pulse from reaching the narrow i.f. filter, Z_1, where it would be stretched in time.

The noise detector should have a threshold effect such that it will not be activated by normal signals. Similarly, the amplifier chain used to process the noise pulse must have a wide bandwidth. A narrow BW would cause pulse stretching to occur. The noise detector should produce a suitable narrow output pulse to ensure that the main receiver is not silenced for too long an interval.

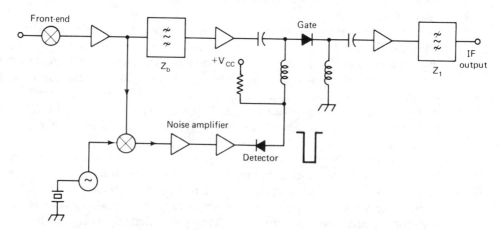

Figure 8.7 Block diagram of a noise blanking system suitable for impulse noise. Wideband information from the front-end is heterodyned to a lower frequency and amplified to produce a pulse which "opens" the noise gate. The filter, Z_b, serves two functions. It provides some selectivity to protect later stages from imd and other nonlinear effects. It also provides a time delay, compensating for the delay encountered in the noise-detecting i.f. amplifier.

The circuitry within the main signal path must be designed with considerable care. It must be "strong" enough that no significant intermodulation distortion occurs to compromise system input intercepts. The gate diode, or whatever type of switch that might be employed, must also meet this linearity constraint.

The noise filter, Z_b, serves two vital purposes. First, it provides some initial selectivity, protecting the following circuitry from strong signals far removed from the receiver frequency. Second, it provides a time delay. This should be matched to that of the noise processing i.f. amplifier chain. Hence, the noise gate is activated at just the instant when the noise pulse arrives.

The noise blanker shown in Fig. 8.7 is typical of those found in many communications receivers. Some systems are more elaborate (12).

Noise blanking is an example of a problem that can occur when a receiver is subjected to pulse-type information. The problem originates primarily from the time domain characteristics of the narrow filters employed. If the filters did not stretch the pulse information into much longer outputs, the blanking problem would be much easier.

The time domain response of filters becomes more significant for specialized receivers. An example would be a receiver designed for the processing of digital information, certainly a design of every increasing interest. The receivers usually operate in the microwave region with an i.f. of perhaps 70 MHz. The i.f. bandwidths are several MHz, needed to process the broad bandwidth digital information transmitted. The i.f. filters should be designed for reasonable pulse response. Moreover, the filter should have a flat group delay. Specifically, the delay of the information envelope through the filter should be constant over the filter width. This may be realized, at least to a desired degree of practical accuracy, by the application of a suitable auxiliary filter. This is an all-pass type, one that has no attenuation for any frequencies. It does, however, have a phase characteristic that is dependent upon frequency. The equalization all-pass filter is designed to have a group delay function which complements that of the i.f. bandpass filter, producing a system with a flat envelope delay (13).

Instrumentation receivers, including the spectrum analyzer, have special requirements for the i.f. systems. First, the time domain response of the filters should be proper, for the instruments are often used to examine pulses. The step and impulse responses should be investigated and designed to offer a reasonable replica of the input information. The filters popular for communications application are often the worst types for instrumentation receivers. A Chebyshev filter would be ideal for a traditional communications system; it would be a poor choice for a spectrum analyzer. Filters designed for a Gaussian or equal phase ripple are much more satisfactory. Often, a spectrum analyzer will use a chain of single resonator filters isolated by amplifiers (synchronously tuned), a scheme that has not been used in communications equipment for decades. That scheme allows the bandwidth to be easily altered and also provides good time domain characteristics. The agc system described earlier (Figs. 8.5 and 8.6) was popular for cw and ssb communications. It employs a fast attack and a slow, controllable rate of decay. This would not be suitable for an instrumentation receiver. Instead, the attack and decay should both be fast. AGC is rarely used in an instrumentation receiver.

The typical i.f. system in a spectrum analyzer consists of two distinct sections. The first is a *variable resolution amplifier,* or VR amp. This is an i.f. chain with a wide variety of available bandwidths. Typically, the widest will exceed 1 MHz while the narrowest will be 10 to 100 Hz. All noise constraints mentioned for a communications system apply equally to the spectrum analyzer VR. In addition, the VR must have a stable gain, often programmable. Amplifiers using extensive negative feedback are usually used. Some were described in Chap. 6. Linearity is also important in an analyzer VR. Stages preceding filters must not generate excess imd; those following the filters must not exhibit gain compression, even when delivering an output of 0 to +10 dBm. Owing to the requirement for gain stability and accuracy, system shielding is even more important than in a communications application.

The shielding problem is further complicated by the general nature of any modern instrumentation. The equipment is compatible with computer control. As such, an instrument will usually contain at least one microprocessor. Numerous other digital systems are also present, serving a variety of functions ranging from digital storage of display information to frequency synthesis. The challenges presented to the rf designer are great.

While not as overwhelming, the problems of digital noise are also increasingly common in communications systems. Most receivers use either frequency synthesis or a frequency counter for readout, or both. These digital systems can cause a great deal of interference if proper shielding and decoupling are not employed.

Following the spectrum analyzer VR amplifier is a logarithmic amplifier, or *log-amp.* This is a system with an output that is directly proportional to the log of the input, producing a vertical display in dB per division. The amplifiers usually depend upon the logarithmic characteristics of a diode for the response.

Log-amps generally fall into one of three types. All use a long chain of identical amplifiers, but differ in the details of implementation. Figure 8.8 shows one type of log-amp cell. A high speed operational amplifier is utilized. The feedback network consists of two arms. One is a resistor that determines the gain when the signals are low enough that diode conduction does not occur. A typical gain would be 10 dB. Ten identical stages would be required to achieve a log amp with a 100-dB

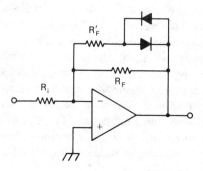

Figure 8.8 A simple form of logarithmic amplifier. The stage has a gain determined by R_f for small signals. Large signals cause the diodes to conduct; gain is then restricted to unity by R_f'. A practical circuit would use a very wide band operational amplifier.

range. As signal levels increase, the diodes in the second arm of the feedback loop begin to conduct. The gain is then determined by the parallel combination of the two arms. The resistors are picked such that the stage has a gain of unity (0 dB) for large signals.

Consider a cascade of several stages like that shown in Fig. 8.8. As signals increase, the output amplifier will first show diode conduction. The logarithmic response will result from that conduction. The output stage saturates at 0 dB gain as the signals grow. However, diode conduction is now present in the preceding stage. The action continues as larger signals are encountered.

The operational amplifier used must have wide bandwidth, usually large compared with the intermediate frequency. A typical spectrum analyzer i.f. is 10 MHz, requiring amplifiers with a 100-MHz bandwidth. Special integrated circuits or op-amps built from discrete transistors are usually used.

The noise problems are severe in a log-amp. The small signal gain may well exceed 100 dB. Hence, it is usually necessary to restrict the noise bandwidth of the log-amp. The restriction cannot be severe though, for wideband information must be amplified. The large gain and the attending noise problems justify the need to drive the log-amp with large signal powers from the VR amp.

A more practical logarithmic amplifier is shown in Fig. 8.9. The input is an emitter follower with a low output impedance. The output signal flows through a pair of conducting diodes. The small-signal diode resistance appears in series with the signal path for small inputs. The current flowing through the two diodes approaches or exceeds the dc bias current as signals increase. The excess signal current is then shunted into resistors R_1 and R_2. The only path to the output stage is through R_4. Small-signal gain is determined primarily by R_3 while that for large signals is controlled

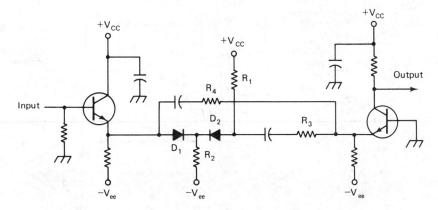

Figure 8.9 One stage of a chain of logarithmic amplifiers. The diodes conduct with a constant bias current. Small signals are passed through the diodes. Large signals cause diode currents in excess of the bias values. Then, the only path for the signals is through R_4, limiting the large signal voltage gain to unity.

by R_4. R_1 is picked so that the bias current in it is half of that in R_2. This ensures equal current in the two diodes.

The complete log-amp is fabricated from a long cascade of identical stages like that shown in Fig. 8.9. The range of an individual stage may be extended through the addition of a second set of diodes with a similar biasing. This circuit is used in the Tektronix 7L18 and 492 spectrum analyzers (14).

Detection occurs at the end of the long-amp chain. Usually, a high gain op-amp is used with diodes in the feedback loop to synthesize a "perfect" rectifier. The detector must have a wide bandwidth. It is usually followed by a low pass filter that is wide enough to pass broadband signal information, but has a sharp enough cutoff to eliminate the i.f. frequency. The "dc" output may be further filtered to "smooth" the noise response. Such additional circuits are termed video filters, for they provide low pass filtering of the video (dc up to a defined limit) band output. Typically, an analyzer will have selectable video filters. In other cases, the video filtering may be realized digitally through averaging methods integrated into the digital storage circuitry. Display storage is virtually mandatory with an analyzer, for the sweep rate may be very slow. It is not uncommon for a trace to require a full minute. Display storage provides a continuously viewable presentation to the user.

A third type of logarithmic amplifier system is shown in Fig. 8.10. Two stages are shown although there are usually many more. Operational amplifiers are again used. Each stage has a well-defined small-signal gain determined by feedback resistors. As the signals increase, the diodes around the feedback element cause the amplifier

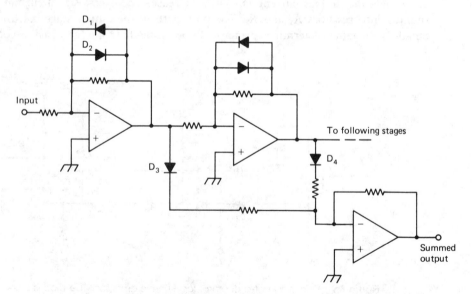

Figure 8.10 A logarithmic amplifier with detection at each stage. The amplifiers have gain for small signals. However, the diodes cause the amplifiers to limit for large signals. The signals at each stage are detected with the results summed to produce a system output.

to limit to an output independent of the input. The saturated gain of a single stage is vanishingly small.

Each stage in the logarithmic amplifier has a detector, shown as D_3 in the figure. The outputs from each stage are all summed in an appropriate additional op-amp.

The design of the log-amps are all similar. The major problem is to obtain diodes with uniform and predictable characteristics.

It should be mentioned that a logarithmic amplifier will process noise differently than would a linear amplifier. Generally, there is about a 2-dB error when measuring noise. The error is in a direction to make the noise appear smaller than the actual value. The literature should be consulted for noise measurement (15).

8.4 FRONT-END DESIGN

The front-end is that part of a receiver that precedes the major selectivity determining filters and is a critical section in determining performance. The front-end must be designed for low noise, for it will determine to a large extent the overall receiver sensitivity. At the same time, it must be designed for the utmost linearity. It is in the front-end where intermodulation distortion occurs to limit two-tone dynamic range. Other nonlinear phenomena are also associated with the front-end including cross-modulation and gain compression. The other major performance limiting phenomenon, so far as adjacent channel interference problems are concerned, is reciprocal mixing. This is determined by the carrier-to-noise ratio of the oscillator or synthesizer and is not significantly affected by the front-end.

Other factors are also controlled by the design of the front-end. The filters used will control the image rejection. Other spurious responses that might occur from harmonic mixing or spurious outputs in the local oscillator system are also determined by the selectivity of the filters. The noise figure may be degraded if the filters are made overly exotic, for a very selective, multiple resonator filter will have a high insertion loss. Compromise is required.

Clearly, the optimum performance will occur when all of the individual elements in the front-end are the best that can be built. The lowest noise figures are desired—reduction of noise figure in any stage will do nothing but improve sensitivity. Similarly, the highest possible output intercept should be sought for each stage. Linearity can only be improved. While this is all perhaps obvious, it is less than practical. First, there are some requirements that are mutually exclusive. For example, the lowest noise figures are usually obtained with amplifiers with relatively low bias current. This, however, will detract from the large-signal performance. The intercept will be degraded.

Other compromises are also required. One might be in the choice of a mixer. A high level diode-ring mixer will probably offer the optimum distortion performance. It, however, usually requires a large amount of lo power. If reciprocal mixing is the major performance limiting factor, it may be wiser to use a more standard diode-ring mixer that requires less lo power. A reduced oscillator drive level will allow the noise sidebands to be suppressed, reducing the reciprocal mixing observed. This is especially significant when a narrow bandwidth receiver with a filter with extremely steep skirt response is utilized.

Cost and complexity are always major considerations. A receiver with a high level switching-mode mixer with no rf amplifier will most certainly offer optimum performance. However, a simple receiver with a dual gate **MOSFET** mixer will often be completely satisfactory, and will be much more economical. Power consumption will usually be less, an additional feature if portability is a consideration.

With all of the compromises, one might wonder how a front-end might be rationally designed. The intended application will generally limit the choice of component types. The designer has one major set of parameters that may still be determined, the gains of the individual front-end elements. Gain distribution may be optimized for a given set of performance requirements and constraints. Much of this section will be devoted to the tradeoffs associated with altering the gain in various parts of the front-end.

Section 8.2 outlined methods for evaluation of receivers. Numerous parameters were mentioned. A vital parameter to optimize is the receiver factor for instrumentation applications and most communications systems. Receiver factor is a bandwidth invariant indicator of the two-tone dynamic range. If a system is to be evaluated with the goal of optimum receiver factor, both system noise figure and input intercept must be calculated.

As an initial example, consider the chain of three amplifiers shown in Fig. 8.11. Each stage is characterized by a stage gain, noise factor, and output intercept. All parameters are algebraic ratios rather than the more common logarithmic representations. The methods for calculating the noise figure and the system input intercept were presented in Chap. 6. We will apply those methods to this example using an approach that is easily used in writing calculator or computer programs.

The system noise figure will depend upon those of the individual stages as well as that of the following system. One might argue that a three-stage amplifier will ensure that system noise figure is dominated by the early amplifiers with little effect from the following system. This is not always accurate in receiver applications. For this example, assume the first stage has a 3-dB noise figure and that the others are 10 dB including that of the system following the chain. Hence, $F_1 = 2$ and $F_2 = F_3 = F_4 = 10$. All gains are 10 dB. The noise figure looking into the third stage is

$$F_3' = F_3 + \frac{F_4 - 1}{G_3} \tag{8.4-1}$$

Similarly, the noise factor looking into the second stage is

$$F_2' = F_2 + \frac{F_3' - 1}{G_2} \tag{8.4-2}$$

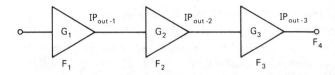

Figure 8.11 Example of a multistage system where net noise figure and intercept are evaluated.

This procedure is repeated for each stage in the system. The example of Fig. 8.11 has a net noise factor of 2.999, or 4.77 dB.

If the system contains lossy elements such as filters they will have a gain equal to the negative of the insertion loss in dB, or the reciprocal of the algebraic loss. The noise figure will equal the insertion loss.

The third order intercept is estimated using the methods derived in Chap. 6. The mechanism is the simple rule that intercepts combine just as resistors in parallel when they are expressed in algebraic units and normalized to the same plane. Assume that the three amplifiers have output intercepts of $+10$, $+20$, and $+40$ dBm, respectively. The algebraic equivalents are then 10, 100, and 10,000 mW. The intercept of the output stage is combined with that of the second with

$$\frac{1}{IP_{22}} = \frac{1}{IP_2} + \frac{1}{IP_{32}} \tag{8.4-3}$$

where the second subscript indicates the plane of normalization. In the example, IP_{32} is the output intercept of the third stage normalized to the output of the second, or 1000 mW owing to the stage gain of 10. IP_{22} is the combined intercept of all stages following and including the second one. The value for the example is $(1/100 + 1/1000)^{-1} = 90.9$ mW, or $+19.58$ dBm.

The procedure is repeated to add the distortions from earlier stages. While closed form equations may be written, they are generally redundant and confusing compared with the simplicity of the actual procedure. In the example, the 90.9 mW combined intercept of the second and third stage is normalized to the plane at the output of the first amplifier to yield 9.09 mW. This is combined with the output intercept of that stage, 10 mW, to yield a net value of 4.762 mW. This is normalized to the input of the chain by reducing the value by the gain of the first stage to yield 0.4762 mW. The overall amplifier has an input intercept then of -3.22 dBm.

The methods may be extended now to a complete receiver analysis. It should be recalled that the resistors-in-parallel analogy represents coherent addition of third order intermodulation distortion powers. As such, it is a worst case. The intercepts may not add coherently, especially when mixers or filters are considered (filters can contribute imd!). However, it is conservative to assume coherent addition.

The equations may be combined in a single program for a handheld calculator or a computer. The programs are surprisingly easy, considering the large number of parameters. Complicated systems may be analyzed with standard features on many hand-held calculators. For example, by using subroutines and indirect addressing, a triple conversion spectrum analyzer front-end may be analyzed with an inexpensive calculator.

A receiver example is shown in Fig. 8.12. The first element is a filter with an insertion loss of 0.5 dB. This is followed by an rf amplifier with 10-dB gain, a 3-dB noise figure, and an output intercept of +30 dBm. The next stage is the major preselector and image stripping filter with an insertion loss of 5 dB. The mixer is assumed to be a diode-ring–type with a 6-dB loss, a comparable noise figure, and an output intercept of +15 dBm. The next stage is an amplifier with 12-dB gain, a 5-dB noise figure, and an output intercept of +40 dBm. This drives the main i.f. filter, Z_3, which is assumed to have an insertion loss of 10 dB. The noise figure of the following i.f. system is assumed to be 8 dB.

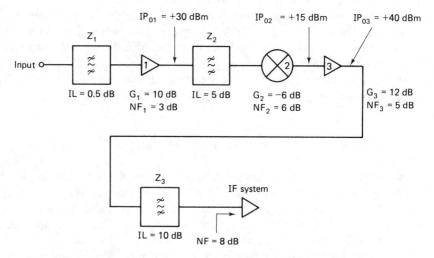

Figure 8.12 Example of a receiver front-end section. The noise figures, gains and intercepts of each stage are known, allowing the overall dynamic range to be evaluated. The gain distribution may be altered for an optimum system dynamic range.

Analysis of the system of Fig. 8.12 shows a system noise figure of 10.5 dB, an input intercept of +14.9 dBm, and an overall receiver factor of +4.4 dBm. The two-tone dynamic range is 100.9 dB if the bandwidth of the main selective filter, Z_3, is assumed to be 500 Hz.

The real value of a program to perform the calculations outlined is that various parameters may be varied to evaluate the effects of changes. A multiple dimensioned optimization may be performed.

A frequent deficiency in many receivers is the utilizaiton of excess gain in the rf amplifier, if indeed one is even used. The rf gain of the first amplifier of Fig. 8.12 may be varied to evaluate the effect. If the gain is dropped to 7.5 dB, system noise figure grows to 12.6 dB while the overall receiver factor improves slightly to +4.8 dBm. Conversely, a 20-dB gain produces a noise figure of 5.0 dB with a receiver factor of −0.1 dBm. The curve of receiver factor versus gain is monotonic. An increase in gain will always yield a slight improvement in noise figure but a degradation in receiver factor. The actual intercepts and dynamic range numbers for an assumed bandwidth may be evaluated with the equations presented in Sec. 8.2.

A more interesting set of curves is produced if the gain of the postmixer i.f. amplifier is varied. The gain shown in Fig. 8.12 is 12 dB. A peak is found if the gain is allowed to vary, assuming that the stage has constant noise figure and output intercept. At a gain of 16 dB, the system noise figure is 9.2 dB while the receiver factor is +5.4 dBm. Further increase in gain will cause a very slight improvement in noise figure, but will cause system imd to be dominated by distortion in the post mixer amplifier. As shown, the system input intercept is limited primarily by the mixer used.

The effect of i.f. system noise figure is easily evaluated. This may be done by either varying the noise figure shown, presently 8 dB, or the insertion loss of the main i.f. filter, presently 10 dB. Choosing the latter, the system noise figure improves to 8.6 dB when the filter loss drops to only 2 dB. The receiver factor is +6.3 dBm. Conversely, with a 14-dB insertion loss for Z_3 the noise figure is 12.6 dB while the receiver factor drops to +2.2 dBm. The noise figure of all parts of the front-end is of vital importance if the receiver is to be optimized for two-tone dynamic range while preserving a reasonable noise figure.

The intercepts may also be varied with the program analysis. For example, increasing the output intercept of the rf amplifier from +30 to +40 dBm shows a slight improvement in receiver factor from +4.4 to +5.6 dBm. There is some contribution to the system intercept from the rf stage. The mixer is still dominant though. A decrease in the rf amplifier output intercept to +20 dBm would drop the receiver factor to −1.0 dBm, causing the amplifier to be the major imd contributing element.

Optimization will sometimes produce an improved receiver factor, but a system noise figure lower than required. Constant receiver factor may be maintained with insertion of an input attenuator. This is common practice in a spectrum analyzer where up to 60 dB of attenuation may be inserted in 10-dB steps. An alternative solution in a communications receiver might be to increase the insertion loss of the image stripping filter, Z_2 of Fig. 8.12. This may allow a more selective filter to be used, improving the immunity to spurious responses.

If the front-end of Fig. 8.12 is altered by changing the output intercept of the rf amplifier to +40 dBm, the output intercept of the mixer to +20 dBm, and the gain of the postmixer amplifier to 16 dB, the overall system is improved measurably. The noise figure becomes 9.2 dB, the input intercept is +19.7 dBm, and the receiver factor is +10.4 dBm. The two-tone dynamic range in a 500-Hz bandwidth is then 104.9 dB.

This analysis is all an approximation. There are several factors which will invalidate it, especially when fairly "strong" receivers are considered, those with a receiver factor exceeding 0 dBm. The mixers may not be well-behaved, making the very concept of using an intercept of questionable validity. The system noise figure may be compromised by an excess noise bandwidth in the i.f. system. Finally, crystal filters may contribute imd. Still, the analysis methods provide a model for design, for they describe the salient features of the system.

REFERENCES

1. Zverev, Anatol I., *Handbook of Filter Synthesis*, John Wiley & Sons, New York, 1967.
2. Manassewitsch, Vadim, *Frequency Synthesizers: Theory and Design*, Chap. 3, John Wiley & Sons, New York, 1976.
3. Hayward, Wes and Bingham, Dick, "Direct Conversion, A Neglected Technique," *QST*, 52, 11, pp. 15–17, November 1968.
4. Lewallen, Roy W., "An Optimized QRP Transceiver," *QST*, 64, 8, pp. 14–19, August 1980.
5. Norgard, Donald E., "The Phase-Shift Method of Single-Sideband Signal Generation," and "The Phase-Shift Method of Single-Sideband Signal Reception," *Proc. IRE*, 44, 12, pp. 1718–1743, December 1956.
6. Weaver, Donald K., Jr., "A Third Method of Generation and Detection of Single-Sideband Signals," *Proc. IRE*, 44, 12, pp. 1703–1705, December 1956.
7. Hayward, Wes, "Defining and Measuring Receiver Dynamic Range," *QST*, 59, 7, pp. 15–21, July 1975.
8. See reference 7.
9. Hayward, Wes, "More Thoughts on Receiver Performance Specification," *QST* (Technical Correspondence), 63, 11, pp. 48–49, November 1979.
10. "The Amplifier Factor," Application Note from Anzac Electronics, Waltham, Mass.
11. Hayward, Wes and DeMaw, Doug, *Solid-State Design for the Radio Amateur*, ARRL, Newington, Conn., 1977.
12. Rohde, Ulrich L., "Increasing Receiver Dynamic Range," *QST*, 64, 5, pp. 16–21, May 1980.
13. Wilkens, Mark; Besser, Les; and Szentirmai, George, "Design Precise IF Filters for Digital Radio Links," *Microwaves*, pp. 65–70, November 1979.
14. "Service Manual for Tektronix Type 492 Spectrum Analyzer," Beaverton, Ore., 1979.
15. Engelson, Morris and Telewski, Fred, *Spectrum Analyzer Theory and Applications*, ARTECH House, Dedham, Mass., 1974.

SUGGESTED ADDITIONAL READINGS

1. Sabin, William, "The Solid-State Receiver," *QST*, 54, 7, pp. 35–43, July 1970.
2. Krauss, Herbert L., Bostian, Charles W., and Raab, Frederick, H., *Solid State Radio Engineering*, John Wiley & Sons, New York, 1980.
3. Miller, Gary M., *Handbook of Electronic Communication*, Prentice-Hall, Englewood Cliffs, N.J., 1979.

Index

A

ABCD matrix, 164
Abrupt junction varactor diode in vco, 331
Additive noise in oscillators, 295
Admittance on Smith Chart, 123
Admittance parameters:
  definitions, 163
  design examples, 183
  design with, 163
Admittance to impedance conversion on Smith Chart, 125
agc:
  in amplifiers, 246
  application to front-end, 361
  control range, 352
  detector, 358
  dynamics, 352
  in FET oscillator, 282
  in an oscillator, 263
  response of receiver, 345, 351
  system design, 358
  threshold, 351

Aging effects in quartz crystals, 288
All-pass filter, 45
All-pole filter, 59
Alpha, 7
Amplifier design:
  with admittance parameters, 176
  with *S*-parameters, 197
Amplifier impedance, relation to voltage gain, 23
Amplifier with multiple feedback paths, 21
Amplitude noise in oscillators, 293
Amplitude-to-phase-noise conversion, 294
Aperture coupling, 100
Attenuation, low pass filters, 66
Attenuation constant, 132
Attenuation slope in filter stopband, 50
Attenuators in i.f. amplifiers, 254
Audio amplifier, receiver, 346
Automatic gain control (*see* agc)
Available power, 19

B

Balanced mixer:
 bandwidth, 245
 bidirectional nature, 244
 bipolar transistor, 245
 FET, 245
Balance in frequency doublers, 312
Balun, 148
Bandpass filter, 45
 coupled resonator, 75
 microstrip, 159
 poles and zeros, 73
 vhf and uhf, 96
Bandstop filter, 45
Bandwidth, effect on noise figure, 207
Bandwidth, receiver measurement, 350
Bandwidth, relation to Q, 58
Base spreading resistance, 9
Beat frequency oscillator, 345
Beta, 2
Beta generator with emitter resistance, 4
bfo, 345
Biasing, bipolar transistor, 10
Bias of Seiler oscillator, 278
Bias resistors, noise from, 214
Bias shift with large signals, 17
Bipolar transistor, 1
 biasing, 10
 i.f. amplifiers, 247
 large signal operation, 14
 models, 3
Blocking, receiver, 355
Blocking capacitor, 13
Bode plots of a PLL, 328
Boltzman's constant, 203
Breakout frequency in oscillator noise, 296
Bridged-T attenuator, 254
Broadband amplifier, example, 188
Broadband balun transformer, 149
Broadband noise figure, 205
Broadband phase noise, 294
Broadband stability, 182, 184, 197
Butterworth filter, 60
 insertion loss, 93
 normalized k and q, 84
 pole locations, 61
Bypass capacitor, 13
Bypassing, 10

C

Capacitive impedance on Smith Chart, 121
Cascaded filters in receivers, 348
Cascade of two-port networks, 169
 with $ABCD$ matrix, 169
 with y-parameters, 171
cb from ce or cc conversion, 168
cc from ce or cb conversion, 168
ce from cb or cc conversion, 167
Circles of constant gain, 198
Characteristic impedance, 112
Chebyshev filter:
 normalized, 61
 pole locations, 64
Clapp factor, 277
Clapp oscillator, 274
 noise in, 296
Close-in phase noise, 295
Coaxial cable, 110, 134
Coherent imd, 229
Collector to base capacitance, bipolar transistor, 9
Colpitts network, 266
Colpitts oscillator, 265
 capacitor size in, 271
 crystal controlled, 288
Common base amplifier, 5
 current gain, 7
Common collector amplifier, 5
Common emitter amplifier, 5
Common terminal in oscillator, 279, 280
Communications receiver, 341
Commutating mixer, 241
COMPACT, 183
Compensation network, i.f. amplifier, 257
Compressed Smith Chart, 305

Index

Conjugate match, 49
Constant current biasing, bipolar transistor, 12
Contours of constant resistance, reactance, 121
Controlled current source with emitter resistance, 4
Conversion loss, diode ring mixer, 245
Converter, frequency, 348
Correlated noise, 204
Coupled resonator filters, 75
Coupler, 156
Coupling coefficient, 82
  measurement, 97
Cross modulation, 222
  in a receiver, 355
Crystal controlled oscillators, 286
  crystal current, 291
Crystal filters, 101
  capacitance in, 106
  coupled resonator, 105
  single element, 105
Current gain, 18
Current limiting in oscillator, 264
Current ratio in ideal transformer, 77
Current robbing for gain control, 249
Cutoff frequency of low pass filter, 50

D

Damped oscillation, 43
Damping factor in a PLL, 330
Damping network, 188
dB, 20
dBm, 20
Decoupling, 13
Degeneration, effect upon stability, 188
Delayed agc, 361
Delayed response in transmission lines, 111
Delta function, 37
Denormalization of filters, 68
Detection of signals, square law, 220
Digital noise in a receiver, 364

Diode mixer, 239
  ring, conversion loss, 245
Diplexer, 145
  in a receiver, 343
Dirac delta function, 37
Direct conversion receiver, 348
Directional coupler, 156
  in amplifiers, 219
  effect of port termination, 157
  stripline, 158
Directivity of bridge, 154
Direct synthesizer, 311
Dishal technique, 95
Distortion (imd)
  in amplifiers, 219
  of filter shape, 87
  in PIN diodes, 255
Divide-by-N synthesizer, 324
  with programmable reference, 335
Double tuned circuit, 90
Doubly balanced mixer, 240
Doubly terminated filters, 47, 75
Dual conversion receiver, 347
Dual forms of low pass filters, 60

E

Ebers-Moll model, 2
Efficiency, Class A amplifier, 26
Emitter degeneration, 7
  in oscillator, 266
  for stability, 182
Emitter follower, 7
Emitter inductance, 187
Exchanges with two-port networks, 167
External Q, 91

F

F_β, 5
Feedback, 20
  effect on imd performance, 231
  in i.f. amplifiers, 256
  for stability, 182

Feedback amplifier, 21
  conditions for impedance match, 25
  noise of, 212
  performance with beta, frequency, terminations, 24
  using junction FET, 30
Feedback capacitor, hybrid-pi, 174
Field effect transistor, 26 (see also Junction FET (JFET))
  noise characteristics, 214
  as voltage controlled resistor, 255
Filter concepts, 32
Filter denormalization, 68
Filtering action, LC filters, 51
Filter tuning, Dishal method, 99
Forward gain control, 248
Forward power port in directional coupler, 157
Fourier frequency in a PLL, 325
Fourier transform, 36
Frequency agile synthesizers, 312
Frequency division, 314
Frequency domain, 32
  network analysis, 36
Frequency multiplier, 221, 222
  effect on phase noise, 297
  methods, 312
Frequency response, 38
  filter, 49
  idealized filter, 44
Frequency synthesis, 311
Front-end, receiver, 343
  design of, 367
F_t, 5, 173 (see also Gain-bandwidth product)

G

G_{max}, 19
  with S-parameters, 196
  with y-parameters, 179
G_T, 19
Gain, types of, 18
Gain-bandwidth product, 5, 173
Gain circles, 198

Gain compression, 226
  in bipolar transistor, 26
  in a receiver, 355
  relation to intercept, 230
Gain control systems, receiver, 355
Gain crossover in PLL, 329
Gain distribution in a receiver front-end, 368
Gain of PLL elements, 326
Gain switching with transistors, 253
GASFET, 26
  in an oscillator, 309
General circuit parameters, 164
General coverage receiver, 347
Generator mismatch with return loss bridge, 153

H

Half wave balun transformer, 149
Half wave filter, 144
Hang agc, 360
Harmonic distortion:
  bipolar transistor amplifiers, 16
  in oscillators, 264
  second, 221
Harmonic mixing, 233
Hartley oscillator:
  bipolar transistor, 283
  crystal controlled, 289
  JFET, 284
Helical resonator, 136
  filters with, 100
Helical transmission line, 135
High level mixers, 243
  imd in, 228
High pass filter, 45
  design, 69
Hot-cold noise source, 208
h-parameters, 164
$H(s)$, transfer function, 38
$H(s)$ filter in PLL, 317, 327
Hybrid combiner, 6-dB, 153
  use in imd test, 233
Hybrid parameters, 164

Index

Hybrid-pi transistor, 5, 172
Hyperabrupt junction varactor diode in vco, 331
Hyperbolic Bessel functions, 16

I

Ideal transformer, 76
i.f. amplifiers, 246
  in a receiver, 345, 355
i.f. feedthrough, measurement, 351
i.f. noise, contribution to receiver noise figure, 357
Image frequency, 233
Image reflection mixer, 244
Image rejection of a receiver, 351
Image response in mixers, 243
Image stripping filter, 233
imd:
  in filter, 369
  measurement in receiver, 352
  in mixers, 242
  multi-tone, 223
  second order, 221
  third order, 222
Immittance, 51
IMPATT diode oscillator, 306
Impedance:
  to admittance conversion on Smith Chart, 125
  frequency domain, 38
  inversion in pi-network, 144
  inverting network with mixers, 236
  mismatching in amplifiers, 183
  normalized, 115
  parameters, 163
  along transmission line, 125
Impedance matching, 48
  to increase gain, 185
  networks, 137
  with Smith Chart, 127
Impedance transformation:
  in ideal transformer, 77

  networks, 75, 137
  by reactive element, 78
Impulse function, 37
Indirect synthesizer, 311
Inductance, emitter lead, 187
Inductive impedance on Smith Chart, 121
Input admittance, amplifier, 177
Input capacitance, bipolar transistor, 5
Input impedance, filter, 50
Input noise, 204
Input resistance, bipolar transistor, 4
Insertion loss, relation to Q, 58, 81
Insertion power gain, 19
Instrumentation receiver, 341
Integrated circuit i.f. amplifiers, 249
Intercept, 219
  of cascaded networks, 228
    in a receiver front-end, 368
  efficiency, 231
  input, 226
    in a receiver, 353
  measurement of, 228
  output, 226
  point, 226
Interconnection of two-port networks, 168
Intermediate frequency amplifier (*see* i.f. amplifiers)
Intermediate frequency from mixer, 232
Intermodulation distortion (*see* imd)
Interstage mismatch in amplifiers, 20
Inverse Fourier transform, 37

J

Junction FET, 26
  biasing, 26
  feedback amplifier, 30
  i.f. amplifier, 248
  large signal operation, 28
  mixer, 234
  oscillator limiting, 29, 281
  small signal model, 27

K

Kuroda identities, 137

L

Ladder method, 51
  with calculator, 53
Laplace frequency, s, 37
Laplace transform, 37
Large signal operation of bipolar transistor:
  with constant current bias, 15
  with constant voltage bias, 14
Lattice filter, crystal, 107
LC networks with Smith Chart, 129
Leakage inductance, 78
Limiting in oscillator, 264
  with bipolar transistor, 17
  with negative resistance, 303
Link coupled inductor, 79
Linvill stability factor, 181
L-network, 79, 138
Loaded Q, 54
Loading to achieve stability, 182
Local oscillator for mixer, 232
Local oscillator modulation in mixer, 242
Logarithmic amplifier in spectrum analyzer, 364
  response to noise, 367
Loop bandwidth of PLL, 329
Loop filter, 317, 327
Loop gain in a PLL, 328
Losses in reactive components, 54
Loss in bandpass filters, 87
Loss in transmission lines, 132
Lower sideband from mixer, 232
Lower sideband ladder, 106
Low noise amplifiers, 215
Low pass filter, 44
  normalized prototype, 45
  effect of finite Q, 87
LR network, 34

M

MAG, 19
Magnetization winding, 150
Matched amplifiers with low noise, 215
Matching impedance with Smith Chart, 127
Matching in predistorted filters, 95
Matrix representation, two-port network, 163
Maximum available gain, 19
Maximum frequency of operation, 182
mds, 349
Mesh realization, bandpass filter, 90
Microphonic effects in PLL, 329
Microstrip, 110, 134
  amplifier, 186
  application in biasing, 188
  bandpass filter, 159
  directional coupler, 158
Microwave oscillators, 301
Minimum detectable signal, 349
Mismatching of impedances in amplifiers, 183
Mixer, 221 (*see also* Balanced mixer)
  bipolar transistor, 237
  diode, 238
  fundamentals, 232
  harmonic, 233
  JFET, 234
  MOSFET, 236
  termination, 244
  types, 234
Models of transistors, 1
Modulation frequency in PLL, 325
Monolithic crystal filter, 108
MOSFET, 26
  i.f. amplifier, 248
  mixer, 236
  switching mode mixer, 245
  as a two-port network, 162
Motional capacitance, crystal, 102
Motional inductance, crystal, 102
Multiple loop synthesizers, 336
Multiplier, 221

N

Natural frequency, 42
  of PLL, 330
Negative feedback, 18 (*see also* Feedback amplifier)
Negative feedback bias in agc amplifier, 251
Negative resistance, 180
  oscillators, 301
  with S-parameters, 195
Network analyzer, 120, 193
Networks, impedance matching, 75, 137
Nodal capacitance, 81, 85
Noise, 202
  bandwidth, 209
  bandwidth of receiver, 350
  blanker, 362
  correlated, 204
  current, 203
  factor, 204 (*see also* Noise figure)
  filter in receiver, 345
  floor of oscillator, 294
  gain, 204
  matching in an amplifier, 210
  models for an amplifier, 210
  in oscillators, 292
  in oscillator starting, 263
  powers, addition of, 204
  in a resistor, 202
  source for measurement, 207
  stripping filter in a receiver, 343
  temperature, 205
  voltage, 203
Noise figure, 205
  of cascaded stages, 206
    in a receiver front-end, 368
  contours, 200, 211
  of an oscillator, 296
  receiver measurement, 351
Nonlinear amplifier in oscillator, 263
Normalized coupling coefficients, 82, 83
Normalized impedance, 115
Normalized k and q, filter design with, 84
Normalized loading coefficients, 82

Normalized low pass filter, 60
Normalized Q of filter, 87
N-port network, 162

O

Offset-divide-by-N PLL, 337
One-on-one PLL, 324
One-port oscillator, 306
Open circuited transmission line, 127
Open circuit impedance parameters, 163
Open stub, 128
Opposite sideband in a mixer, 233
Optical devices for gain control, 255
Optimization of two-port network, 183
Optimized dynamic range, receiver, 371
Oscillators:
  basic concepts, 261
  limiting in, 17, 29
  noise, 292
    measurement, 298
  operating level, 268
    Colpitts, 271
Output admittance of amplifier, 177
Output capacitance, bipolar transistor, 9
Output intercept, 226
Output resistance, bipolar transistor, 9
Output resistance, common collector amplifier, 8

P

Parallel two-port networks, 169
Parallel wire line, 110, 134
Passband, 32
Passband ripple, 60
Passive network, two-port parameters, 166
Phase constant, 118, 132
Phase detector, sampling, 319
Phase detector using diodes, 318
Phase determination with return loss bridge, 153
Phase-frequency detector, digital, 320, 321
Phase-locked loop, 311, 317, 323

Phase margin in a PLL, 328
Phase modulation in a PLL, 327
Phase noise in oscillators, 293
Phase noise modulation in Colpitts oscillator, 268
Phasing method of ssb, 349
Pierce oscillator, 287
Piezoelectric effect, 102
PIN diode attenuator, 254
PIN diode for gain controlled amplifier, 252
Pi-network, 139
Planes in ladder filter, 52
PLL, 311, 317, 323
Polarity reversing transformer, 148
Poles, 39, 42
  bandpass filter, 73
  series tuned circuit, 42
Pole-zero plot, 44
Pole-zero spacing of crystal, 104
Port, 161
Pot cores, 80
Power, 18
Power gain, 19
Power limitation of crystal oscillator, 297
Power output, bipolar transistor, 25
Predistorted filters, k and q values, 90
Preselector filter in a receiver, 233, 343
Primary inductance, 77
Probe coupling, 97
Product detector, 345
Propagation constant, 132
Propagation velocity, 114
Proportional-to-absolute-temperature, 13
Prototype filter, 45
Pseudobalun, 148
PTAT, 13
Push-push doubler, 313

Q

Q, 40, 54
  of combined elements, 57
  of filter poles, 68
  in impedance matching networks, 139
  model, 54, 55
  of an oscillator, 296
  quartz crystal, 103
  relation to bandwidth, 58
  of resonator, measurement, 96
q, end section of low pass filter, 82
Q_e, external Q, 91
Quadrature coupler, 155
Quality factor (see Q)
Quartz crystal, 102
  filters, 101
  in oscillators, 286

R

RC network, 33
Reactive feedback in low noise amplifiers, 212
Receiver design, 341
Receiver factor, 355
Receiver measurements, 349
Reciprocal mixing, 300
  in a receiver, 343, 350
    measurement, 355
Reentrant modes on transmission lines, 129
Reference frequency in PLL, 326
Reference noise in a PLL, 330
Reflected power in directional coupler, 157
Reflected wave, 117
Reflection coefficient, 116, 121
  of amplifier, 193
  reciprocal, 195
Reflection gain, 305
Resistors-in-parallel rule for imd, 230
Resonant frequency, 40
Resonator, transmission line, 128
Resonator Q, 54
  measurement of, 96
Return loss, 120, 133
Return loss bridge, 151
Reverse gain control, 248
Reverse transmission parameters, 164
rf amplifier in a superhetrodyne, 233, 343
rf signal as mixer input, 232
Richards' transformation, 137

Index

Ringing, 44
Ripple in passband, 60

S

s, Laplace frequency, 37
Saturation of transistor in oscillator, 273
Scattering parameters, 191 (*see also* S-parameters)
Seiler factor, 277
Seiler oscillator, 276
  noise, 296
Selective mismatching in broadband amplifiers, 190
Sensitivity, filter components, 67
Sensitivity of receiver (*see* mds)
Series connection, two-port networks, 169
Series-current feedback, 20
Series resistance, quartz crystal, 102
Series tuned circuit, 39
Series-voltage feedback, 20
Shape factor, filter, 344
Shift register as frequency divider, 316
Short circuit admittance parameters, 163
Short circuited transmission line, 127
Shorted stub, 128
Shunt-current feedback, 20
Shunt-voltage feedback, 20
Signal-plus-noise-to-noise ratio, measurement, 349
Signal-to-noise ratio, 204
Single conversion receiver, 342
Single tone input, 220
Single tuned circuit, 81
Singly balanced mixer, 240
Skin effect, 110
Slotted line, 120
Slow wave structure, 136
Small signal limit, bipolar transistor, 18
Small signal model, bipolar transistor, 3
Small signal operation, bipolar transistor, 17
Smith Chart, 121
  construction, 123
  equation, 119
  simulating program, 130
  spiral motion on, 132
  two-color, 125
  use with S-parameters, 193
Solenoidal inductors, 80
Source impedance, effect on noise, 211
S-parameters, 191
  oscillator design with, 304
Spectrum analyzer as a receiver, 342
Spectrum analyzer for imd measurements, 224
Spot noise figure, 205
Spurious modes in quartz crystals, 108
Spurious oscillations in oscillators, 284
Spurious outputs from mixers, 233
Spurious responses in a receiver, 346
Square law detector, 220
Square law mixer, 234
Squegging in an oscillator, 283
Stability contours, 200
Stability factor:
  S-parameters, 196
  y-parameters, 181
Stability of two-port network, 175
Stabilizing an amplifier, methods, 182
Standing waves, 119
Starting gain of oscillator, 267
Starting in an oscillator, 263
  Colpitts, 266
Step function, 38
Step recovery diode multiplier, 314
Stern stability factor, 181
Stopband, 32
  attenuation in crystal filters, 106
  attenuation of receiver, 350
  of bandpass filters, 94
  of filter, 46
Strip line, 110
Stripline directional coupler, 158
Stub, open or shorted line, 128
Stub, Q of resonator, 134
Superheterodyne receiver, 221, 342
Surface acoustic wave resonators, 312
Switched reactances in vco, 333
Switching mode mixer, 234, 238

Symmetrical network, two-port parameters, 166
Symmetry, bandpass filter shape, 73, 95
Synchronous counter divider, 315
Synchronously tuned filters in receiver i.f., 363

T

Tack hammer, 51
Tapped inductor, 79
Temperature compensation, oscillators, 286
Temperature stability of oscillators, 285
Termination of mixers, 244
Thermal noise, 203
Thermal runaway, 10
Three port y matrix, 167
Three terminal oscillator, 308
Time domain network analysis, 33
T-network, 144
Toroid cores, 80
Tracking PLL, 324
Transcent response of filter, effects in receiver, 344
Transconductance, small signal, 3
Transducer gain, 19
  of filter, 49
  with S-parameters, 193
  with y-parameters, 178
Transfer admittance, 36
Transfer function, 36
  frequency domain, 38
  generalized nonlinear, 219
  polynomial, 42
Transformer, ideal, 76
Transformer feedback:
  bipolar transistor amplifier, 218
  FET amplifier, 216
Transmission line, 109, 110
  measurements, 151
  practical, 132
  transformer, 147
    construction, 151
Transmission parameters, 164
Traveling waves, 114

Two-port parameters, 161
  interconnections, 168
  matrix representation, 163
  relations between, 165
  stability with, 175
Two stage oscillator, 291
Two-terminal oscillator, 306
Two-tone dynamic range, receiver, 353
Two-tone input, 220
Tuned circuit:
  parallel, 56
  series, 39
Tunnel diode oscillator, 306

U

uhf amplifier construction, 187
uhf bandpass filters, 96
Unilateral amplifier, 20, 178
Unilateral approximation, 198
Unilateral figure-of-merit, 198
Unit impulse, 37
Unit step function, 37
Unloaded Q, 54
Up-conversion receiver, 347
Upper sideband ladder, 106
Upper sideband from mixer, 232

V

Varactor diode:
  effects in mixers, 242
  multiplier, 314
  offset voltage, 332
  in oscillator, 318
  Q, 56, 332
  temperature characteristic, 332
Variable crystal oscillator, 288
Variable resolution amplifier, 364
vco in a PLL, 326, 330
Velocity of propagation, 114
VFET, 26
vhf bandpass filters, 96
Voltage controlled oscillators in PLL, 326, 330

Index

Voltage gain, 18
Voltage limiting in an oscillator, 264, 274
Voltage reflection coefficient, 116
Voltage standing wave ratio, 119
Voltage waves:
  with S-parameters, 191
  on transmission lines, 115
vswr, 119
VXO, 288

W

Warm-up drift, oscillators, 285
Wave equation, 114
Waves on transmission lines, 112
Well-behaved imd, 226
Well-behaved mixer in a receiver, 353

Y

Y-factor, noise measurements, 207
YIG, 308
Yitrium-iron garnet, 308
y-parameters:
  of beta generator with r_e, 166
  experimental viewpoint, 163
  for oscillator design, 302

Z

z, normalized impedance, 115
Z_o, characteristic impedance, 115
Zero-degree hybrid, 154
Zeros, 39, 42
z-parameters, 163

ARRL MEMBERS

This proof of purchase may be used as a $3.00 credit on your next ARRL purchase. Limit one coupon per new membership, renewal or publication ordered from ARRL Headquarters. No other coupon may be used with this coupon. Validate by entering your membership number from your *QST* label below:

INTRO TO RADIO
FREQUENCY DESIGN

PROOF OF
PURCHASE

FEEDBACK

Please use this form to give us your comments on this book and what you'd like to see in future editions, or e-mail us at **pubsfdbk@arrl.org** (publications feedback).

Where did you purchase this book?
☐ From ARRL directly ☐ From an ARRL dealer

Is there a dealer who carries ARRL publications within:
☐ 5 miles ☐ 15 miles ☐ 30 miles of your location? ☐ Not sure.

License class:
☐ Novice ☐ Technician ☐ Technician Plus
☐ General ☐ Advanced ☐ Extra

Name _____

ARRL member? ☐ Yes ☐ No
Call sign _____

Daytime Phone (___) _____ Age _____

Address _____

City, State/Province, ZIP/Postal Code _____

If licensed, how long? _____

Other hobbies _____

_____

Occupation _____

For ARRL use only   INTRO TO RF
Edition   1 2 3 4 5 6 7 8 9 10 11 12
Printing   2 3 4 5 6 7 8 9 10 11 12

From _____
    _____
    _____

> Please affix postage. Post Office will not deliver without postage.

EDITOR, INTRO TO RADIO FREQUENCY DESIGN
AMERICAN RADIO RELAY LEAGUE
225 MAIN ST
NEWINGTON CT 06111-1494

.... please fold and tape ....